Building Intelligent Agents

An Apprenticeship Multistrategy Learning Theory, Methodology, Tool and Case Studies

Building Intelligent Agents

An Apprenticeship Multistrategy Learning Theory,
Methodology, Tool and Case Studies

Gheorghe Tecuci

Department of Computer Science
George Mason University
Fairfax, VA
USA

Contributing Writers

Tomasz Dybala, Michael Hieb, Harry Keeling, Kathryn Wright,
Philippe Loustaunau, David Hille, Seok Won Lee

ACADEMIC PRESS
San Diego   London   Boston
New York   Sydney   Tokyo   Toronto

Academic Press
525 B Street, Suite 1900, San Diego, California 92101-4495, USA
http://www.apnet.com

Academic Press Limited
24–28 Oval Road, London NW1 7DX, UK
http://www.hbuk.co.uk/ap/

ISBN 0–12–685125–5

Library of Congress Cataloging-in-Publication Data

Tecuci, Gheorghe.
    Building intelligent agents: an apprenticeship multistrategy
learning theory, methodology, tool and case studies / by Gheorghe
Tecuci.
        p.   cm.
    Includes index.
    ISBN 0–12–685125–5 (alk. paper)
    1. Intelligent agents (Computer software)   I. Title.
QA76.76 I58T43   1998
006.3'3—dc21   98-14478
                                                                             CIP

A catalogue record for this book is available from the British Library

Typeset by Blackpool Typesetting Services Ltd, Blackpool, Lancashire

Printed and bound by Antony Rowe Ltd, Eastbourne

Transferred to digital printing 2005

Contents

Preface

This book presents a theory and methodology of building intelligent agents, together with a tool and case studies. The defining feature of this approach to building agents is that a person teaches the agent how to perform domain-specific tasks. This teaching of the agent is done in much the same way as teaching a student or apprentice, by giving the agent examples and explanations, as well as supervising and correcting its behavior. This approach, called Disciple, in which the agent learns its behavior from its teacher, integrates many machine-learning and knowledge acquisition techniques, taking advantage of their complementary strengths to compensate for each other's weaknesses. As a consequence, the Disciple approach significantly reduces (or even eliminates) the involvement of the knowledge engineer in the process of building an intelligent agent. The agent can be directly taught and trained by a user to assist him/her or other users in various ways: by helping the user to perform his/her tasks, by performing tasks on the user's behalf, by monitoring events or procedures for the user, by advising other users on how to perform a task, by training or teaching other users, or by helping different users collaborate. This work is part of a long-term vision where personal computer users will no longer be simply consumers of ready-made software, as they are today, but also developers of their own software assistants.

Most of the current approaches to building intelligent knowledge-based agents can be divided into two broad categories: knowledge acquisition approaches and machine-learning approaches. In the knowledge acquisition approaches, the agent is actually built by a knowledge engineer who manually encodes the expertise acquired from the domain expert into the agent's knowledge base. This indirect transfer of knowledge, from the domain expert through the knowledge engineer to the agent, leads to a long, painful and inefficient agent development process. In the machine-learning approaches, the agent employs automated learning methods to learn all of its knowledge from data or experience. However, the difficulty of this task, and the capabilities of the current autonomous learning methods, limit the application of this approach to building simple prototype agents. As knowledge acquisition attempts to automate more of the agent development process and machine learning attempts to automatically develop more complex agents, there is an increasing realization within these two research communities that a better approach to building intelligent agents is through an integration of different knowledge acquisition and learning methods. In Chapter 1 we briefly review and contrast the main knowledge acquisition, machine-learning and integrated approaches to building intelligent agents.

In Chapter 2 we present, at an intuitive level, the Disciple approach for building intelligent agents, illustrating it with two examples: building a manufacturing assistant and building an educational assessment agent. In the Disciple approach, a person (who is an expert in some domain, but may not have prior experience with computers or knowledge bases), receiving limited assistance from a knowledge engineer, can teach the agent in a way that resembles the way the expert would teach a student or his/her assistant. For instance, the expert may teach the agent how to solve a certain type of problem by providing a concrete example, helping the agent to understand the solution, supervising the agent as it attempts to solve analogous problems, and correcting its mistakes. Through only a few such natural interactions, the agent will be guided into learning complex problem solving rules, and in extending and correcting its knowledge base. This process is based on a cooperation between the agent and the human expert in which the agent helps the expert to express his/her knowledge using the agent's representation language, and the expert guides the learning actions of the agent. The central idea is the use of synergism at several levels. At an upper level is the synergism between teaching (of the agent by the expert) and learning (from the expert by the agent). For instance, the expert may select representative examples to teach the agent, may provide explanations, and may answer the agent's questions. The agent, on the other hand, will learn general rules that are difficult to be defined by the expert, and will consistently integrate them into the knowledge base. At a lower level is the synergism among different learning methods employed by the agent. By integrating complementary learning methods (such as inductive learning from examples, explanation-based learning, learning by analogy, learning by experimentation) in a dynamic, task-dependent way, the agent is able to learn from the human expert in situations in which no single-strategy learning method would be sufficient.

The next two chapters of the book present in detail the concept-based knowledge representation of the Disciple approach, as well as the problem-solving, knowledge acquisition and learning methods. They assume no prior knowledge from the reader. Chapter 3 covers knowledge representation, through semantic networks and rules, and the main inference methods used with these representations. Chapter 4 presents the methods of systematic elicitation of knowledge for developing an initial knowledge base, the rule-learning method based on explanations and analogy, and the rule refinement method based on analogy, experimentation, and explanation-based and example-based induction. It also presents the exception-based refinement of the knowledge base that employs knowledge discovery and elicitation. These two chapters cover many of the topics introduced in artificial intelligence, machine-learning, and knowledge acquisition courses, in a unified way.

Chapter 5 presents the methodology for building intelligent agents by using the Disciple learning agent shell that implements the methods from the previous chapters. It shows in detail how a user interacts with the Disciple learning agent shell to teach and train it to become a domain-specific agent. This chapter also makes the transition to the remaining chapters of the book. Each of the last four chapters presents a case study of using the Disciple methodology and shell to develop intelligent agents for complex domains. While all four case studies illustrate the same methodology, they are different in many ways, not only because the domains are different, but more importantly because the issues that we have stressed in developing these agents are different, yet complementary. Therefore, each case study will bring much more to the reader than yet another application of the same approach.

Chapter 6 presents in detail an application of the Disciple methodology to the building, training and using of two agents that generate history tests to assist in the assessment of students' understanding and use of higher-order thinking skills. These two agents are representative of the class of agents built by an expert (in education and history, in this case) to assist other users (history teachers and students). One of the assessment agents is integrated with the MMTS (Multimedia and Thinking Skills) educational system, creating a system with expanded capabilities, called Intelligent MMTS (IMMTS). The other agent is a stand-alone agent that can be used independently of the MMTS software. The IMMTS system has been field-tested in American history classes in several middle schools on American installations in Germany and Italy. This case study covers all the phases of the agent development process: analyzing the problem domain and defining agent requirements, defining the top-level ontology of the knowledge base, developing domain dependent modules, building the initial knowledge base, teaching the agent how to generate tests, developing the assessment engine and the graphical user interface, and finally, verifying, validating and maintaining the agent. The presentations of the other three case studies will only contain an overview of the applied methodology, stressing more those aspects that make them different from this case study.

Chapter 7 presents the case study of developing a statistical analysis assessment and support agent. This agent is integrated in a university-level introductory science course and is accessed on the Internet through a web browser. The course, The Natural World, introduces students to the world of science using collaborative assignments and problem-centered group projects that look at scientific issues which underlie public policy making and stimulate the development of students' analytic skills. This web-based course focuses on three issues (diseases, evolutionary history, and nutrition) and is taught at George Mason University as part of an innovative curriculum designed for the New Century College, GMU's newest college. It responds to a growing belief that the goal of education should no longer be dominated by the need to transfer information, but by the need to help students locate, retrieve, understand, analyze, construct, and use information, and by the need to train life-long learners. The agent supports two aspects of students' learning in this course: students' knowledge and understanding of statistics and students' analyses of issues related to statistics. It does this in three ways. First, it can be used as a traditional test generator, as in the case of the assessment agent discussed in Chapter 6. Second, it integrates the documents accessed by students on the web and interacts with the students during the learning process. Finally, it can be used as an assistant by the students as they work through their assigned projects. For most of these projects, students are required to include some data analysis to support their argument, and so students can use the statistical analysis assessment and support agent to help them extract all possible relevant information from a particular data set and support their analysis of the data. This particular function, that of an assistant, is similar to the one of the design assistant presented in Chapter 8. An interesting feature of Chapter 7 is that it is written from the perspective of the domain expert who has trained the agent and therefore it gives a first-hand account of how an expert interacts with a Disciple agent.

Chapter 8 presents a case study of using the Disciple approach to develop an agent that behaves as a personal assistant to the user. As opposed to the agents presented in Chapters 6 and 7, which are built by an expert for other users, this agent is developed by a user to be an assistant to him/herself. The agent is continuously supervised and customized by the user according to the changing practices in the user's domain, as well as the needs and the preferences of the user. In this case study we will focus on an engineering design domain; more specifically, computer configuration tasks. The growing complexity of contemporary engineering designs requires design tools that play a more active role over the whole design process. One way to make the design tools more active is to have them behave as assistants and partners in the design process. The human designer and the assistant create designs together. The level of interaction between them depends on the quantity and quality of the assistant's knowledge. Initially, the agent will behave as a novice that is unable to compose the majority of designs. In this case the designer takes initiative and creates the designs, and the assistant learns from the designer. Gradually, the assistant learns to perform most of the routine (but usually more labor intensive) designs within an application domain. Because of its plausible reasoning capabilities, the assistant can also propose innovative designs that are corrected and finalized by the designer. Creative designs are specified by the designer. As a result of learning, designs that were innovative for the assistant become routine, and designs that were creative become first innovative and later routine ones. This process constantly increases the capabilities of the design team consisting of the human expert and the computer assistant.

Chapter 9 presents a case study of building and training a Disciple agent to behave as a military commander in a virtual environment. This is a type of agent that is trained by a user to perform tasks on user's behalf. The virtual military environment is the ModSAF (Modular Semi-Automated Forces) distributed interactive simulation that enables human participants at various locations to enter a synthetic world containing the essential elements of a military operation. There are many applications of such interactive simulations, among the most important being that of training individuals that interact with each other, such as collective training of military forces in large-scale exercises, peace-keeping operations or disaster relief. ModSAF is a very complex real-time application which simulates military operations. It provides facilities for creating and controlling virtual agents (such as virtual tanks, planes, ships, or infantry) acting in a realistic representation of the physical environment that utilizes digital terrain databases. In the ModSAF environment, human participants may cooperate with, command or compete against virtual agents. In the case study presented, an officer trains the Disciple agent to behave as a virtual company commander in the ModSAF environment. The officer may train the agent to perform various missions (such as a road march, bounding overwatch, or occupying a position) using the graphical interface of ModSAF, which consists of a topographical map and various editors. The officer instructs the agent by giving an initial example of how to perform the mission that is to be taught. Then the officer guides and helps the agent to understand why the example given is a good solution to the mission problem. Next the agent will engage the officer in a dialog, showing the officer examples of how it would solve other similar missions, and asking for explanations if its solution is not correct. Through such simple and natural interactions, the agent learns complex rules for accomplishing its missions.

Finally, the bibliography contains, in addition to the works cited in the book, basic titles from the fields of machine learning, knowledge acquisition and intelligent agents.

This book addresses issues of interest to a large spectrum of readers from research, industry, and academia. Researchers interested in building intelligent agents, whether through knowledge acquisition methods or through machine-learning methods, will find in this book an integrated approach that uses the complementariness of many basic knowledge acquisition and machine-learning techniques in order to take advantage of their strengths and to compensate for each others weaknesses. We believe that through such an approach it will some day be possible to develop learning agent shells that will be customized, taught and trained by normal users as easily as they now use personal computers for text processing or email.

Practitioners who develop various types of systems (educational systems, simulation systems, database systems, knowledge-based systems, etc.) will find in this book a detailed, yet intuitive presentation of an agent development methodology and tool, as well as several case studies of developing intelligent agents that illustrate different types of agents that are relevant to a wide variety of application domains. Such an agent could either enhance the capability, generality and usefulness of another computer system (like the history assessment agent), could be integrated within and enhance a larger multisystem environment (like the virtual military commander or the statistical analysis assessment and support agent), or could be used as an assistant to the user (like the engineering design assistant or the manufacturing assistant) or as an agent that performs various tasks at the user's request (like the stand-alone history assessment agent). These are, of course, only the few examples that are treated in detail in this book, but they show the generality of the proposed approach and may inspire the practitioners to use the Disciple methodology and shell to develop other types of agents for their particular needs.

This volume could also be used as a textbook for an introductory course in artificial intelligence, machine learning or knowledge acquisition. Chapters 3 and 4 cover, in a unified way, many of the topics introduced in these courses, both at the level of general principles and with detailed methods and many examples. Then Chapter 5 presents a practical implementation of a learning agent shell that will allow the students a hands-on experience. Finally, Chapters 6–9 describe detailed case studies on how to apply the principles, techniques, methods and the learning shell presented in the previous chapters to build practical intelligent agents. These case studies are good examples of collaborative student projects, and provide many ideas for simpler projects.

The curriculum vitae of an idea*

The origin of the ideas that have evolved into the apprenticeship multistrategy approach for building intelligent agents, presented in this book, can be traced back to 1984, when I first addressed the problem of how to build an instructable system. At that time I was a researcher at the Romanian Research Institute for Informatics, headed by Mihai Drăgănescu, who has supported me in many ways and more than I can express during my entire research career. Soon after that I had the chance to meet and work with Zani Bodnaru, who enthusiastically reacted to my proposal of building an expert system that would assist him in designing plans for loudspeaker manufacturing and could learn the knowledge it needed to design these plans. While investigating how our expert system could learn, I realized that none of the learning methods I knew were applicable in that domain. There were not many examples from which to learn rules through empirical induction, and there was not the strong domain theory needed to use the newer explanation-based learning approach. But there was a cooperative expert that was willing to interact with the system and to help it learn. It is by trying to solve this concrete learning problem that led to my development of a new kind of learning method that would integrate a form of explanation-based learning (to understand an example, with the expert's help, and to generalize it), a kind of learning by analogy and experimentation (to automatically generate additional examples to be shown to the expert), and a kind of empirical induction (to learn from these examples). Because the system was trying to learn from the expert in an apprentice style (Mitchell *et al.*, 1985), I called it Disciple. This basic method turned out to be one of the first multistrategy approaches to learning that later grew into the subfield of multistrategy machine learning (Tecuci, 1993b; Michalski and Tecuci, 1994). It also turned out to be one of the first attempts to integrate machine learning and knowledge acquisition (Tecuci *et al.*, 1994; Tecuci and Kodratoff,

---

* This phrase was introduced by Mihai Drăgănescu.

1995). Yves Kodratoff reacted very enthusiastically to this new learning approach and invited me to present it at the First European Working Session on Learning that he was organizing in Orsay, in February 1986. We presented a joint paper at that workshop, and this marked the beginning of a long cooperation between us on this subject. Yves supported me in many ways and contributed many ideas to the Disciple approach, being also the adviser of my Ph.D. thesis, 'Disciple: a theory, methodology and system for learning expert knowledge', completed in 1988. Our joint work was supported by the Romanian Academy and the French National Center for Scientific Research.

After the completion of the Ph.D. thesis, I continued the development of the Disciple approach, improving the learning methods and extending them with guided knowledge elicitation (Tecuci, 1991, 1992a and b). This work, done at George Mason University, was supported by NSF, ONR and DARPA. The more recent developments to the Disciple approach have been done in collaboration with my Ph.D. students at George Mason University. Tomasz Dybala developed the interface modules of Disciple, modularized the learning method, and developed a scaled-up version of the Disciple learning agent shell. He also developed the design assistant presented in Chapter 8 and contributed to the assessment agent presented in Chapter 6 and the virtual commander agent presented in Chapter 9. Michael Hieb enhanced the interactions between Disciple and the user, and contributed to the development of Disciple's knowledge elicitation methods. He is also the main developer of the agent presented in Chapter 9 to which David Hille and J. Mark Pullen have also contributed. Harry Keeling significantly contributed to the development of the assessment agent presented in Chapter 6 and also contributed to the assessment and support agent presented in Chapter 7. Kathryn Wright ported the Disciple shell on Macintosh, further developed it and contributed to the assessment and support agent presented in Chapter 7. My colleague, Philippe Loustanou, was the domain expert and the driving force behind the development of the assessment and support agent presented in Chapter 7. Seok Won Lee worked on the bibliography and helped significantly with finalizing the book. Hadi Rezazad and Ping Shyr read several drafts of the book and made valuable suggestions. Tom Mitchell contributed indirectly to the Disciple approach through his influential work in machine learning, especially the version space representation, apprenticeship learning and explanation-based learning.

Andrew Sage, Murray Black and Peter Denning have helped and supported me in many ways. Henry Hamburger has been a wonderful colleague and friend that was always available when I needed his help. I would also like to acknowledge the more recent support from AFOSR, DARPA, and NSF, and especially David Gunning and David Luginbuhl. This support has facilitated the latest developments and applications of the Disciple approach and the writing of this book. Kate Brewin, Bridget Shine and Tamsin Cousins from Academic Press have been very patient with us and are a great team to work with.

Finally, my wife Sanda and our daughter Miruna Gabriela, to whom I am dedicating my work, have an invaluable contribution to what I am and do.

Gheorghe Tecuci

About the Contributing Writers

Gheorghe Tecuci (*all chapters*) is a Professor of Computer Science at George Mason University and a member of the Romanian Academy. He received the M.S. degree in Computer Science from the Polytechnic Institute of Bucharest in 1979, graduating first among all the Computer Science students at the Polytechnic Universities of Romania. He received two Ph.D. degrees in Computer Science, one from the University of Paris-Sud and the other from the Polytechnic Institute of Bucharest, both in 1988. He has directed the Learning Agents Laboratory of the Computer Science Department of George Mason University since 1995, and the Center for Machine Learning, Natural Language Processing and Conceptual Modeling of the Romanian Academy since 1992. He has also held research and/or teaching positions at Research Institute for Informatics in Bucharest, University of Paris-Sud, and Polytechnic Institute of Bucharest. Dr Tecuci has published around 100 scientific papers, most of them in artificial intelligence. He contributed to the development of two important research areas of artificial intelligence: multistrategy learning, and the integration of machine learning and knowledge acquisition. Both these areas are viewed as key technologies for the development of intelligent systems and for the generalized application of artificial intelligence to complex real-world problems. Dr Tecuci co-edited (with R. S. Michalski) *Machine Learning: a Multistrategy Approach* (Morgan Kaufmann, 1994) and (with Y. Kodratoff) *Machine Learning and Knowledge Acquisition: Integrated Approaches* (Academic Press, 1995). He was the program chairman of the first international workshops on multistrategy learning (MSL-91 and MSL-93, Harpers Ferry, WV), and of the first workshop on integrated knowledge acquisition and machine learning (at IJCAI-93, Chambery, France). He also gave the first tutorials on these topics (at IJCAI-93, AAAI-93, and IJCAI-95). The research of Dr Tecuci has been sponsored by Air Force Office of Scientific Research, National Science Foundation, Defense Advanced Research Project Agency, National Research Council, Office of Naval Research, Romanian Academy, Romanian Research Institute for Informatics, European Economic Community, and French National Research Center. His current address is Department of Computer Science, George Mason University, 4400 University Drive, Fairfax, VA 22030, USA. Email: tecuci@gmu.edu.

Tomasz Dybala (*co-author, Chapters 4, 5, and 8*) is a Lead Software Engineer and Analyst at NASDR, the company responsible for regulating the securities industry in the USA. He is currently involved in the development and deployment of the break detection and knowledge discovery system for the Nasdaq stock market. He received his Ph.D. in Information Technology (1996) from George Mason University, USA, and his MS in Computer Science (1985) from Stanislaw Staszic University, Poland. Prior to coming to the USA, he worked as a researcher/instructor, and a software engineer for several leading Polish research and education institutions and electronic corporations. From 1991 to 1996, Tomasz was associated with the School of Information Technology and Engineering at GMU where he worked on development of machine-learning algorithms and on their industrial applications. Since

1994 he has been a research associate with the Learning Agents Laboratory at GMU where he participated in the development of the Disciple shell for building intelligent agents with learning capabilities. He also applied the shell to build interactive design assistants, to develop intelligent tutors for multimedia systems, and to train virtual commanders of computer generated forces. He supervised and taught graduate students. Tomasz has published over a dozen journal and conference papers in the areas of machine learning, applications of knowledge-based technologies to engineering design, and control engineering. His current interest is in intelligent agents with learning capabilities, integrated knowledge and software-engineering environments, and knowledge discovery and data mining systems.

Michael R. Hieb (*co-author, Chapters 4 and 9*) is a Senior Research Scientist with the Modeling and Simulation Group of AB Technologies in Alexandria, Virginia. He is leading a research program to develop interfaces between virtual military simulations and real-world command and control systems. He received his Ph.D. in Information Technology (1996) from George Mason University, USA. His Ph.D. thesis was on instruction techniques for training virtual agents. This work advanced and adapted Disciple to teach new tasks to Modular Semi-Automated Forces (ModSAF) simulation agents. He received his MS in Engineering Management specializing in Artificial Intelligence and Human Factors (1990) from George Washington University, USA. His master's thesis was on machine discovery in large scale engineering systems. He received his BS in Nuclear Engineering (1980) from the University of California, Santa Barbara, USA. Prior to his MS degree, he worked as a Nuclear Engineer, directing startup testing at three nuclear power stations and receiving a Senior Reactor Operator certification. Michael worked on AI Blackboard systems for NASA at IntelliTek, Inc. and served as a consultant to Computer Sciences Corporation on AI projects. Michael conducted his doctoral research with Dr Gheorghe Tecuci in the Learning Agents Lab in the Computer Science Department of GMU. He has published over 30 papers in the area of instructable agents, knowledge acquisition and computer-generated forces. His research interest is in methods for applying advanced simulation technologies to specific real-world domains and developing scaleable behavior for virtual agents.

Kathryn Wright (*co-author, Chapter 5*) is a Ph.D. student in the School of Information Technology and Engineering at George Mason University and is a researcher in the Learning Agents Laboratory where she is currently involved in the application of Disciple to the High Performance Knowledge Base program, a DARPA project. She received her M.S. in Operations Research (1993) from George Mason University. From 1985 to 1996, she worked as a communications software engineer at Data Systems Analysts, Inc., a company dedicated to providing customized software and information technology solutions to the telecommunications industry. Her area of involvement was in Government-owned and -operated message-switching systems. Prior to that, she spent 11 years as a communications officer in the United States Air Force. Her current interest is in intelligent agents with learning capabilities, methods for reasoning under uncertainty, and adaptive system management for distributed messaging systems.

Harry Keeling (*co-author, Chapter 6*) is a Ph.D. candidate in the School of Information Technology and Engineering at George Mason University and is a researcher in the Learning Agents Laboratory. His doctoral research explores the application of machine learning and knowledge acquisition methods to the development of intelligent educational agents that support intelligent tutoring and learning environments. As an AI researcher, Harry has significantly contributed to the design and development of several intelligent learning agents. With an emphasis on the design and development of problem-solving components, Harry has participated in the development of the Disciple shell and its machine-learning algorithms. Before joining this research team in 1995, he spent seven years as a researcher and lecturer in the Information Systems and Analysis Department at Howard University in Washington DC. There he taught systems analysis and design methodologies and his research efforts resulted in the development of computer-based training systems for first-year students in computer science courses. He received his B.S. in Computer Science from George Washington University (1980). His Masters thesis was on the use of inductive assertions in the verification of computer systems.

Philippe Loustaunau (*co-author, Chapter 7*) is a senior analyst at Systems Planning and Analysis (SPA), Inc., located in Alexandria, Virginia. SPA is a small business whose mission is to provide top-level decision makers with timely assessments that integrate the technical, operational, and programmatic aspects of strategic, Navy and other national security issues. Dr Loustaunau's current projects include the analysis of the strategic deterrent force structure and the characteristics and capabilities of Trident submarines, weapon systems and missiles; and the development of innovative solutions and enabling technologies to improve the utility and performance of defense and commercial systems (in particular the design, modeling and analysis of electro-mechanical materials, optical systems, active/adaptive control strategies, and smart structural systems). He is also collaborating with Dr Tecuci and the *Learning Agent Laboratory* on the applications of AI techniques, and in particular the applications of the *Disciple* approach, to the military domain. Prior to his current position, Dr Loustaunau spent nine years at George Mason University (GMU) where he was an Associate Professor of Mathematics, directed an active research program in applied mathematics and provided leadership in wide-ranging department and university projects. While at GMU, Dr Loustaunau's research focused on applied and computational mathematics, in particular symbolic and algebraic computation, artificial intelligence, algorithms, coding theory, mathematical physics, and software development. He has published 25 articles and two books and has been a frequently invited speaker at professional conferences. He was Director of Graduate Studies and acting Chair of the Department of Mathematics. He has also been involved in a number of innovative educational projects. The work described in Chapter 7 was done while Dr Loustaunau was at GMU and is continuing as this book goes to press.

David Hille (*co-author, Chapter 9*) is a Ph.D. candidate in Information Technology at George Mason University. He works as a senior computer scientist for ANSER, a public service research institute, and as a researcher at the Computer Science Department at George Mason University. His research focuses on the integration of apprenticeship learning and adversarial planning in building automated agents. He has worked as a professional game designer, creating tactical war games, such as Combat Leader, published by Strategic Simulations, Inc., that included simple automated agents acting entities such as tank commanders, section leaders, squad leaders, platoon leaders, and company commanders. He has developed database applications, including Budget Analysis and Tracking System to monitor defense program funding and budget execution and Electronic Source Selection System used to assist in the Government source selection decision-making process. He has worked on a number of projects in artificial intelligence. These projects include Internal Control Expert, an expert system used to evaluate the internal controls of an organization; Cut Drill Expert System, used to evaluate candidates for budget cuts among Government programs; and Captain, a project for developing automated commanders for use in Modular Semi-Automated Forces (ModSAF) simulations. He is currently working on Disciple, extending it to permit planning and problem solving in complex domains.

Seok Won Lee (*co-author, Bibliography*) is a Ph.D. student in the School of Information Technology and Engineering at George Mason University and is a researcher in the Learning Agents Laboratory, working with Professor Gheorghe Tecuci. He received his B.S. in Computer Science from Dongguk University at Seoul, Korea in 1992 and M.S. in Computer Science with specialization in Artificial Intelligence from University of Pittsburgh at Pittsburgh in 1995. Prior to joining to the School of Information Technology and Engineering at GMU, he was a Ph.D. researcher in the Intelligent Systems Laboratory working with Professor Bruce G. Buchanan and a teaching assistant for the Department of Computer Science at University of Pittsburgh. His M.S. thesis concerned the application of Inductive Rule Learning System (RL) to multiple large databases in order to discover new patterns and relationships among distributed heterogeneous large data and contributed a new methodology for high-quality knowledge acquisition. He has published over 10 papers in the areas of constructive induction, multistrategy learning, knowledge discovery, knowledge management, and machine learning applications to medical diagnosis. His primary research interest is in intelligent agents with learning capabilities, and in data analysis using machine learning, knowledge acquisition, and knowledge discovery and data-mining techniques. His current research focuses on collaborative intelligent knowledge acquisition agents for high-performance knowledge bases. He is a member of the Korean-American Scientists and Engineers Association (KSEA), American Association for Artificial Intelligence (AAAI), IEEE Systems, Man and Cybernetics Society, and Association for Computing Machinery (ACM).

Notation

Below is a summary of symbols and notation used in this book.

Acronyms

ADS	Assistant design space
AFOSR	Air Force Office of Scientific Research
CDC	Center for Disease Control
CGFs	Computer-generated forces
DARPA	Defense Advanced Research Projects Agency
D_C	Creative design
D_I	Innovative design
DIMM	Dual in-line memory module
DIS	Distributed interactive simulation
D_R	Routine design
GMU	George Mason University
IDS	Innovative design space
IMMTS	Intelligent MMTS
KB	Knowledge base
KE	Knowledge element
LGG	Least general generalization
LR	Plausible lower bound rule
mGG	Minimally general generalization
MGS	Maximally general specialization
ML & KA	Machine learning and knowledge acquisition
MMTS	Multimedia and thinking skills
ModSAF	Modular Semi-Automated Forces
NCC	New Century College
NSF	National Science Foundation
PVS	Plausible version space
PVSR	Plausible version space rule
RAM	Random access memory
RDS	Routine design space
SAFORs	Semi-Automated Forces
SIMM	Single interface memory module
TDS	Target design space
TS	Testing set
UR	Plausible upper bound rule

General symbols and abbreviations

$\{A\ B\ C\}$ — In a clausal representation, denotes a set of objects or constants from which another object may take one or more values. For example, the expression 'concept$_k$ ISA $\{$concept$_i$ concept$_j\}$' expresses the fact that any instance of concept$_k$ is an instance of concept$_i$ or an instance of concept$_j$ or an instance of both

C — Concept

Cl_j — The class of the variable $?j$ in the plausible lower bound condition of a rule

Cu_j — The class of the variable $?j$ in the plausible upper bound condition of a rule

D — Domain

D_f — The set of objects from the domain of the feature f

F — Fact

KE_i — The ith knowledge element

K_g — A general description of an object which is part of a general task defined in the conditional part of a rule or part of one or more general operations defined in the outcome of the same rule

$\mathscr{L}$ — A knowledge representation language, consisting of the tuple (V, C, F, S, H, T, O, L) where

V is a set of variables
C is a set of constants
F is a set of features
S is a semantic network of objects and instances
H is a set of theorems and properties of the elements of V, C, F and S
T is a set of generic tasks
O is a set of problem solving operations, and
L is a set of logical connectors

$=_\mathscr{L}$ — Equals in the representation language $\mathscr{L}$

L_x — The class of the object concept $?x$ from the plausible lower bound condition of a rule, such that $L_x = (l_1 \ldots l_n)$. Each concept l_i is a minimal generalization of the known positive examples of $?x$ that does not cover any of the known negative examples of $?x$ and is less general than (or as general as) at least one concept u_j from U_x, and is not covered by at least one element l_j of L_x

O_g — A general operation to be performed which is referenced in the outcome of a rule

$O_{g1}, \ldots, O_{gn}$ — A sequence of general operations to be performed

$(P\ v)$ — A property–value pair, where the property P has the value v

R — Rule

R_f — The set of objects from the range of the feature f

$REL(r_i, r_j)$ — An entity r_i is related to an entity r_j by the relation REL (e.g. 'explains(r_i, r_j)' means that the entity r_i explains the entity r_j). Also written as $(r_i$ REL $r_j)$

σ — A function which defines a substitution of the form $(x_1 \leftarrow t_1, \ldots, x_n \leftarrow t_n)$, where each x_i $(i = 1, \ldots, n)$ is a variable and each t_i $(i = 1, \ldots, n)$ is a term

σl_i	If l_i is an expression in the representation language $\mathcal{L}$, then σl_i is the expression obtained by substituting each x_i from l_i with t_i
T_g	A general task to be accomplished which is referenced in the conditional part of a rule
U_x	The class of the object concept ?x from the plausible upper bound condition of a rule, such that $U_x = (u_1 \dots u_m)$. Each concept u_i from U_x is a maximal generalization of all the known positive examples of ?x that does not cover any of the known negative examples of ?x and is more general than (or as general as) at least one concept l_j from L_x and is not covered by any element u_j of U_x
?X	The variable '?X' represents some unknown entity in the knowledge base
(?Z ?X ?M)	A tuple of three objects

Knowledge elements

ASSERT	A general operation which adds new elements to the knowledge base
DECOMPOSE	A general operation which decomposes a task into several subtasks
DIF-FROM	A relationship used to indicate that an object x is different from an object y
FEATURE$_n$	The nth feature where feature is a general term denoting either a property or a relation of an object
INSTANCE-OF	A relationship used to relate an instance with its immediate ancestor in a generalization hierarchy
IS	A relationship used to relate a concept or an instance with any one of its ancestors in a generalization hierarchy
ISA	A relationship used to relate a concept with its immediate ancestor in a generalization hierarchy
SOMETHING	The concept at the top of the concept generalization hierarchy. The most general concept in the semantic network of the knowledge base
TYPE-OF	A relationship used to relate a concept or an instance with all of its ancestors in a generalization hierarchy

Logical symbols

$\forall$	Universal quantifier ('for all')
$\exists$	Existential quantifier ('there exists')
$\wedge$	Conjunction ('and')
$\vee$	Disjunction ('or')
$\rightarrow$	Implication ('if … then …')
$=$	Equals
$\neq$	Does not equal

Sets

$\in$	Is a member of
$\cap$	Intersection
$\cup$	Union
$\{1, 2, \ldots, n\}$	Set consisting of the numbers from 1 to n
$[0; 10]$	Interval containing all the real numbers between 0 and 10, inclusive

1
Intelligent Agents

This chapter presents the definition of an intelligent agent and discusses some of the general issues in developing such an agent. The key issue is how to represent, capture, and maintain the knowledge necessary for the agent to successfully perform its functions. There are two complementary approaches to this issue: knowledge acquisition and machine learning. We briefly review and contrast them. Then we review the main integrated approaches that use the complementariness of knowledge acquisition and machine learning.

1.1 What is an intelligent agent?

Encouraged by many small demonstrations of 'intelligent' behavior, early artificial intelligence (AI) papers predicted the development of 'complete' systems that would exhibit most of the characteristics that we associate with intelligence in human behavior. Such systems would perceive the environment through artificial eyes and ears, would communicate in natural language, would reason, solve problems and formulate plans, would act on the environment to achieve their goals, and would learn from their experience. However, after a period of great enthusiasm and expectations, it was soon realized that building such 'complete' intelligent systems that are able to act, not just in a 'toy' environment but in a real-world one, is far more difficult than anticipated. As a result, the AI research shifted toward studying each function of an intelligent system in isolation from each other. This led to parallel development of several branches of AI, such as natural language processing, theorem proving, planning and problem solving, learning, and vision. Progress achieved in each of these areas has brought the state of the art in AI to a point where building systems which exhibit several cognitive functions associated with human intelligence is now feasible. Such systems are called intelligent agents. There is, as one would expect, a great variety of such systems, each with different capabilities. This diversity caused a heated debate on which of them should be called agents. Many AI researchers have advanced their own definitions that, for most part, characterize the type of agents they are developing (Brustoloni, 1991; Maes, 1994; Smith *et al.*, 1994; Hayes-Roth, 1995; Russell and Norvig, 1995; Wooldridge and Jennings, 1995b; Franklin and Graesser, 1996). The following is a general characterization of an intelligent agent that encompasses many aspects of these definitions and indicates what we mean by an intelligent agent in the context of this book.

*An **intelligent agent** is a knowledge-based system that perceives its environment (which may be the physical world, a user via a graphical user interface, a collection of other agents, the Internet, or other complex environment); reasons to interpret perceptions, draw inferences, solve problems, and determine actions; and acts upon that environment to realize a set of* *goals or tasks for which it was designed. The agent interacts with a human or some other agents via some kind of agent-communication language and may not blindly obey commands, but may have the ability to modify requests, ask clarification questions, or even refuse to satisfy certain requests. It can accept high-level requests indicating what the user wants and can decide how to satisfy each request with some degree of independence or autonomy, exhibiting goal-directed behavior and dynamically choosing which actions to take, and in what sequence. It can collaborate with its user to improve the accomplishment of his/her tasks or can carry out such tasks on user's behalf, and in so doing employs some knowledge or representation of the user's goals or desires. It can monitor events or procedures for the user, can advise the user on how to perform a task, can train or teach the user, or can help different users collaborate.*

The behavior of the agent is based on a correspondence between the external application domain of the agent and an internal model of this domain consisting of a knowledge base and an inference engine (see Figure 1.1). The knowledge base contains the data structures representing the entities from the agent's application domain such as objects, relations between objects, classes of objects, laws and actions. The inference engine consists of the programs that manipulate the data structures in the knowledge base in order to solve the problems for which the agent was designed.

1.2   What is a learning agent?

By far the most difficult problems in developing and using an intelligent agent are the encoding of knowledge in the knowledge base (known as 'the knowledge acquisition bottleneck') and the modification of this knowledge in response to changes in the application domain or in the requirements of the agent ('the knowledge maintenance bottleneck'). *An agent that is able by itself to acquire and maintain its knowledge is called a **learning agent**.* In addition to the knowledge base and the inference engine, it contains a learning engine consisting of the programs that create and update the data structures in the knowledge base (see Figure 1.2).

The learning agent could learn from a variety of information sources in the environment. It could learn from its user or from other agents (either by being directly instructed by them

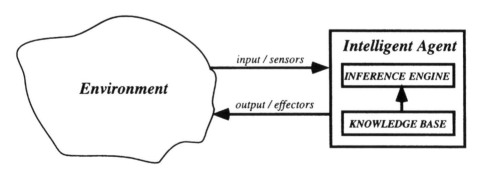

Figure 1.1   The overall architecture of an intelligent agent.

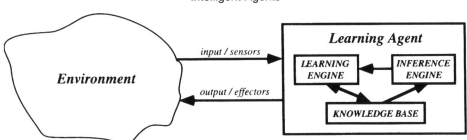

Figure 1.2   The overall architecture of a learning agent.

or just by observing and imitating their behavior), it could learn from a repository of information (such as a database), or it could learn from its own experience.

If an agent could learn, then the process of building and maintaining its knowledge base (KB) would be significantly simplified. However, building a learning engine is a very difficult task because we do not yet understand well enough the cognitive process of learning. But we know that knowledge-free learning is a hopeless enterprise in this case. Therefore, even a learning agent should be provided with some background knowledge to bootstrap learning. However, the developer of the agent need only build an initial incomplete and perhaps incorrect knowledge base because the agent will be able to extend and correct it through learning. The agent will also be able to customize the knowledge for its user and to update it to reflect changes in the environment.

Because building the agent's learning engine is such a complex task, most of the current intelligent agents are not learning agents. However, learning methods may still be used to build their knowledge bases, as will be presented in the following sections.

In the next section we will briefly present the main stages of building the knowledge base of an intelligent agent.

1.3   Phases in the development of an intelligent agent's knowledge base

As has been mentioned in the previous section, the most difficult part of developing an intelligent agent is building its knowledge base. One could distinguish three main phases in the process of building the knowledge base: *systematic elicitation of expert knowledge*, *knowledge base refinement*, and *knowledge base reformulation* (Bareiss *et al.*, 1989; Tecuci, 1992a, b; Gunning, 1996).

- During *systematic elicitation of expert knowledge* one creates the foundation knowledge of the agent by selecting the knowledge representation scheme, and developing the basic terminology and the conceptual structure of the knowledge base. The result of this elicitation is an initial incomplete and possibly partially incorrect knowledge base that is refined and improved during the next phases of knowledge base development.
- In the *knowledge base refinement* phase the knowledge base is extended and debugged. The result of the knowledge refinement phase should be a knowledge base that is

complete and correct enough to provide correct solutions to most problems to be solved by the agent.

- During the *knowledge base reformulation* phase the knowledge base is reorganized for improving the efficiency of problem solving.

The development of knowledge bases has been a central goal in two areas of AI, namely knowledge acquisition and machine learning, and has resulted in a variety of approaches in each discipline. In the next section we will review these approaches.

1.4   Approaches to knowledge base development

In the following sections we will briefly review and contrast the main knowledge acquisition, machine learning and integrated approaches to building intelligent agents.

1.4.1   Knowledge acquisition approaches

Knowledge acquisition has focused on improving and partially automating the acquisition of knowledge from human experts by knowledge engineers. The goal is to construct an accurate and efficient formal representation of the expert's knowledge. As illustrated in Figure 1.3, a knowledge engineer interacts with a domain expert to understand how the expert solves problems and what knowledge he or she uses. Then the knowledge engineer chooses the representation of knowledge, builds the inference engine, elicits knowledge from the expert, conceptualizes it and represents it into the knowledge base. This knowledge elicitation and representation process is particularly difficult because the form in which the expert expresses his or her knowledge is significantly different from how it should be represented in the knowledge base. Moreover, the expert typically fails to specify the knowledge that is common sense or implicit in human communication, but that needs to be explicitly represented in the knowledge base. After the knowledge is elicited it has to be verified by the expert and the knowledge engineer needs to make corrections in the knowledge base. This indirect transfer of knowledge, between the domain expert and the knowledge base, through the knowledge engineer, leads to a long, painful and inefficient knowledge base development process.

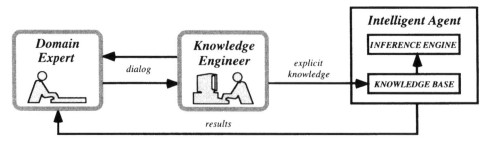

Figure 1.3   Overview of the general knowledge acquisition approach.

Much of the work in knowledge acquisition has concentrated on *systematic elicitation of expert knowledge*. It has produced techniques such as interviewing (Gammack, 1987; LaFrance, 1987), protocol analysis (Ericsson and Simon, 1984), elicitation of repertory grids (Shaw and Gaines, 1987; Boose and Bradshaw, 1988; Gaines and Shaw, 1992a, b), domain modeling (Wielinga *et al.*, 1992; Breuker and Van de Velde, 1994), and guided elicitation for role-limiting problem-solving methods (Marcus, 1988).

Knowledge acquisition research on *knowledge base refinement* has been oriented toward building refinement tools for an expert and a knowledge engineer. In general, the expert uses the problem-solving abilities of the agent to solve typical problems for which the solutions are already known in order to identify the need for additional knowledge (e.g., when the agent cannot solve a problem) or errors in the knowledge base (e.g., when the agent proposes an incorrect solution to some problem). Many tools developed in mainstream knowledge acquisition research provide an inference engine, a representation formalism in which the knowledge base can be encoded, and mechanisms for eliciting, verifying and revising knowledge expressed in that formalism. These tools trade power (i.e., the assistance given to the expert) against generality (i.e., their domain of applicability), covering a large spectrum, with power at one end and generality at the other end. At the power end of the spectrum there are tools customized to a problem-solving method and a particular domain (Musen, 1993). At the generality end are the tools applicable to a wide range of tasks or domains (Davis, 1979; Gruber, 1993a, b; Farquhar *et al.*, 1996). In between are tools that are method-specific and domain independent (Chandrasekaran, 1986; McDermott, 1988).

1.4.2 Machine learning approaches

Machine learning has focused on developing autonomous algorithms for acquiring knowledge from data and for knowledge compilation and organization. The goal is to improve a knowledge base agent's competence and/or efficiency in problem solving. Verification of the learned knowledge is through experimental testing on data sets different from those on which learning was performed.

Figure 1.4 gives an overview of how an agent's knowledge base would be learned from a database of examples using a general machine-learning approach. Notice that the learning system may or may not be part of the agent. In the former case, the agent would be a learning one.

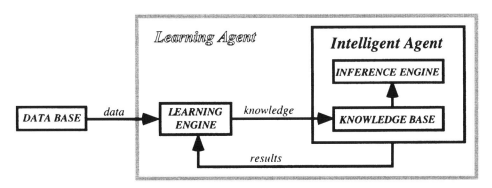

Figure 1.4 Overview of the general machine learning approach.

While *systematic elicitation of expert knowledge* is a main concern of knowledge acquisition research, it is largely ignored in machine-learning research. Machine-learning research has assumed that there already exists a representation language and background knowledge before any learning will take place.

Machine-learning research has produced several basic learning strategies such as:

- *empirical inductive learning from examples* – learning the definition of a concept by comparing positive and negative examples of the concept in terms of their similarities and differences, and inductively creating a generalized description of the similarities of the positive examples (Mitchell, 1978; Michalski, 1983; Quinlan 1986a);
- *explanation-based learning* – learning an operational definition of a concept by proving that an example is an instance of the concept and by deductively generalizing the proof (Mitchell *et al.*, 1986; DeJong and Mooney, 1986; Mooney and Bennet, 1986);
- *analogical learning* – learning new knowledge about an entity by transferring it from a similar entity (Winston, 1980; Gentner, 1983; Prieditis, 1988; Falkenhainer *et al.*, 1989; Russell, 1989);
- *abductive learning* – hypothesizing causes based on observed effects (Josephson *et al.*, 1987; O'Rorke *et al.*, 1990; Josephson and Josephson, 1994);
- *conceptual clustering* – classifying a set of objects into different classes/concepts and learning a description of each such class/concept (Michalski and Stepp, 1983; Fisher and Langley, 1985; Fisher, 1987; Kodratoff and Tecuci, 1988b);
- *quantitative discovery* – discovering a quantitative law relating values of variables characterizing an object or a system (Langley *et al.*, 1987);
- *reinforcement learning* – updating the knowledge, based on feedback from the environment (Barto *et al.*, 1981; Sutton, 1988; Watkins, 1989; Kaelbling, 1990);
- *genetic algorithm-based learning* – evolving a population of individuals over a sequence of generations, based on models of heredity and evolution (Goldberg, 1989; DeJong, 1998);
- *neural network learning* – evolving a network of simple units to achieve an input–output behavior based on a simplified model of the brain's dendrites and axons (Rumelhart and McClelland, 1986).

Most of these methods extend the knowledge base with new concept descriptions or facts, or refine some elements of the knowledge base. Explanation-based learning is intended to optimize some elements of the knowledge base to increase the problem-solving efficiency of the agent. While each of these single-strategy learning methods performs a very limited refinement or optimization operation on the knowledge base, these types of operations are complementary and many of them are necessary to build the knowledge base. The next section presents more advanced machine-learning approaches that combine several such single-strategy learning methods and can perform a more comprehensive refinement or optimization of the knowledge base.

1.4.3   Multistrategy learning approaches

Each of the learning methods mentioned in the previous section can be characterized in terms of the input from which the agent learns, the *a priori* knowledge the agent needs to have in order to learn, the inferences performed by the agent during learning, what is actually learned, and the effect learning has on agent's knowledge base. Table 1.1 contains a brief description of some of the basic learning methods in terms of these features.

Table 1.1 Complementary nature of some basic machine learning strategies

Type of learning	Short intuitive definition	Input information needed	A priori knowledge needed	Type of inference	Learned knowledge	Result of learning
Empirical inductive	Learning a concept from a set of positive (and negative) examples, by comparing the examples in terms of their similarities and differences, and creating a generalized description of the similarities of the positive examples	A set of positive and negative examples of the concept to be learned	Little or none	Induction	Concept definition	KB refinement
Explanation-based	Learning a concept description from a single example E by explaining why E is an example of the concept	One example of the concept to be learned	Complete KB to explain the input example	Deduction	Operational concept definition	KB optimization
Analogical	Acquiring new knowledge about a target entity by transferring knowledge from a known similar entity	A target entity	A source entity similar to the target	Analogy	New knowledge about target	KB refinement
Abductive	Adopting a hypothesis that explains an observation	An observation which does not follow deductively from the KB	Incomplete KB to partially explain the observation (most useful if it is causal knowledge)	Abduction	New knowledge in KB	KB refinement
Multistrategy	Integrating two or more learning strategies in order to perform learning tasks that are beyond the capabilities of the individual learning strategies which are integrated	Any combination of the above	No knowledge or incomplete KB or incorrect KB or complete KB or any combination of these features	Any combination of the above	Any combination of the above	KB refinement and/or optimization

For instance, in the case of empirical inductive learning from examples, in which the primary type of inference is inductive generalization/specialization, the input may consist of *many (positive and/or negative) examples* of some concept C, the knowledge base usually contains only a *small amount of knowledge* related to the input, and the goal is to *learn a description of C* in the form of an inductive generalization of the positive examples which does not cover the negative examples. This description extends or refines the knowledge base and may improve the competence of the agent, that is, improve its ability to solve a larger class of problems and to make fewer mistakes.

In the case of explanation-based learning, in which the primary type of inference is deduction, the input may consist of only *one example* of a concept C, the knowledge base should contain *complete knowledge about the input*, and the goal is to *learn an operational description of C* in the form of a deductive generalization of the input example. This description is a reorganization of some knowledge pieces from the knowledge base and may improve the problem solving efficiency of the agent.

Both analogical learning and abductive learning extend the knowledge base with new pieces of knowledge and usually improve the competence of the agent. In the case of analogical learning, the input may consist of a *new entity I*, the knowledge base should contain an *entity S which is similar to I*, and the goal is to *learn new knowledge about the input I* by transferring it from the known entity *S*. In abductive learning, the input may be *a fact F*, the knowledge base should contain *causal knowledge related to the input* and the goal is to *learn a new piece of knowledge* that would account for the input.

This brief characterization of the learning tasks of different single-strategy learning methods shows that these methods have a limited applicability because each requires a special type of input and background knowledge and learns a specific type of knowledge. On the other hand, the complementary nature of these requirements and results naturally suggests that by properly integrating the single-strategy methods, one can obtain a synergistic effect in which different strategies mutually support each other and compensate for each other's weaknesses.

A large number of multistrategy learning agents that integrate various learning strategies have already been developed (Michalski, 1993; Tecuci, 1993b; Michalski and Tecuci, 1994). Some integrate empirical induction with explanation-based learning, while others integrate symbolic and neural net learning, or deduction with abduction and analogy, or quantitative and qualitative discovery, or symbolic and genetic algorithm-based learning, and so on.

There are several general frameworks for the design of multistrategy learning agents. One such framework consists of a global control module and a toolbox of single-strategy learning modules, all using the same knowledge base (Pazzani, 1988; Morik, 1994; Veloso *et al.*, 1995). The control module analyzes the relationship between the input and the knowledge base and decides which learning module to activate.

Another framework consists of a cascade of single-strategy learning modules, in which the output of one module is an input to the next module (Lebowitz, 1986; Flann and Dietterich, 1989; Hirsh, 1989; Shavlik and Towell, 1990; Whitehall, 1990; Gordon and Subramanian, 1993; Danyluk, 1994; Mooney and Ourston, 1994).

Yet another framework consists of integrating the elementary inferences (like deduction, analogy, abduction, generalization, specialization, abstraction, concretion, etc.) that characterize the individual learning strategies, and thus achieving a deep integration of these strategies (Bergadano and Giordana, 1990; Genest *et al.*, 1990; Tecuci, 1991, 1994; Widmer, 1994).

Most of the developed multistrategy learning methods refine the knowledge base of a simple classification agent to correctly classify a given set of positive and negative examples (Cohen, 1991; Baffes and Mooney, 1993; Saitta and Botta, 1993; Towell and Shavlik, 1994).

1.4.4 Complementary nature of machine learning and knowledge acquisition

As shown in the previous sections, knowledge acquisition and machine learning share the same goal of developing a knowledge base but their approaches are complementary in many respects, as summarized in Table 1.2. Together they cover all three phases of the knowledge base development process.

As mentioned above, knowledge acquisition research has mainly addressed the phases of knowledge elicitation and knowledge base refinement, while machine-learning research has mainly addressed knowledge base refinement and reformulation.

The primary source of knowledge for the knowledge acquisition methods is the domain expert. The knowledge is elicited using a variety of interactive techniques. On the other hand, machine-learning methods assume a pre-existing repository of examples or data and use autonomous methods to refine and/or optimize it.

The most important aspect of these complementary approaches is that many of the hard problems in one are significantly easier in the other, and can therefore be solved more easily by employing the techniques from the other area. Representative examples of hard problems in machine learning are the problem of new terms, the credit/blame assignment problem, and the definition of the learner's representation language, background knowledge, input examples, and bias. The difficulty of these problems can be alleviated by employing

Table 1.2 Complementary nature of machine learning and knowledge acquisition

Type of research	Main phases of KB development addressed	Primary source of knowledge	Main feature of the method	Hardest problems
Knowledge acquisition	Knowledge elicitation KB refinement	Domain expert	Interactive	Definition of general knowledge that characterizes specific examples Verification of KB consistency and reduction of any inconsistencies Reorganization of KB to improve problem-solving efficiency
Machine learning	KB refinement KB reformulation	Data	Autonomous	The problem of new terms The credit/blame assignment problem Definition of the learner's representation language, background knowledge, input examples, and bias

methods for systematic elicitation of knowledge and by involving the expert in the learning loop.

Representative examples of hard problems in knowledge acquisition are the definition of general problem-solving knowledge that characterizes specific examples, the verification of knowledge base consistency and reduction of any inconsistencies, and the reorganization of the knowledge base to improve the problem-solving efficiency of the agent. These are precisely the kinds of problems that have been successfully addressed by machine learning research.

As knowledge acquisition attempts to automate more of the knowledge base development process and machine learning attempts to automatically develop more complex knowledge bases, there is an increasing realization in these two research communities that a better approach to building knowledge bases for intelligent agents is through an integration of different knowledge acquisition and learning methods (Buchanan and Wilkins, 1993; Tecuci et al., 1994; Tecuci and Kodratoff, 1995). Some of the most representative integrated approaches are presented in the next section.

1.4.5   Integrated machine learning and knowledge acquisition approaches

An integrated machine-learning and knowledge acquisition (ML & KA) approach to the problem of building the knowledge base of an intelligent agent can take advantage of their complementary natures, using machine learning techniques to automate the knowledge acquisition process and knowledge acquisition techniques to enhance the power of the learning methods. As illustrated in Figure 1.5, in such an approach the expert interacts with the knowledge-based system via an ML & KA component which performs most of the functions of the knowledge engineer. It allows the expert to communicate expertise in a way familiar to him/her and is responsible for building, updating and reorganizing the knowledge base.

Notice that, as in the case of the learning approaches (see Figure 1.4), the learning component may or may not be part of the agent. In the former case, the agent would be a learning one.

Some of the most representative approaches to the integration of machine learning and knowledge acquisition are:

• *Methods integrating elicitation of repertory grids with empirical inductive learning.* Examples of systems or learning agents implementing such methods are AQUINAS

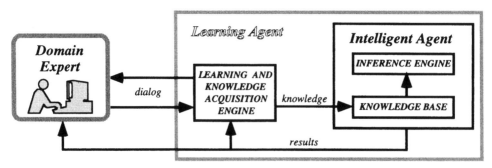

Figure 1.5 Overview of the integrated machine-learning and knowledge acquisition approach.

(Boose and Bradshaw, 1988) and KSS0 (Gaines and Shaw, 1992a and b) that elicit repertory grids from experts and then learn concepts and rules from these grids through conceptual clustering and empirical induction. The main strength of this approach is that the repertory grids can be easily elicited from an expert, through a very natural dialog. Also, other concepts and inference rules can be automatically learned from repertory grids. The main weakness is that more complex knowledge structures are difficult to generate from repertory grids since the grids are oriented toward representing declarative attribute-based knowledge.

- *Case-based reasoning and learning.* These are methods that integrate knowledge elicitation and refinement of cases with inductive generalization and automatic (re-)organization of cases. Examples of case-based reasoning agents are CABINS (Sycara and Miyashita, 1995), and CREEK (Aamodt, 1995). PROTOS (Porter *et al.*, 1990) is a case-based agent that also integrates explanation-based learning. The main strengths of this approach are that cases can be easily elicited from experts, that elicitation and learning are naturally integrated with problem solving, and that it can be used in domains that are more difficult to formalize. The main weakness, however, is its dependence on some similarity measure that is not easy to define.

- *Example-guided knowledge base revision.* These are methods to improve a knowledge base to become consistent with a given set of examples. Knowledge base revision may be explicitly controlled by the expert, as in the case of SEEK (Ginsberg *et al.*, 1989), or may be implicitly controlled by the expert (through predefined biases), as in the case of PTR (Koppel *et al.*, 1995). In principle, this approach shows a natural integration of knowledge acquisition and machine learning where knowledge acquisition methods would be used to define an initial imperfect knowledge base and machine learning methods would be used to improve the knowledge base by using the information contained in the examples. However, the current methods do not, in fact, address the problem of building the initial knowledge base. The main advantage of this approach is that it provides a way to integrate knowledge from the examples into the knowledge base and such examples are easily available in some applications.

- *Learning apprentices.* These are interactive knowledge-based consultants able to assimilate new knowledge by observing and analyzing the problem-solving steps contributed by their expert users through their normal use of the agents. Examples of such agents are LEAP (Mitchell *et al.*, 1990), Odysseus (Wilkins, 1990), CLINT (De Raedt, 1993), and CAP (Mitchell *et al.*, 1994). These agents exploit the complementary characteristics of ML & KA. The expert provides concrete examples from which the apprentice learns general rules and concepts. Because the expert verifies the knowledge base agent's performance, the resultant knowledge base is already verified. Learning and problem solving are naturally integrated. This approach is very useful for building personal assistants. A weakness of this approach is that it requires the agents to have an initial knowledge base in order to be able to learn.

- *Programming by Demonstration* (Cypher, 1993; Schlimmer and Hermens, 1993; Lieberman, 1994; Smith *et al.*, 1994; Maulsby and Witten, 1995; Atkeson and Shaal, 1997). This refers to a type of programming in which an agent is taught how to perform a task by watching an end user as he/she demonstrates the task. It emphasizes the interaction with the end user within a graphical user interface, as well as interactive methods of teaching and learning. Programming-by-demonstration agents typically learn from only a few examples and attempt to minimize the instruction effort of the user. However, the current methods place more emphasis upon the graphical user interface, while the ML & KA methods are *ad hoc*.

The next chapter will give a general description of the Disciple approach to building intelligent agents, an apprenticeship multistrategy learning approach that synergistically integrates many basic ML & KA techniques in order to take advantage of their strengths and to compensate for each other's weaknesses.

2
General Presentation of the Disciple Approach for Building Intelligent Agents

This chapter gives a general description of the Disciple approach for building intelligent agents and briefly illustrates it with two cases: an agent developed by an expert to be his/her assistant and an agent developed by an expert to be used by non-expert users.

2.1   An overview of the Disciple approach

Disciple is an apprenticeship, multistrategy learning approach for developing intelligent agents where an expert teaches the agent how to perform domain-specific tasks in a way that resembles the way the expert would teach an apprentice, by giving the agent examples and explanations as well as by supervising and correcting its behavior. The Disciple approach eliminates, or significantly reduces, the involvement of the knowledge engineer in the process of building the agent. It also creates a natural environment for knowledge acquisition from the human expert.

The Disciple approach is implemented in the Disciple learning agent shell. We define a *learning agent shell* as consisting of a learning engine and an inference engine that support a representation formalism in which a knowledge base can be encoded, as well as a methodology for automatically building the knowledge base. The architecture of the Disciple learning agent shell is presented in Figure 2.1.

The Disciple shell is a system with advanced knowledge acquisition and learning capabilities, with some elementary problem-solving capabilities, and an empty knowledge base. The knowledge acquisition and learning component implements a wide range of complementary knowledge acquisition and learning techniques that allow an expert to teach the agent in a natural way, as mentioned above. The problem-solving component performs only some domain-independent operations like network matching or rule matching. The shell also has a domain-independent graphical user interface, and a knowledge base manager to access and modify the knowledge in the knowledge base.

Typically, the Disciple shell is customized for a specific application domain and then taught by the expert to solve problems in that domain. To customize the Disciple shell the software/knowledge engineer develops a domain-specific interface on top of the domain-independent graphical user interface of the shell. This domain-specific interface will allow the domain experts to communicate with the agent as close as possible to the way they communicate in their environment. In addition to this, the software/knowledge engineer may also need to develop a domain-specific problem-solving component by extending the existing problem-solving component of the agent. The knowledge elicitation and learning component of the Disciple shell is general and the only customization that may be necessary will be the setting of certain parameters.

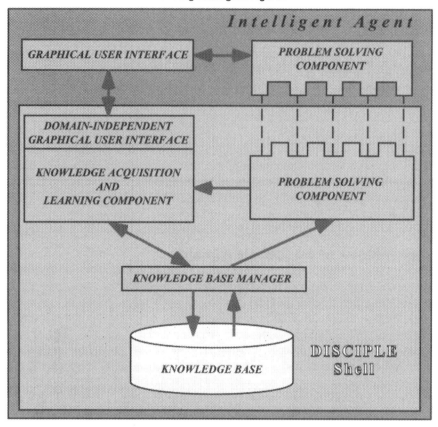

Figure 2.1   General architecture of the Disciple learning agent shell.

Once the customized agent has been built, the expert interacts directly with the agent to define an initial knowledge base by engaging in a knowledge elicitation dialog in which the agent will guide the expert to provide whatever knowledge he or she can easily express. Alternatively, the agent's knowledge base could be initialized from a library of reusable ontologies, common domain theories, and generic problem-solving strategies (Neches *et al.*, 1991; Genesereth and Ketchpel, 1994) or it could be initialized from an existing knowledge base, such as the CYC knowledge base (Lenat, 1995b). In any case, this initial knowledge base of the agent, built mainly through knowledge elicitation techniques, is assumed to be incomplete and partially incorrect.

After the initial knowledge base has been developed, the agent is taught by the expert how to perform domain-specific tasks, through apprenticeship and multistrategy learning. For instance, the expert may teach the agent how to solve a certain type of problem by providing a concrete example, helping the agent to understand the solution, supervising the agent as it attempts to solve analogous problems, and correcting its mistakes. Through only a few such natural interactions, the agent will be guided into learning complex problem-solving rules, and into extending and correcting its knowledge base. This process is based on a cooperation between the agent and the human expert in which the agent helps the expert to express his/her knowledge using the agent's representation language, and the expert guides the learning actions of the agent.

The central idea of the Disciple approach is to facilitate the agent-building process by the use of synergism at three different levels. First, there is synergism between different learning methods employed by the agent. By integrating complementary learning methods (such as inductive learning from examples, explanation-based learning, learning by analogy, learning by experimentation) in a dynamic, task-dependent way, the agent is able to learn from the human expert in situations in which no single-strategy learning method would be sufficient. Second, there is synergism between teaching (of the agent by the expert) and learning (from the expert by the agent). For instance, the expert may select representative examples to teach the agent, may provide explanations, and may answer agent's questions. The agent, on the other hand, will learn general rules that are difficult to be defined by the expert, and will consistently integrate them into its knowledge base. Third, there is synergism between the expert and the agent in solving a problem. They form a team in which the agent solves the more routine but labor-intensive parts of the problem and the expert solves the more creative ones. In the process, the agent learns from the expert, gradually evolving toward an 'intelligent' agent.

There are two main types of agents that could be developed: an assistant to the expert or an agent to be used by non-expert users. An example of an assistant to the expert is a manufacturing assistant that collaborates with the expert in designing plans for loud-speaker manufacturing. In such a case the expert is both the developer and the user of the agent. After an initial phase of training where the expert teaches the agent basic knowledge about the application domain, the expert and the agent start solving problems in collaboration and the agent continues to improve itself by learning from the expert's contributions to the joint problem-solving process. In this way the agent becomes a better and better assistant, taking over more of the problem-solving process and leaving to the expert the task of overseeing and of making creative decisions. An informal description of the process of developing this agent is given in Section 2.2. More details are given in the following chapters.

An example of an agent developed by an expert for non-expert users is an assessment agent that is taught by an educational and history expert to interact with history students (its non-expert users). The agent will generate tests to be solved by students, and could also provide them with hints, answers and explanations. In the case of such an agent, the expert develops the agent and then delivers it to its non-expert users. An informal description of the process of developing this agent is given in Section 2.3. A detailed description is given in Chapter 6.

During its lifetime the agent may need to be retrained and possibly redeveloped. The retraining phase is not distinct from initial training in the case where the agent is an assistant to its expert developer. Indeed, training is a continuous process. However, if changes are needed in the way the agent learns and solves problems, then it has to be redeveloped by the software/knowledge engineer. In the case of an agent used by non-expert users, training and retraining are distinct phases. For instance, if the history curriculum taught in schools has been updated, then the agent would have to be retrained by the educational and history expert.

In the following sections we will briefly characterize the development of the two types of agents mentioned above. Both of them will be described in greater detail in later chapters.

2.2 Building a manufacturing assistant

In this section we will intuitively present the process of building, training and using an agent that acts as an assistant to an expert in designing plans for loudspeaker manufacturing. This process follows the stages presented in the previous section.

The first stage of building an agent is to customize the general Disciple shell into an expert and learning shell for loudspeaker manufacturing, by building a specialized problem solver. The problem solver of the manufacturing assistant implements a problem reduction approach to problem solving. In this approach a problem is solved by successively reducing it to simpler subproblems. This process continues until the initial problem is reduced to a set of elementary problems, that is, problems with known solutions. In this way, the task of manufacturing a loudspeaker, for instance, is reduced to a set of elementary actions which represents precisely the plan to manufacture the loudspeaker. This approach is similar to those of PLANX10 (Sridharan and Bresina, 1983), NOAH (Sacerdoti, 1977), NONLIN (Tate, 1977), SIPE (Wilkins, 1988), TWEAK (Chapman, 1987) and others (Allen *et al.*, 1990). This problem solver requires that the knowledge base of the manufacturing assistant contains two main types of knowledge. The first is knowledge about the object instances and concepts from the application domain. They include descriptions of the various components of a loudspeaker, as well as of the materials and instruments used in loudspeaker manufacturing. These instances and concepts are organized in a semantic network. The second type of knowledge is rules for loudspeaker manufacturing. In general, such a rule indicates the conditions under which a general manufacturing operation could be decomposed into several, lower level operations.

After the Disciple shell is customized into a manufacturing assistant shell, the expert develops an initial knowledge base by expressing whatever useful knowledge he/she can easily express. To this purpose the expert interacts with the agent through the editing and browsing modules of the knowledge acquisition and learning component. Since object descriptions are much easier to express than problem-solving rules, one would expect that the initial knowledge base will consist mainly of a semantic network of object instances and concepts. Moreover, this semantic network will most likely be incomplete and possibly even partially incorrect. However, it is guaranteed to be consistent, because the use of the agent's editing modules will not allow inconsistent knowledge to be introduced.

A sample of the initial knowledge base developed by the expert is presented in Figure 2.2 (each unlabelled link is a link that relates a concept with a more general concept).

Next, the agent is used to interactively solve problems, according to the following scenario: The expert gives the agent the problem to solve and the problem-solving component starts solving this problem by showing the expert all the problem-solving steps. The expert may agree with them or reject them. Therefore, during the course of its functioning as a problem solver, the agent may encounter two situations:

- The current problem-solving step (which we shall call the partial solution) is accepted by the expert. In this case, no learning will take place.
- The agent is unable to propose any partial solution to the current problem (or the solution it proposes is rejected by the expert). Then the agent requests a solution from the expert. Once this solution is given, a learning process will take place. The agent will try to learn a general rule so that, when faced with problems similar to the current one (which it has been unable to solve), it will become able to propose a solution similar to the solution given by the expert to the current problem. In this way, the agent progressively evolves from a helpful assistant in problem solving to an expert.

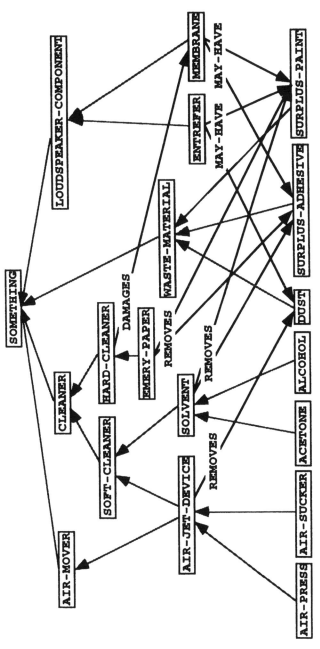

Figure 2.2 A sample of the initial semantic network developed by the manufacturing expert.

```
Perform the task
MANUFACTURE OBJECT LOUDSPEAKER

by performing the tasks

   1. MAKE OBJECT CHASSIS-ASSEMBLY

      Perform this task by performing the tasks
         1.1 FIX OBJECT CONTACTS ON CHASSIS
         1.2 MAKE OBJECT MECHANICAL-CHASSIS
         1.3 FINISH-OPERATIONS ON ENTREFER

         Perform this task by performing the tasks
            1.3.1 CLEAN OBJECT ENTREFER OF DUST
            1.3.2 VERIFY OBJECT ENTREFER

   2. MAKE OBJECT MEMBRANE

   3. ASSEMBLE OBJECT CHASSIS-ASSEMBLY WITH MEMBRANE

      Perform this task by performing the tasks
         3.1 ATTACH OBJECT MEMBRANE TO CHASSIS-ASSEMBLY
         3.2 ATTACH OBJECT RING TO CHASSIS-MEMBRANE-ASSEMBLY

         Perform this task by performing the tasks
            3.2.1 APPLY OBJECT MOWICOLL ON RING
            3.2.2 PRESS OBJECT RING ON CHASSIS-MEMBRANE-ASSEMBLY

   4. FINISH-OPERATIONS ON LOUDSPEAKER
```

Figure 2.3 Problem-solving operations: decompositions of tasks into simpler subtasks.

Let us suppose that the agent receives the problem of designing a plan for manufacturing a loudspeaker:

MANUFACTURE OBJECT LOUDSPEAKER

It will try to solve this problem by successive decompositions and specializations of the complex operation of manufacturing the loudspeaker into simpler operations, then better defining these simpler operations by choosing tools, materials, or verifiers, which are in turn successively refined, as illustrated in Figures 2.3 and 2.4. The agent will combine such decompositions and specializations, building a problem-solving tree like the one in Figure 2.5. This process continues until all the leaves of the tree are elementary tasks. Elementary tasks are tasks that will be performed by the entity manufacturing the loudspeaker.

The above tree is a standard AND tree, the solution to the problem from the top of this tree consisting of the leaves of the tree. That is, to manufacture the loudspeaker, one has to perform the sequence of operations indicated in Figure 2.6.

When the agent is unable to solve a problem, it asks for a solution from the expert. Once this solution is given, a learning process takes place in which the agent will learn a general problem-solving rule covering the present problem-solving episode. To illustrate this situation let us consider again the problem-solving tree in Figure 2.5 and let us assume that the agent was not able to further reduce the task 'CLEAN OBJECT ENTREFER OF DUST'. In this case the reduction from the problem-solving tree had to be given by the expert, as shown in Figure 2.4. Starting from this particular problem-solving episode the agent learns the general reduction rule in Figure 2.7. This rule allows the agent to reduce new 'CLEAN' tasks. For instance, it is now able to reduce the task

CLEAN OBJECT LOUDSPEAKER OF SURPLUS-ADHESIVE

to the task

CLEAN OBJECT LOUDSPEAKER OF SURPLUS-ADHESIVE WITH ACETONE

In this way the agent learns new knowledge and improves its problem-solving abilities.

In the following we will illustrate the process of learning the rule in Figure 2.7. The agent balances its limited knowledge by using a multistrategy learning method whose power comes from the synergism between different learning strategies: explanation-based learning, learning by analogy and experimentation, learning from examples, and learning by questioning the user.

First the agent tries to understand why the problem solving episode in Figure 2.4 is correct. It uses heuristics to propose a set of plausible partial explanations from which the expert has to select the correct ones, as indicated in Figure 2.8. The explanations selected by the expert are marked by an '*'. The expert may also indicate other pieces of explanations.

```
Perform  the  task
      CLEAN  OBJECT  ENTREFER  OF  DUST
by performing the  task
      CLEAN  OBJECT  ENTREFER  OF  DUST  WITH  AIR-SUCKER
```

Figure 2.4 Problem-solving operation: specialization of a task.

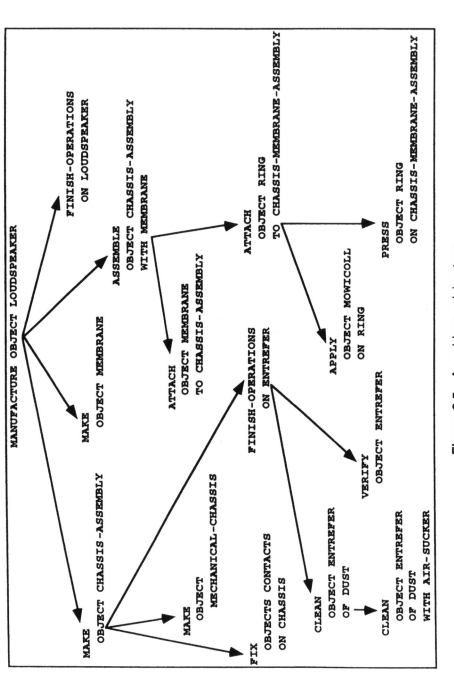

Figure 2.5 A problem-solving tree.

Perform the task
    MANUFACTURE   OBJECT   LOUDSPEAKER

by performing the tasks
    FIX OBJECTS CONTACTS ON CHASSIS
    MAKE OBJECT MECHANICAL-CHASSIS
    CLEAN OBJECT ENTREFER OF DUST WITH AIR-SUCKER
    VERIFY OBJECT ENTREFER
    MAKE OBJECT MEMBRANE
    ATTACH OBJECT MEMBRANE TO CHASSIS-ASSEMBLY
    APPLY OBJECT MOWICOLL ON RING
    PRESS OBJECT RING ON CHASSIS-MEMBRANE-ASSEMBLY
    FINISH-OPERATIONS ON LOUDSPEAKER

Figure 2.6 A solution obtained through problem reduction.

Next the agent generalizes this explanation to an analogy criterion which will allow it to generate problem-solving episodes analogous to the problem-solving episode in Figure 2.4. The agent also generates an initial form of the general problem-solving rule to be learned. This rule is shown in Figure 2.9. When compared with the rule in Figure 2.7, one can notice that it has two conditions instead of a single applicability condition. The plausible lower-bound condition corresponds to the explanation from Figure 2.8 and is a very specific condition. The plausible upper bound condition is an analogy criterion.

During the next stages of learning the agent uses analogical reasoning to generate examples that are similar to the one in Figure 2.4 and shows them to the expert who has to accept or to reject them, thus characterizing them as positive or negative examples of the rule to be learned. Then, the agent changes the conditions of the rule to become consistent with these examples. In the end, both conditions will converge toward the single condition shown in Figure 2.7.

To generate an example, the agent looks in the knowledge base for objects satisfying an explanation similar to the explanation of the problem-solving episode indicated by the expert (see Figure 2.8) and finds that ALCOHOL REMOVES SURPLUS-ADHESIVE and MEM-BRANE MAY-HAVE SURPLUS-ADHESIVE (see Figure 2.2). Based on this, the agent generates the problem-solving episode in Figure 2.10. Because the generated example is accepted by the expert the plausible lower bound condition of the rule in Figure 2.9 is generalized to cover it.

Similarly, the agent finds that EMERY-PAPER REMOVES SURPLUS-ADHESIVE and MEMBRANE MAY-HAVE SURPLUS-ADHESIVE, and generates the problem-solving episode from the top of Figure 2.11. Because this problem-solving episode is rejected by the expert, the agent attempts to 'understand' why it is wrong. Then it tries to elicit the corresponding explanation of why the initial problem-solving episode in Figure 2.4 is correct. This additional piece of explanation (AIR-SUCKER NOT-DAMAGE ENTREFER) is used to update the rule being learned. The learning process decreases the distance between the two conditions of the rule until the agent learns the rule in Figure 2.7.

Details about the processes behind these interactions are provided in the following chapters. What is important here is that through easy and natural interactions with the human expert, the agent is able to learn problem-solving rules which would be difficult to be manually defined by the expert.

```
IF
?C          IS          CLEAN           ; If the task to perform is
            OBJECT                      ; clean ?X of ?Y and
            OF          ?X
                        ?Y

?X          IS          SOMETHING       ; ?X is something
            MAY-HAVE    ?Y              ; that may have ?Y and
            MADE-OF     ?M              ; is made of ?M and

?Y          IS          WASTE-MATERIAL  ; ?Y is some waste material and

?M          IS          MATERIAL        ; ?M is some material and

?Z          IS          CLEANER         ; ?Z is a cleaner
            REMOVES     ?Y              ; that removes ?Y
            NOT-DAMAGE  ?M              ; without damaging ?M

THEN
SPECIALIZE  TASK        ?C              ; Then perform the task
            TO          ?C1

?C1         IS          CLEAN           ; clean ?X of ?Y with ?Z
            OBJECT
            OF          ?X
                        ?Y
            WITH        ?Z
```

Figure 2.7   A reduction rule learned from an example indicated by the expert.

```
Select   the   explanations   of   the   reduction

     Perform   the   task
             CLEAN   OBJECT   ENTREFER   OF   DUST
     by  performing  the  task
             CLEAN   OBJECT   ENTREFER   OF   DUST   WITH   AIR-SUCKER

   *  ENTREFER   MAY-HAVE   DUST   ∧   AIR-SUCKER   REMOVES   DUST
      ENTREFER   MAY-HAVE   SURPLUS-PAINT
```

Figure 2.8 Agent–expert dialog for understanding a problem-solving episode.

2.3 Building an assessment agent for higher-order thinking skills

In this section we will intuitively present the process of building, training and using an educational agent. This agent generates history test questions to assess student's understanding and use of the higher-order thinking skills of evaluating historical sources for relevance, credibility, consistency, ambiguity, bias, and fact *vs* opinion. This agent is representative of the class of agents built by an expert to be used by non-experts. The agent can dynamically generate a test question, based on a student model, together with the answer, hints and explanations. It can be used by teachers to test the students, or it can be used by students to learn and test themselves.

Figure 2.12 shows a test question generated by the assessment agent. The student is asked to imagine that he or she is a reporter and has been assigned the task to write an article for *Christian Recorder* during the pre-Civil War period on slave life. He or she has to analyze the historical source 'Slave Cultivating Sugar Cane' in order to determine whether it is relevant to this task. In the situation illustrated in Figure 2.12 the student answered correctly. Therefore the agent confirms the answer and provides an explanation for it, as indicated in the lower right pane of the window.

There are several other types of test questions that the agent can generate. They will be presented in Chapter 6, together with a detailed description of the process of building this agent. In the following we will only intuitively illustrate this process.

First, the knowledge engineer customizes the general Disciple learning agent shell into a specific assessment agent learning shell, by defining several interface modules. One is the *source viewer* that allows the agent to access and display historical sources from a multimedia database. Another is a *customized example editor* that allows the expert to give examples of test questions using natural language templates. Yet another is a *test interface* that allows the agent to display test questions as shown in Figure 2.12. The knowledge engineer also builds an assessment engine that relies on example generation, a basic problem solving operation supported by the Disciple shell.

After the assessment agent learning shell has been built, the history expert or teacher interacts with it to develop its initial knowledge base and to teach it higher-order thinking skills such as how to judge the relevance of a historical source to a given task. The expert starts by choosing a historical theme (such as slavery in America) for which the agent will generate test questions. Then the expert identifies a set of historical concepts that are appropriate and necessary to be learned by the students. The expert also identifies a set of historical sources

```
Plausible Upper Bound IF
?C          IS       CLEAN           ; If the task to perform is
            OBJECT   ?X              ; clean ?X of ?Y and
            OF       ?Y

?X          IS       SOMETHING       ; ?X is something
            MAY-HAVE ?Y              ; that may have ?Y and

?Y          IS       SOMETHING       ; ?Y is something and

?Z          IS       SOMETHING       ; ?Z is something
            REMOVES  ?Y              ; that removes ?Y


Plausible Lower Bound IF
?C          IS       CLEAN           ; If the task to perform is
            OBJECT   ?X              ; clean ?X of ?Y and
            OF       ?Y

?X          IS       ENTREFER        ; ?X is entrefer
            MAY-HAVE ?Y              ; that may have ?Y and

?Y          IS       DUST            ; ?Y is dust and

?Z          IS       AIR-SUCKER      ; ?Z is air-sucker
            REMOVES  ?Y              ; that removes ?Y

THEN
SPECIALIZE  TASK     ?C
            TO       ?C1

?C1         IS       CLEAN           ; Then perform the task
            OBJECT   ?X              ; clean ?X of ?Y with ?Z
            OF       ?Y
            WITH     ?Z
```

Figure 2.9   The initial form of the rule to be learned.

```
May I perform the task
    CLEAN OBJECT MEMBRANE OF SURPLUS-ADHESIVE
by performing the task
    CLEAN OBJECT MEMBRANE OF SURPLUS-ADHESIVE WITH ALCOHOL?  Yes
```

Figure 2.10   An example generated by analogy with the one provided by the expert.

```
May I perform the task
    CLEAN OBJECT MEMBRANE OF SURPLUS-ADHESIVE
by performing the task
    CLEAN OBJECT MEMBRANE OF SURPLUS-ADHESIVE WITH EMERY-PAPER?  NO

Select the explanations of why this problem solving episode
is wrong

*   EMERY-PAPER DAMAGES MEMBRANE
    MEMBRANE MAY-HAVE SURPLUS-PAINT

Look again to the problem solving episode you have indicated
to me.

Perform the task
    CLEAN OBJECT ENTREFER OF DUST
by performing the task
    CLEAN OBJECT ENTREFER OF DUST WITH AIR-SUCKER

Could you explain why this is correct?
    AIR-SUCKER NOT-DAMAGE ENTREFER
```

Figure 2.11   Another sample dialog between the agent and the expert.

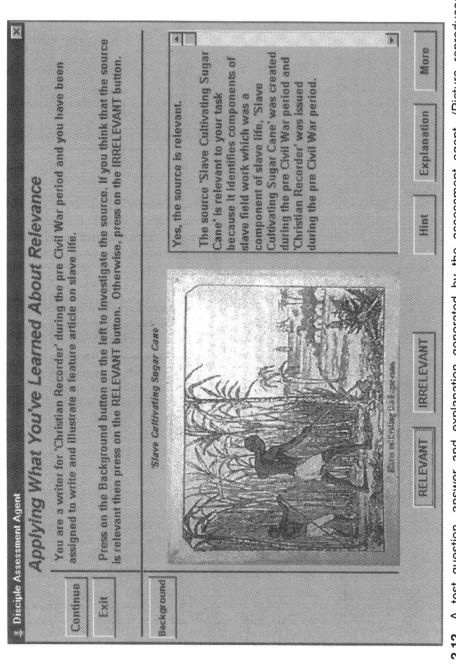

Figure 2.12 A test question, answer and explanation generated by the assessment agent. (Picture reproduced from LC-USZ62-37809, Library of Congress, Prints and Photographs Division, Civil War Photographs.)

that illustrate these concepts and will be used in test questions. All these concepts and the historical sources are represented by the history expert/teacher in the knowledge base of the assessment agent, by using the various editors and browsers of the customized learning agent shell. For instance, in the top right-hand side of Figure 2.13 one can see the source viewer that displays the historical source 'Slave Cultivating Sugar Cane'. Under it is the *concept editor* that is used to describe the historical source. 'Slave Cultivating Sugar Cane' is defined as being a painting. It illustrates the concepts slave work, farm crop, male field slave and plantation. It also identifies components of plantation and slave field work. Other information is also provided, such as the audience for which this historical source is appropriate and when it was created. The concepts used to describe the historical sources are hierarchically organized in a semantic network, as indicated in the *concept browser* window shown in the top left of Figure 2.13. For instance, slave cotton planting is defined as being a type of slave field work which, in turn, is a type of slave work. This initial knowledge base of the agent is assumed to be incomplete and even possibly partially incorrect and will be improved during the next stages of knowledge base development.

Once the initial knowledge base is built, the expert teaches the agent higher-order thinking skills such as judging the relevance of a source to a given task. During this process the agent learns reasoning rules and also extends and improves the semantic network of objects and concepts.

To teach the agent how to judge the relevance of a source to a given task, the history expert interacts with the customized example editor to provide an example of a task and a source relevant to this task, as shown in Figure 2.14. He/she does this by simply selecting values for the variables in the task description.

The agent tries to understand why the source 'Slave Cultivating Sugar Cane' is relevant to the given task by using various heuristics to propose plausible explanations from which the expert has to choose the correct ones. The expert may also provide additional explanations.

The explanations generated by the agent and accepted by the expert are shown in the bottom pane of Figure 2.15. According to these explanations, the source 'Slave Cultivating Sugar Cane' is relevant because it identifies components of slave field work which was a component of slave life, and it was created during the pre-Civil War period.

Based on the selected explanations and on the initial example shown in the top left pane of Figure 2.15, the agent automatically generates the initial rule shown in the top right pane of Figure 2.15. Next, the agent generates examples analogous to the initial example, as shown in Figure 2.16.

The example from the right-hand side of Figure 2.16 is accepted by the expert. Consequently, the plausible lower-bound condition of the rule is generalized to cover it by making the generalizations indicated in the bottom pane of the window in Figure 2.16. The generalized rule is shown in the top part of the *refine rule* window from Figure 2.16.

Continuing with such interactions, the agent finally learns the rule in Figure 2.17. Having such rules in its knowledge base, the agent can generate test questions like the one from Figure 2.13. That test question is simply an instance of the rule in Figure 2.17. The task description, hint, and explanations are the corresponding instances of the patterns associated with the rule. Other types of test questions require more complex processing, involving the use of several such rules, as will be presented in detail in Chapter 6.

The following chapters present in detail the knowledge representation, reasoning and learning mechanisms used in the Disciple shell that allow the agents to exhibit the behavior illustrated above.

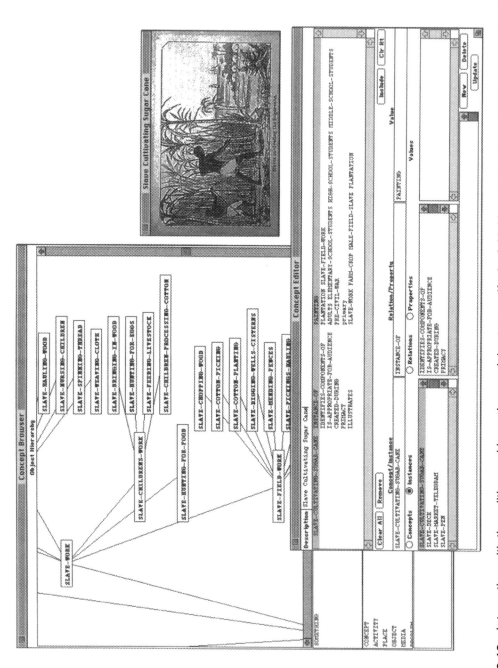

Figure 2.13 Interaction with the editing and browsing modules to specify the initial knowledge base. (Picture reproduced from LC-USZ62-37809, Library of Congress, Prints and Photographs Division, Civil War Photographs.)

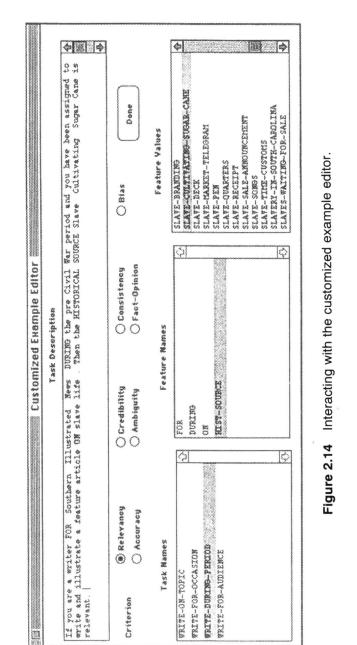

Figure 2.14 Interacting with the customized example editor.

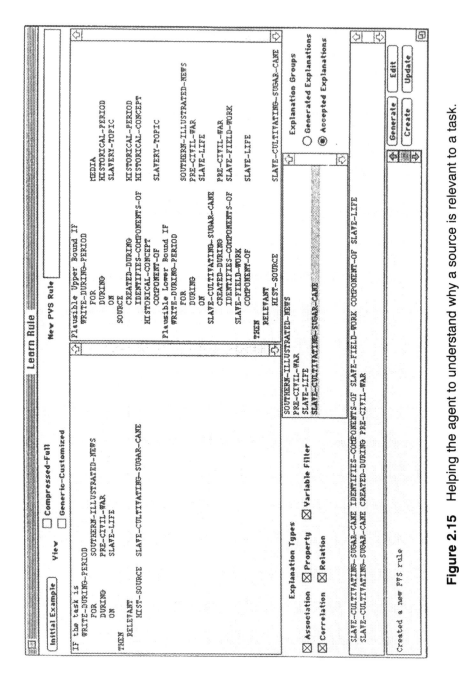

Figure 2.15 Helping the agent to understand why a source is relevant to a task.

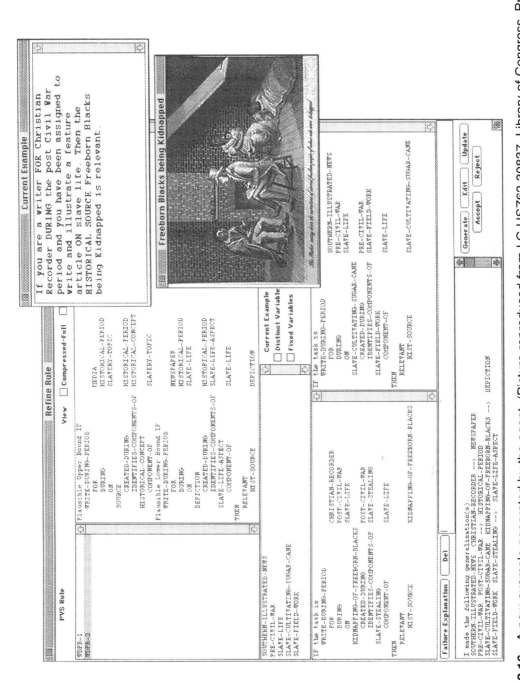

Figure 2.16 A new example generated by the agent. (Picture reproduced from LC-USZ62-30837, Library of Congress, Prints and Photographs Division, Civil War Photographs.)

```
IF
  ?W1    IS          WRITE-DURING-PERIOD
         FOR         ?S1
         DURING      ?P1
         ON          ?S2

  ?P1    IS          HISTORICAL-PERIOD

  ?S1    IS          MEDIA
         ISSUED-DURING   ?P1

  ?S2    IS          SLAVERY-TOPIC

  ?S3    IS          SOURCE
         IDENTIFIES-COMPONENTS-OF   ?S4
         CREATED-DURING   ?P1

  ?S4    IS  SLAVE-LIFE
         COMPONENT-OF   ?S2
THEN
  RELEVANT  HIST-SOURCE ?S3

Task  Description
  You  are  a  writer  for  ?S1  during  the  ?P1  period  and  you
  have  been  assigned  to  write  and  illustrate  a  feature
  article  on  ?S2.

Operation  Description
  ?S3  is  relevant

Explanation
  ?S3  identifies  components  of  ?S4  which  was  a  component  of
  ?S2,  ?S3  was  created  during  the  ?P1  period  and  ?S1  was
  issued  during  the  ?P1  period.

Hint
  To  determine  if  this  source  is  relevant  to  your  task
  investigate  if  it  identifies  components  of  something  that
  was  a  component  of  ?S4,  check  when  was  it  created,  and
  when  ?S1  was  issued.

Right  Answer
  The  source  ?S3  is  relevant  to  your  task  because  it
  identifies  components  of  ?S4  which  was  a  component  of
  ?S2,  ?S3  was  created  during  the  ?P1  period  and  ?S1  was
  issued  during  the  ?P1  period.

Wrong  Answer
  Investigate  this  source  further  and  analyze  the  hints  and
  explanations  to  improve  your  understanding  of  relevance.
  You  may  consider  reviewing  the  material  on  relevance.
  Then  continue  testing  yourself.
```

Figure 2.17   The refined rule.

3
Knowledge Representation and Reasoning

In this chapter we will describe in detail the knowledge representation and reasoning mechanisms of the Disciple agents. We will first describe the knowledge representation. Then we will describe the basic learning operations supported by this representation. Finally we will present the elementary problem-solving methods which can be used with Disciple's representation to develop a large variety of problem solvers. Although the purpose of this chapter is to provide the general representation capabilities of the Disciple agents independent of an application domain, to facilitate the understanding of these capabilities, most of the examples given are from the domain of loudspeaker manufacturing that was introduced in Chapter 2.

3.1 Knowledge representation

In this section we will describe Disciple's knowledge representation, a hybrid representation that was specially designed to support both learning and problem solving. We will start with a brief description of what is a knowledge representation and which are its main features. Then we will describe in detail the two main components of Disciple's knowledge representation, the semantic network for representing domain objects, and the rules for representing agent's problem-solving operations.

3.1.1 Introduction

3.1.1.1 *What is a knowledge representation?*

The knowledge representation of an intelligent agent is a correspondence between the external application domain and an internal symbolic reasoning system. The symbolic reasoning system is the agent's model of the external world and consists of data structures for storing information and procedures for manipulating these data structures. For each relevant element E from the agent's application domain, such as an object, a relation between objects, a distinct class of objects, a law, or an action, there is an expression $R(E)$ in the agent's domain model that represents that entity. This mapping between the elements of the application domain and those of the domain model allows the agent to reason about the application domain, by performing reasoning processes in the domain model, and transferring the conclusions back into the application domain. For instance, in order to find a solution to a problem P in the application domain, this problem is first represented as P_m in the agent's domain model (see Figure 3.1). Next the agent looks for a solution S_m of P_m in its domain model. Then the obtained solution S_m is reverse-mapped into S, which is the solution of the problem P .

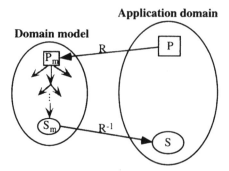

Figure 3.1 The use of the domain model for solving problems in an application domain.

A basic feature of a representation is its meaning or semantics which, in the case of Disciple, is based on linguistics. Each element in the application domain is represented in the agent's domain model by the word used in natural language to designate it. A complex entity is represented in terms of its elements and its meaning is given by the meaning of the words representing those elements.

3.1.1.2 *General features of a knowledge representation*

Figure 2.1 shows the overall architecture of the Disciple learning agent shell, identifying the main components of a Disciple agent: the knowledge base (containing the data structures that represents the entities from the agent's application domain), the problem-solving component (consisting of the programs that manipulate the data structures in the knowledge base in order to solve the problems for which the agent was designed), and the knowledge acquisition and learning component (consisting of the programs that create and update the data structures in the knowledge base). Therefore, when defining a knowledge representation for an intelligent agent, one has to consider four important features related to these components: representational adequacy, inferential adequacy, problem-solving efficiency and learning efficiency (Rich and Knight, 1991).

- ● **Representational adequacy** *is the ability to represent all of the kinds of knowledge that are needed in a certain application domain.*
- ● **Inferential adequacy** *is the ability to represent all of the kinds of inferential procedures needed in the application domain.* Inferential procedures manipulate the representational structures in such a way as to derive new structures corresponding to new knowledge inferred from old knowledge. They are used both during problem solving and during learning.
- ● **Problem-solving efficiency** *is the ability to represent efficient problem-solving procedures.* One could do this, for instance, by incorporating into the knowledge structure additional information that can be used to focus the attention of the inference mechanisms in the most promising directions.
- ● **Learning efficiency** *is the ability to easily acquire and learn new information and to integrate it within the agent's knowledge structures, as well as to modify the existing knowledge structures to better represent the application domain.*

3.1.1.3 Hybrid knowledge representation

The knowledge representation formalisms such as predicate calculus, production rules, semantic networks or frames (Brachman and Levesque, 1985), can be compared in terms of the four features described in the previous section. For instance, predicate calculus (McCarthy, 1968; Kowalski, 1979; Genesereth and Nilsson, 1987) has a high representational and inferential adequacy, but a low problem-solving efficiency. The complexity of first-order predicate calculus representation makes it very difficult to implement learning methods and they are not efficient. Therefore, most of the existing learning methods are based on restricted forms of first-order logic or even on propositional logic. However, new knowledge can be easily integrated into the existing knowledge due to the modularity of the representation. Thus, the learning efficiency of predicate calculus is moderate.

Production rules (Waterman and Hayes-Roth, 1978; Anderson, 1983; Brownston *et al.*, 1985; Laird *et al.*, 1987) possess similar features. They are particularly well-suited for representing knowledge about what to do in predetermined situations. Therefore many agents use this type of representation. However, they are less adequate for representing knowledge about objects.

Semantic networks and frames (Quillian, 1968; Minski, 1975; Bobrow and Winograd, 1977; Sowa, 1984; Brachman and Schmolze, 1985; Lenat and Guha, 1990) are, to a large extent, complementary to production systems. They are particularly well-suited for representing objects and states, but have difficulty in representing processes. As opposed to production systems, their inferential efficiency is very high because the structure used for representing knowledge is also a guide for the retrieval of the knowledge. However, their learning efficiency is low because the knowledge that is added or deleted affects the rest of the knowledge. Therefore, new knowledge has to be carefully integrated into the existing knowledge.

In defining the knowledge representation for the Disciple agents, we have considered all the four features of a representation defined above, but we have paid special attention to the learning efficiency. That is, we have defined a type of knowledge representation that is particularly well-suited for knowledge acquisition and learning, although it may be less powerful from the point of view of the other features. The knowledge representation is based on the notion of concept and supports the fundamental operations with the concepts, such as comparing the generality of concepts, generalizing concepts, and specializing concepts. All the problem-solving and learning mechanisms of the Disciple agents are based on these elementary operations.

Disciple agents use a hybrid knowledge representation, where domain objects are represented using semantic networks and problem-solving strategies are represented using rules. These two different knowledge representation formalisms are combined in such a way as to compensate for each other's weaknesses and to take advantage of their strengths and complementariness. The advantage of a hybrid knowledge representation over a uniform one is that it allows a more natural representation of the diverse types of pieces of knowledge characterizing a given application domain. The objects are described in terms of their properties and relations, and are hierarchically organized according to the 'more-general-than' relation, thus forming a hierarchical semantic network. The next section describes the representation of domain objects and their organization into a semantic network.

The rules are expressed in terms of the object names, properties and relations. The meaning of the rules depends on the application domain. These rules may be inference rules for inferring new properties and relations of objects from other properties and relations

(Tecuci, 1992b); general problem-solving rules, for instance, rules that indicate the decomposition of complex problems into simpler subproblems (Tecuci and Kodratoff, 1990); or even action models that describe the actions that could be performed by an agent (for instance, a robot), in terms of their preconditions, effects and involved objects (Tecuci, 1991). The representation of rules is described in Section 3.1.3.

3.1.2   Semantic network representation of objects

The underlying idea of semantic networks is to represent knowledge in the form of a graph in which the nodes represent objects, situations, or events, and the arcs represent the relations between them. In defining the semantic network representation of the objects from the agent's application domain, we will take a bottom-up approach, presenting first how elementary entities are represented and then how more complex representations are built from the elementary ones. The elementary entities are the instances, the elementary concepts, the properties, and the relations.

3.1.2.1   Characterization of instances and concepts

- *An **instance** is a representation of a particular entity in the application domain.* An example of an instance is **DESK1** in Figure 3.2.
- *A **concept** is a representation of a set of instances.* For example, the concept **DESK** represents the set of all desks from the application domain (see Figure 3.2).

We indicate that an instance belongs to a concept by using the relation **INSTANCE–OF**:

DESK1   INSTANCE–OF   DESK

Elementary concepts are denoted by natural language (English) words (like desk, membrane, adhesive, container, etc.) which have a clear meaning for the user with respect to what entities they represent.

- *An entity belonging to the set of instances represented by a concept is called a **positive example** of the concept.*

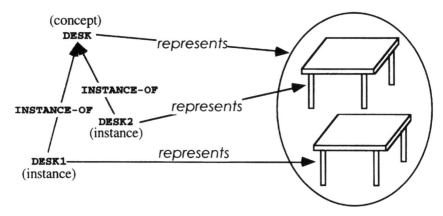

Figure 3.2   Representation of the domain objects.

- *An entity that does not belong to the set of instances represented by a concept is called a* **negative example** *of the concept.*

To illustrate, **DESK2** is a positive example of the concept **DESK** and a negative example of the concept **MEMBRANE**.

3.1.2.2 Intuitive definition of generalization

Generalization is a fundamental relation between concepts. *Intuitively, a concept P is said to be **more general than** (or a generalization of) another concept Q if and only if the set of instances represented by P includes the set of instances represented by Q.* For example, let **ACETONE** be the concept representing the set of all substances of type acetone, and **SOLVENT** be the concept representing all substances that can dissolve other substances. Because any acetone is also a solvent, the concept **SOLVENT** is said to be more general than the concept **ACETONE**. Figure 3.3 shows seven concepts of different generality, each concept denoting a set of instances.

The concept **SOMETHING** represents all the objects from the application domain. Therefore, the concept **SOMETHING** is more general than any other concept that represents a set of concepts from the application domain.

Let us notice that the above definition of generalization is extensional, based upon the instance sets of concepts. In order to show that P is more general than Q, this definition would require the computation of the (possibly infinite) sets of the instances of P and Q. Therefore, it is useful in practice only for showing that P is not more general than Q. Indeed, according to this definition, it is enough to find an instance of Q that is not an instance of P because this shows that the set represented by Q is not a subset of the set represented by P. In Section 3.2.1, we will give an intentional definition of generalization that will allow one to compare the generality of the concepts.

One may express the generality relation between two concepts by using the relation **ISA**:

ACETONE   ISA   SOLVENT

Consequently, one may represent the generality relations between the concepts in Figure 3.3 in the form of a partially ordered network that is usually called a generalization hierarchy (see Figure 3.4). The leaves of these hierarchies are instances of the concepts which are represented by upper-level nodes.

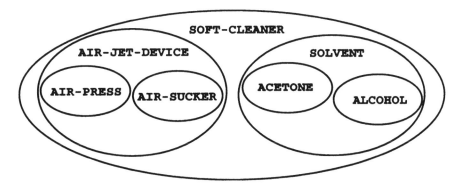

Figure 3.3   Concepts of different generality.

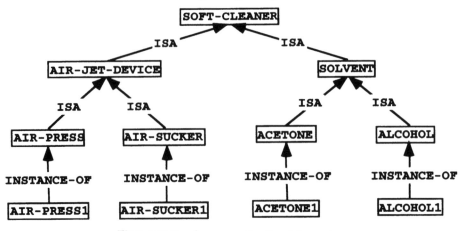

Figure 3.4   A generalization hierarchy.

3.1.2.3   Properties and relations

The objects in the application domain may be described in terms of their characteristics and relationships with each other. These characteristics, collectively referred to as features, are represented as properties and relations in Disciple.

A **property** *P* is defined by a *name*, a *description* (which is the natural language word or phrase denoting that property in the application domain), a *domain* (which is the set of objects that may have that property), and a *range* (which is the set of possible values of the property). The property values are constants in the application domain such as numbers or strings.

Based on their range, one could distinguish between several types of properties, such as, set properties, number properties, and string properties, as illustrated in Table 3.1. For instance, **HAS–COLOR** is a set property (see Table 3.1). Its description, which is used in the communication with the user, is 'has the color'. The domain of **HAS–COLOR** is **SOME-THING**, which means that any object may have a color. The range of **HAS–COLOR** is the set **{red yellow blue white clear}**, which means that its value could be an element from this set. **STATE** is another example of a set property. An example of a number property is **COST**, the range of which is defined by the interval of allowable values. Also, an example of a string property is **SURNAME**, the range of which is a string.

Similarly, a **relation** *R* is defined by a *name*, a *description* (which is the natural language word or phrase denoting that relation in the application domain), a *domain* (the set of objects

Table 3.1   Definitions of properties of objects

Name	Description	Domain	Range
STATE	state	SOMETHING	{solid fluid gas}
HAS–COLOR	has the color	SOMETHING	{red yellow blue white clear}
COST	cost	SOMETHING	[0, 9999999999.99]
SURNAME	surname	PERSON	string

Table 3.2 Definitions of relations between objects

Name	Description	Domain	Range
ISA	is a	SOMETHING	SOMETHING
INSTANCE-OF	is an instance of	SOMETHING	SOMETHING
IS	is	SOMETHING	SOMETHING
TYPE-OF	is a type of	SOMETHING	SOMETHING
PART-OF	is a part of	SOMETHING	SOMETHING
DIF-FROM	is different from	SOMETHING	SOMETHING
GLUES	glues	ADHESIVE	MATERIAL

that may have that relation), and a *range* (the set of objects that may be related by R). Table 3.2 contains some examples of relations.

The relations **ISA** and **INSTANCE-OF** have already been introduced in the previous sections. Two other relations derived from them are **IS** and **TYPE-OF**. The following statements define the meaning of the **IS** relation:

$$\forall \ ?X, \ (?X \ IS \ ?X)$$

$$\forall \ ?X, \ \forall \ ?Y, \ [(?X \ INSTANCE-OF \ ?Y) \rightarrow (?X \ IS \ ?Y)]$$

$$\forall \ ?X, \ \forall \ ?Y, \ [(?X \ ISA \ ?Y) \rightarrow (?X \ IS \ ?Y)]$$

That is, an instance or a concept is in the relation **IS** with itself and with any of its superconcepts from the generalization hierarchies to which it belongs. We have introduced this relation as an abstraction of the **ISA** and **INSTANCE-OF** relations because in many situations one could use either **ISA** or **INSTANCE-OF**. In such situations we will just use the **IS** relation.

The relation **TYPE-OF** is another abstraction of **ISA** and **INSTANCE-OF**. While **ISA** and **INSTANCE-OF** refers to one superconcept of a concept or instance, **TYPE-OF** refers to all the superconcepts of this entity. For example, with respect to the generalization hierarchy in Figure 3.4, the value of the relation **TYPE-OF** for the concept **ACETONE** is {**SOLVENT SOFT-CLEANER**}. Also, the value the relation **TYPE-OF** for the instance **ACETONE1** is {**ACETONE SOLVENT SOFT-CLEANER**}.

Other general relations that are useful in a great variety of domains are **PART-OF** (used to indicate that an object is part of another object) and **DIF-FROM** (used to indicate that an object x is different from an object y).

Besides these domain-independent relations there are also domain-dependent relations between the objects from the application domain that need to be represented in the agent's domain model. An example of such a relation is **GLUES** (see Table 3.2) which relates an adhesive to the material it glues.

3.1.2.4 Definition of instances and concepts

When designing a knowledge base, one has to first specify some elementary concepts (like **ACETONE**, **ALCOHOL**, **SOLVENT**, **SOFT-CLEANER**), as well as properties and relations that may characterize instances and concepts. Once elementary concepts, properties and relations are specified, one can define new concepts and instances as logical expressions of the known concepts.

*The basic representation unit for defining new concepts is called a **clause** and has the following form:*

$$\text{concept}_k \quad \text{ISA} \quad\quad \text{concept}_i$$

$$\text{FEATURE}_1 \quad \text{value}_1$$

$$\vdots$$

$$\text{FEATURE}_n \quad \text{value}_n$$

A *feature* is defined as either a relation between two concepts or a property of a concept. For instance, if 'value$_1$' is a concept name (e.g. 'concept$_j$') then FEATURE$_1$ is a relation between 'concept$_k$' and 'concept$_j$'. Alternatively, if 'value$_1$' is a constant (e.g. **5** or **fluid**), then FEATURE$_1$ is a property of concept$_k$ the value of which is the constant.

The above clause is a necessary definition of 'concept$_k$'. It defines 'concept$_k$' as being a subconcept of 'concept$_i$' and having additional features. This means that if 'concept$_i$' represents the set C_i of instances, then 'concept$_k$' represents a subset C_k of C_i. The elements of C_k have the features 'FEATURE$_1$', ..., 'FEATURE$_n$' with values that are less general than or as general as 'value$_1$', ..., 'value$_n$', respectively.

Using this clausal representation one can define a concept as being a subconcept of a known concept and having additional properties and relations. For instance, the concept **SOLVENT** can be defined as being a soft-cleaner that removes some paint and some adhesive, and is fluid:

$$\text{SOLVENT} \quad \text{ISA} \quad\quad \text{SOFT–CLEANER}$$

$$\text{REMOVES} \quad \text{PAINT}$$

$$\text{REMOVES} \quad \text{ADHESIVE}$$

$$\text{STATE} \quad\quad \text{fluid}$$

As a convenience, the definition of **SOLVENT** can also be written in the following, more condensed, form:

$$\text{SOLVENT} \quad \text{ISA} \quad\quad \text{SOFT–CLEANER}$$

$$\text{REMOVES} \quad \text{PAINT ADHESIVE}$$

$$\text{STATE} \quad\quad \text{fluid}$$

Notice that this is a possibly incomplete description of a solvent. Therefore it should only be interpreted as a necessary but not sufficient definition of a solvent. That is, any solvent is necessarily a soft-cleaner, removes some paint, removes some adhesive, and is fluid. There might be, however, other objects having all these properties without being solvents.

The clause for defining concepts can be rewritten as a conjunctive predicate calculus expression, where 'IS', 'FEATURE$_1$', ..., 'FEATURE$_n$' are the names of the predicates. For instance, the above necessary definition of solvent can be rewritten as follows:

∀ ?X [(?X IS SOLVENT)

⇒ ∃ ?Y ∃ ?Z, (?X IS SOFT–CLEANER) ∧ (?X REMOVES ?Y) ∧ (?Y IS PAINT) ∧

(?X REMOVES ?Z) ∧ (?Z IS ADHESIVE) ∧ (?X STATE fluid)]

Therefore, the clause for describing concepts corresponds to a conjunctive, function-free predicate calculus expression.

In the clause for defining concepts, instead of 'concept$_i$', there could be a list of concepts 'concept$_{i1}$ ... concept$_{im}$'. In such a case, 'concept$_k$' is defined as a subconcept of each of the concepts from the list. That is, the list represents the instances occurring in the intersection of the included concepts. To illustrate this, the following expression describes **AIR–JET–DEVICE** as being both a **SOFT–CLEANER** and an **AIR–MOVER**:

AIR–JET–DEVICE   ISA   SOFT–CLEANER AIR–MOVER

One could also represent the fact that a 'concept$_k$' is a subconcept of a 'concept$_i$' or a sub-concept of a 'concept$_j$' or of both, by using the notation:

concept$_k$ ISA   {concept$_i$ concept$_j$}

This expresses the fact that any instance of 'concept$_k$' is an instance of 'concept$_i$' or an instance of 'concept$_j$' or both.

Using the same notation we can express the fact that the value of a property is one of the values from a set. For instance, we can represent the fact that the state of an adhesive is fluid or solid as follows:

ADHESIVE   ISA   SOMETHING

STATE   {fluid solid}

The clause for defining an instance of a concept is similar to that of defining a concept except that instead of the 'ISA' relation one uses the 'INSTANCE-OF' relation:

instance$_k$ INSTANCE-OF   concept$_i$

FEATURE$_1$ value$_1$

$\vdots$

FEATURE$_n$ value$_n$

To illustrate, the following is the definition of **MEMBRANE1**:

MEMBRANE1   INSTANCE–OF   MEMBRANE

A concept defined by a clause is represented by a semantic network fragment, as shown in Figure 3.5.

Let us consider the concept and instance definitions from Table 3.3. The semantic network representation of these concepts and instances is shown in Figure 3.6. This semantic network is a part of the semantic network from the knowledge base of the loudspeaker manufacturing assistant.

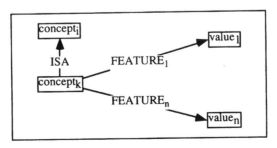

Figure 3.5   Semantic network representation of a concept.

Table 3.3   Sample concept and instance descriptions

AIR-JET-DEVICE	**ISA**	**SOFT-CLEANER AIR-MOVER**
	REMOVES	**DUST**
AIR-PRESS	**ISA**	**AIR-JET DEVICE**
AIR-PRESS1	**INSTANCE-OF**	**AIR-PRESS**
AIR-SUCKER	**ISA**	**AIR-JET-DEVICE**
AIR-SUCKER1	**INSTANCE-OF**	**AIR-SUCKER**
SOLVENT	**ISA**	**SOFT-CLEANER**
	REMOVES	**PAINT**
	REMOVES	**ADHESIVE**
	STATE	**fluid**
ACETONE	**ISA**	**SOLVENT**
ACETONE1	**INSTANCE-OF**	**ACETONE**
ALCOHOL	**ISA**	**SOLVENT**
ALCOHOL1	**INSTANCE-OF**	**ALCOHOL**

3.1.2.5   General characterization of the semantic network

As has been shown, the semantic network of a Disciple agent consists of instances and concepts describing objects or groups of objects from the application domain. There are three main features of this semantic network:

- It may be incomplete;
- It may be partially incorrect;
- It does not contain variables.

The semantic network may be incomplete in the sense that it may not contain all the relevant concept and instance descriptions. It may also be incomplete because the representations may not specify all the properties and relations of the represented objects. The semantic network may even be partially incorrect in the sense that some descriptions may be wrong. However, the Disciple agents are able to learn and therefore to complete and improve this semantic network, as will be presented in Chapter 4.

As indicated in Section 3.1.2.4, the definitions of instances and concepts do not contain any variables. This may clearly appear as a significant limitation. However, as will be shown in the next section, the elements of this semantic network are used as terms for the definition of concepts and rules which do contain variables.

3.1.3   General concepts and rules

The semantic network presented in the previous section defines the instances and the basic concepts from the application domain in terms of their properties and relations. Using these instances, concepts, properties and relations one can define more complex entities such as more complex concepts, tasks, and reasoning rules. We will start by defining the representation language for these entities and then we will describe their representation.

3.1.3.1   Representation language for general concepts and rules

A knowledge representation language defines a syntax and semantics for expressing knowledge in a form that an agent can use. We define a representation language for general

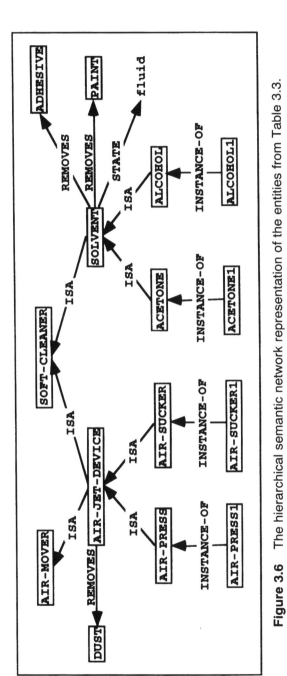

Figure 3.6 The hierarchical semantic network representation of the entities from Table 3.3.

concepts and rules as follows:

- Let V be a set of *variables*. For convenience in identifying variables, their names start with '?', for instance, **?X**. Variables will be used to denote unspecified instances of concepts.
- Let C be a set of *constants*. Examples of constants are the numbers (such as '**5**'), the values of the properties (such as **blue**), and the boolean constants 'true' and 'false'. We define a *term* to be either a variable or a constant.
- Let F be a set of *features*, where a feature may be a property or a relation. The set F includes the domain independent features **ISA**, **INSTANCE-OF**, **IS**, **TYPE-OF**, and **DIF-FROM**, as well as other domain specific features, such as **GLUES**.
- Let S be the semantic network defined in Section 3.1.2. It consists of a set of concepts and instances defined using the clause representation presented in Section 3.1.2.4, where the feature values are either constants or other concepts and instances. That is, there are no variables in the definition of a concept or an instance from S.

  The concepts and the instances from S are related by the generalization relations **ISA** and **INSTANCE-OF**, as indicated by their definitions. S includes the concept **SOME-THING** which represents all the instances from the application domain and is therefore more general than any other concept.
- Let H be the set of theorems and properties of the features, variables, and constants. Two properties of any feature are its domain and its range. Other features may have special properties. For instance, the relation ISA is transitive (see Section 3.3.1). Also, a concept or an instance inherits the features of the concepts that are more general than it (see Section 3.3.1).
- Let T be the set of generic tasks that could be performed by the agent. The representation of the tasks will be described in Section 3.1.3.3.
- Let O be the set of generic operations used by the agent to accomplish its tasks. This may include domain independent operations such as **DECOMPOSE** (that decompose a task into simpler subtasks) or **ASSERT** (that asserts a new piece of knowledge into the knowledge base of the agent), as well as domain specific operations. The representation of the operations will be described in Section 3.1.3.3.
- Let L be a set of connectors. L includes the logical connectors AND ($\wedge$) and OR ($\vee$), the connectors '{' and '}' for defining alternative values of a feature, the connectors '[' and ']' for defining a numeric interval, the delimiter ',', and the **symbols 'IF'**, **'Plausible Upper Bound IF'**, **'Plausible Lower Bound IF'** and **'THEN'**.

We call the tuple $\mathcal{L} = (V, C, F, S, H, T, O, L)$ a *representation language* for general concepts and rules.

In the following sections we will show how the various knowledge elements from the agent's knowledge base are represented using this language.

3.1.3.2   General concepts
In the representation language $\mathcal{L}$ one can define a general concept by using the following clause:

$$\begin{array}{lll} variable_k & \text{IS} & entity_i \\ & \text{FEATURE}_1 & value_1 \\ & \vdots & \\ & \text{FEATURE}_n & value_n \end{array}$$

where 'variable$_k$' is a variable denoting a generic instance of the defined concept, 'entity$_i$' is a concept or an instance from S, 'FEATURE$_1$', ..., 'FEATURE$_n$' are features from F, and 'value$_1$', ..., 'value$_n$' are constants or variables from C or V. If 'value$_1$' is a variable, then it should be described by another clause.

The defined concept represents the set of instances satisfying the clausal representation. Sometimes we will refer to this concept by using the name of the variable. We will also refer to entity$_i$ as the *class* of variable$_k$.

For instance, the following clause represents the concept 'fluid adhesive':

?X	IS	ADHESIVE
	STATE	fluid

It represents all the instances that satisfy the description of **?X**, that is, it represents all the instances that are adhesives and are fluid. We refer to this concept as the concept **?X** and we refer to **ADHESIVE** as being the class of **?X**.

Similarly, one could define the concept 'fluid adhesive that glues paper' as follows:

?X	IS	ADHESIVE
	STATE	fluid
	GLUES	?M
?M	IS	PAPER

One should notice that the value of the relation **GLUES** is another generic instance **?M** that is defined by another clause.

3.1.3.3   Rules

The agent's representation language contains a set T of generic tasks to be performed by the agent. A generic task is defined by three elements: name, features and description. For instance, Table 3.4 contains the definition of three generic tasks of the loudspeaker manufacturing assistant.

ATTACH is a generic task that expresses the attachment of an object to another object. A specific instance of this task, such as **Attach membrane1 to chassis-assembly1** is represented by using the clausal representation as follows:

?A	IS	ATTACH
	OBJECT	MEMBRANE1
	TO	CHASSIS-ASSEMBLY1

Table 3.4   Definitions of some generic tasks for the loudspeaker manufacturing assistant

Name	Features	Description
ATTACH	OBJECT, TO	Attach /OBJECT to /TO
APPLY	OBJECT, ON	Apply /OBJECT on /ON
PRESS	OBJECT, ON	Press /OBJECT on /ON

OBJECT and **TO** are features from the set F of the representation language. They are defined in the same way as the object feature (see Section 3.1.2.3). For instance, the description of **OBJECT** is **object**, its domain is **TASK** and its range is **SOMETHING**.

The description of the above task is obtained from the description of **ATTACH** (i.e. **Attach /OBJECT to /TO**) by replacing /**OBJECT** with the description of the value of the feature **OBJECT** (which is **membrane1**), and by replacing /**TO** with the description of the value of the feature **TO** (which is **chassis-assembly1**).

The following is the representation of a task of attaching a membrane **?M** to a chassis assembly **?C**, where both **?M** and **?C** are variables denoting unspecified instances of membrane and chassis assembly, respectively:

?A	IS	ATTACH
	OBJECT	?M
	TO	?C
?M	IS	MEMBRANE
?C	IS	CHASSIS-ASSEMBLY

Notice that the above expression consists of a sequence of three clauses which are to be considered as connected by the logical AND operator. Therefore, the description of the above task consists of a conjunction of three clauses.

The generic operations used by the agent to accomplish its tasks are also defined by three elements: name, features and description. A specific operation is an instance of a generic operation.

The following is an instance of the **DECOMPOSE** operation:

DECOMPOSE	TASK	?A
	INTO	?AP ?P
?A	IS	ATTACH
	OBJECT	MEMBRANE1
	TO	CHASSIS-ASSEMBLY1
?AP	IS	APPLY
	OBJECT	CONTACT-ADHESIVE1
	ON	MEMBRANE1
?P	IS	PRESS
	OBJECT	MEMBRANE1
	ON	CHASSIS-ASSEMBLY1

This operation decomposes the task **?A** into the simpler tasks **?AP** and **?P**. It is by successive application of this operation that the manufacturing assistant builds the problem-solving tree in Figure 2.5. Because **DECOMPOSE** operates on tasks, its description must include features which take tasks rather than concepts, instances or constants, as their values. These special features are called arguments.

```
IF                          ; IF
        Tg                  ; Tg is a general task to be accomplished and
        Kg                  ; there exist the objects Kg
   THEN                     ; THEN
        Og                  ; apply the general operation Og
```

Figure 3.7   The general structure of a rule.

Besides general operations, like **DECOMPOSE** or **ASSERT**, which are useful in a variety of domains, the user of the Disciple agent shell may define other type of operations that are necessary for the particular problem solver of the agent to be built.

The general structure of a rule is the one from Figure 3.7. Usually there is only one operation, but there could also be a sequence $O_{g1}, ..., O_{gn}$ of operations. If the operation O_g involves some new tasks to be performed by the agent, then the description of these tasks follow the description of the operation. However, if O_g and the new tasks contain some general object concepts, then their descriptions are part of K_g.

```
IF
   ?A          IS          ATTACH          ; the task is to attach
               OBJECT      ?X              ; object ?X
               TO          ?Y              ; to ?Y and

   ?X          IS          SOMETHING       ; ?X is made of ?M and
               MADE-OF     ?M

   ?Y          IS          SOMETHING       ; ?Y is made of ?N and
               MADE-OF     ?N

   ?Z          IS          ADHESIVE        ; ?Z is an adhesive
               GLUES       ?M              ; that glues ?M and
               GLUES       ?N              ; glues ?N and
               STATE       fluid           ; is fluid and

   ?M          IS          MATERIAL        ; ?M is a material and

   ?N          IS          MATERIAL        ; ?N is a material

THEN

   DECOMPOSE   TASK        ?A              ; decompose the task ?A
               INTO        ?AP    ?P       ; into the tasks ?AP and ?P

   ?AP         IS          APPLY           ; ?AP is the task to apply
               OBJECT      ?Z              ; object ?Z
               ON          ?X              ; on ?X

   ?P          IS          PRESS           ; ?P is the task to press
               OBJECT      ?X              ; object ?X
               ON          ?Y              ; on ?Y
```

Figure 3.8   An exemplary rule.

```
IF
    ?A     ATTACH,       OBJECT   ?X,   TO    ?Y
    ?X     SOMETHING,    MADE-OF  ?M
    ?Y     SOMETHING,    MADE-OF  ?N
    ?Z     ADHESIVE,     GLUES    ?M,   GLUES ?N,   STATE   fluid
    ?M     MATERIAL
    ?N     MATERIAL
THEN
    DECOMPOSE    TASK   ?A,   INTO  ?AP   ?P
    ?AP    APPLY, OBJECT ?Z,    ON     ?X
    ?P     PRESS, OBJECT ?X,    ON     ?Y
```

Figure 3.9   An exemplary rule in condensed form.

Figure 3.8 shows a rule which states that if the task to be accomplished is to attach one object to another object and there is a fluid adhesive that glues the materials of the two objects, then one should decompose this task into two simpler tasks: the task of applying the adhesive on the first object and the task of pressing the first object on the second one. Notice that the variable **?Z** represents a concept contained in the operation but its clause appears in the condition of the rule.

A more condensed form of the rule in Figure 3.8 is presented in Figure 3.9. In this case all the features of a concept are written on the same line, separated by ','.

3.1.3.4   Plausible version space rules

Many of the rules in the agent's knowledge base are plausible version space (PVS) rules learned by the agent, as will be shown in Chapter 4. These rules have two conditions that represent a PVS for a hypothetical exact condition. An example of such a rule is shown in Figure 3.10.

- *The **plausible lower bound condition** is an expression that, as an approximation, is less general than the hypothetical exact condition of the rule.* By this we mean that most of the instances of the plausible lower-bound condition are also instances of the hypothetical exact condition.
- *The **plausible upper bound condition** is an expression that, as an approximation, is more general than the hypothetical exact condition.*

Figure 3.10 shows graphically the relations between these three concepts: the hypothetical exact condition of the rule, the plausible lower bound condition, and the plausible upper bound condition. The universe of instances is an n-dimensional space, where n is the number of object variables from the rule's conditions (n is 6 in the case of the rule from Figure 3.10). The set of instances of each of these three conditions is represented as an elliptical region. As can be seen from Figure 3.10, the plausible lower bound ellipse is strictly included in the plausible upper bound ellipse. The ellipse corresponding to the hypothetical exact condition almost covers the plausible lower bound ellipse, and is almost covered by the plausible upper bound ellipse.

During learning, the plausible lower bound and the plausible upper bound are both generalized (i.e. extended) and specialized (i.e. contracted) to better approximate the hypothetical exact condition. Notice, however, that this learning process takes place in an incomplete representation language based on the incomplete semantic network S (see Section 3.1.2)

Plausible Upper Bound IF

?A	IS	ATTACH
	OBJECT	?X
	TO	?Y
?X	IS	SOMETHING
	MADE-OF	?M
?Y	IS	SOMETHING
	MADE-OF	?N
?Z	IS	ADHESIVE
	GLUES	?M
	GLUES	?N
	STATE	{fluid gas}
?M	IS	MATERIAL
?N	IS	MATERIAL

Plausible Lower Bound IF

?A	IS	ATTACH
	OBJECT	?X
	TO	?Y
?X	IS	LOUDSPEAKER-COMPONENT
	MADE-OF	?M
?Y	IS	LOUDSPEAKER-COMPONENT
	MADE-OF	?N
?Z	IS	ADHESIVE
	GLUES	?M
	GLUES	?N
	STATE	fluid
?M	IS	MATERIAL
?N	IS	MATERIAL

THEN

DECOMPOSE	TASK	?A
	INTO	?AP ?P
?AP	IS	APPLY
	OBJECT	?Z
	ON	?X
?P	IS	PRESS
	OBJECT	?X
	ON	?Y

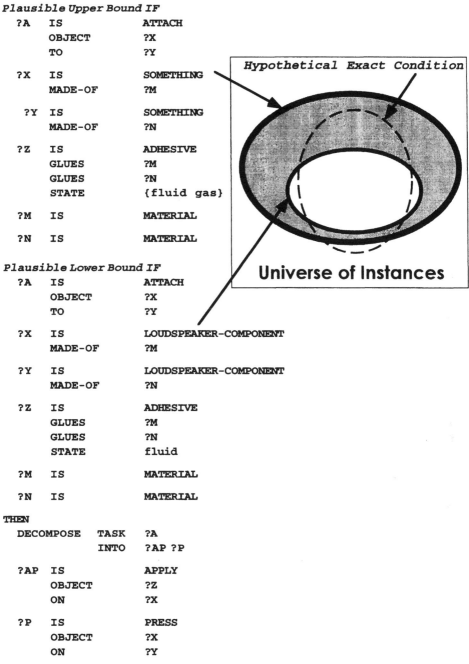

Figure 3.10 A PVS rule.

which may not even contain all the concepts needed to learn an exact rule. This space, however, may also change during learning.

3.2 Generalization in the representation language of the agent

The knowledge representation of the Disciple agents was defined in such a way as to facilitate the basic operations involved in learning. This led to the concept-based representation presented in the previous section. In this section we describe in detail the basic learning operations dealing with concepts: comparing the generality of concepts, generalizing concepts, and specializing concepts.

3.2.1 Formal definition of generalization

In Section 3.1.2.2 a concept is defined as a representation of a set of instances. In order to show that a concept P is more general than a concept Q, this definition would require the computation of the (possibly infinite) sets of the instances of P and Q. In this section we will give an intentional definition of generalization based on a representation language $\mathcal{L}$ (see Section 3.1.3.1) that allows one to determine whether a concept P is *more general than* another concept Q by examining the descriptions of P and Q, without computing the sets of instances that they represent. The definition of generalization is based on substitutions.

A **substitution** is a function $\sigma = (x_1 \leftarrow t_1, ..., x_n \leftarrow t_n)$, where each x_i $(i = 1, ..., n)$ is a variable and each t_i $(i = 1, ..., n)$ is a term. If l_i is an expression in the representation language $\mathcal{L}$, then σl_i is the expression obtained by substituting each x_i from l_i with t_i.

3.2.1.1 Term generalization
In the representation language $\mathcal{L}$, defined in Section 3.1.3.1, a term is a constant or a variable. An unrestricted variable **?X** is *more general than* any constant (such as **fluid**) and is *as general as* any other unrestricted variable (such as **?Y**).

3.2.1.2 Clause generalization
Let us consider the concepts described by the following two clauses, C_1 and C_2:

$$
\begin{array}{lll}
C_1 = v_1 & \text{IS} & c_1 \\
 & f_{11} & v_{11} \\
 & \ldots & \\
 & f_{1m} & v_{1m} \\
C_2 = v_2 & \text{IS} & c_2 \\
 & f_{21} & v_{21} \\
 & \ldots & \\
 & f_{2n} & v_{2n}
\end{array}
$$

We say that the clause C_1 is *more general than* the clause C_2 if there exists a substitution σ such that

- $\sigma v_1 = v_2$;
- $c_1 = c_2$;
- $\forall\, i \in \{1, ..., m\}, \exists\, j \in \{1, ..., n\}$ such that $f_{1i} = f_{2j}$ and $\sigma v_{1i} = v_{2j}$.

For instance, the concept

$$C_1 = \text{?x} \quad \textbf{IS} \qquad \textbf{ADHESIVE}$$

$$\textbf{STATE} \qquad \textbf{fluid}$$

is more general than the concept

$$C_2 = \text{?y} \quad \textbf{IS} \qquad \textbf{ADHESIVE}$$

$$\textbf{STATE} \qquad \textbf{fluid}$$

$$\textbf{COLOR} \qquad \textbf{white}$$

Let $\sigma = (\textbf{?X} \leftarrow \textbf{?Y})$. As one can see, σC_1 is a part of C_2, that is, each feature of C_1 is also a feature of C_2. The first concept represents the set of all fluid adhesives, while the second one represents the set of all fluid adhesives that are also white. Obviously the first set includes the second one, and therefore the first concept is more general than the second one.

Let us notice, however, that this definition of generalization does not take into account the theorems and properties of the representation language $\mathcal{L}$. In general one needs to use these theorems and properties to transform the clauses C_1 and C_2 into equivalent clauses C_1' and C_2' respectively. Then one shows that C_1' is *more general than* C_2'. We use the notation '$=_{\mathcal{L}}$' to indicate that two entities are equal in the representation language $\mathcal{L}$. Therefore, the definition of the *more general than* relation in $\mathcal{L}$ is as follows. A clause C_1 is *more general than* another clause C_2 if and only if there exists C_1', C_2', and a substitution σ, such that:

- $C_1' =_{\mathcal{L}} C_1$;
- $C_2' =_{\mathcal{L}} C_2$;
- $\sigma v_1 =_{\mathcal{L}} v_2$;
- c_1 is more general than c_2 in $\mathcal{L}$;
- $\forall i \in \{1, ..., m\}, \exists j \in \{1, ..., n\}$ such that $f_{1i}' =_{\mathcal{L}} f_{2j}'$ and $\sigma v_{1i}' =_{\mathcal{L}} v_{2j}'$.

For instance, let us consider that the semantic network of instances and concepts from the representation language $\mathcal{L}$ contains the following descriptions of **SOLVENT** and **ALCOHOL**:

$$\textbf{SOLVENT} \quad \textbf{ISA} \qquad \textbf{SOFT-CLEANER}$$

$$\textbf{REMOVES} \quad \textbf{PAINT}$$

$$\textbf{REMOVES} \quad \textbf{ADHESIVE}$$

$$\textbf{STATE} \qquad \textbf{fluid}$$

$$\textbf{ALCOHOL} \quad \textbf{ISA} \qquad \textbf{SOLVENT} \quad \textbf{INFLAMMABLE-OBJECT}$$

Let us now show that in this representation language the concept

$$C_1 = \text{?X} \quad \textbf{IS} \qquad \textbf{SOLVENT}$$

$$\textbf{REMOVES} \quad \textbf{?M}$$

$$\textbf{STATE} \qquad \textbf{fluid}$$

is more general than the concept

$$C_2 = \text{?Y} \quad \textbf{IS} \qquad \textbf{ALCOHOL}$$

$$\textbf{REMOVES} \quad \textbf{?N}$$

$$\textbf{COLOR} \qquad \textbf{blue}$$

In this example C_1 does not need to be transformed. However, we need to transform C_2 into an equivalent clause C'_2 such that for every feature of C_1 there is a corresponding feature of C'_2. By using the inheritance of properties from **SOLVENT** to **ALCOHOL**, one can conclude that any **ALCOHOL** is **fluid** from the fact that every **SOLVENT** is **fluid**. Therefore

$C_2 =_{\mathcal{L}} C'_2 = $?Y	IS	ALCOHOL
	REMOVES	?N
	COLOR	blue
	STATE	fluid

Now let us consider the substitution is $\sigma = ($?X $\leftarrow$?Y, ?M $\leftarrow$?N$)$ and let us apply it to C_1:

$\sigma C_1 = $?Y	IS	SOLVENT
	REMOVES	?N
	STATE	fluid

As one can see, the class of σC_1 (i.e. **SOLVENT**) is more general than the class of C'_2 (i.e. **ALCOHOL**). Also, any feature of σC_1 is also a feature of C'_2 and the corresponding values are equal. Therefore C_1 is more general than C'_2, which is equal to C_2.

In the following we will always assume that the equality is in $\mathcal{L}$ and we will no longer indicate this.

3.2.1.3   Conjunctive formula generalization

By a conjunctive formula we mean a conjunction of clauses, as follows:

?X	IS	ADHESIVE
	STATE	fluid
	GLUES	?M
?M	IS	PAPER

where, for notation convenience, we have dropped the AND connector. Therefore, any time there is a sequence of clauses, they are to be considered as being connected by AND.

Let us consider two concepts, A and B, defined by the following conjunctive formulas:

$$A = A_1 \wedge A_2 \wedge \cdots \wedge A_n$$
$$B = B_1 \wedge B_2 \wedge \cdots \wedge B_m$$

where each A_i ($i = 1, ..., n$) and each B_j ($j = 1, ..., m$) is a clause.

A is *more general than* B if and only if there exist A', B', and σ such that:

- $A' = A$, $A' = A'_1 \wedge A'_2 \wedge \cdots \wedge A'_p$;
- $B' = B$, $B' = B'_1 \wedge B'_2 \wedge \cdots \wedge B'_q$;
- $\forall i \in \{1, ..., p\}, \exists j \in \{1, ..., q\}$ such that $\sigma A'_i = B'_j$.

Otherwise stated, one transforms the concepts A and B, using the theorems and the properties of the representation language, so as to make each clause from A' more general than a corresponding clause from B'. Notice that some clauses from B' may be 'left-over', that is, they are matched by no clause of A', as in the following example.

Let the two concepts A and B be defined as follows:

$A = {?}\mathbf{U}$	IS	ADHESIVE
	GLUES	?V
	GLUES	?W
?V	IS	SOMETHING
?W	IS	SOMETHING
$B = {?}\mathbf{X}$	IS	ADHESIVE
	GLUES	?Y
	STATE	fluid
?Y	IS	SOMETHING

One may transform B by using the idempotence of the AND operator ($P = P \wedge P$), thus obtaining:

$B' = {?}\mathbf{X}$	IS	ADHESIVE
	GLUES	?Y
	GLUES	?Y
	STATE	fluid
?Y	IS	SOMETHING
?Y	IS	SOMETHING

Let us now consider $\sigma = ({?}\mathbf{U} \leftarrow {?}\mathbf{X}, {?}\mathbf{V} \leftarrow {?}\mathbf{Y}, {?}\mathbf{W} \leftarrow {?}\mathbf{Y})$ and apply it to A:

$\sigma A = {?}\mathbf{X}$	IS	ADHESIVE
	GLUES	?Y
	GLUES	?Y
?Y	IS	SOMETHING
?Y	IS	SOMETHING

One can see that the first clause of σA is a part of the first clause of B' and that the second and the third clauses in σA and B' are identical. Therefore, A is more general than B'. Because $B = B'$, one can conclude that A is more general than B.

3.2.1.4 Disjunctive formula generalization

Obviously $P \vee Q$ is more general than P, but let us now consider the more general case of two expressions A and B described by two disjunctive formulas:

$$A = A_1 \vee A_2 \vee \cdots \vee A_n$$

$$B = B_1 \vee B_2 \vee \cdots \vee B_m$$

where each A_i ($i = 1, \dots, n$) and B_j ($j = 1, \dots, m$) is a clause.

A is *more general than* B if and only if there exist A', B', and σ such that:

- $A' = A$, $A' = A'_1 \vee A'_2 \vee \cdots \vee A'_p$
- $B' = B$, $B' = B'_1 \vee B'_2 \vee \cdots \vee B'_q$
- $\forall j \in \{1, \dots, q\}$, $\exists i \in \{1, \dots, p\}$ such that $\sigma A'_i = B'_j$.

Otherwise stated, one transforms A and B, using the theorems and the properties of the representation language, so as to make each literal from B' less general than a corresponding literal from A'.

One can naturally extend the above definition to apply to concepts represented in *disjunctive normal form* such as the following one:

$$A = (A_{11} \wedge A_{12} \wedge \cdots \wedge A_{1k}) \vee \cdots \vee (A_{n1} \wedge A_{n2} \wedge \cdots \wedge A_{nl})$$

3.2.2 Generalization rules

The definition of generalization in terms of substitutions provides a way of formally characterizing the *more-general-than* relation. It can be used to prove that a concept is more general than another concept. However, this definition does not show how one can generalize a concept. One way to make this definition operational is to define generalization rules which can be used to generalize a concept:

- A **generalization rule** *is a rule that, when applied to a concept, transforms it into a more general concept. The generalization rules are usually inductive transformations. The inductive transformations are not truth-preserving but falsity-preserving. That is, if A is true and is inductively generalized to B, then the truth of B is not guaranteed. However, if A is false then B is also false.*

There are two other types of transformation rules:

- A **specialization rule**, *which transforms a concept into a less general one. The reverse of any generalization rule is a specialization rule. Specialization rules are deductive, truth-preserving transformations.*
- A **reformulation rule**, *which transforms a concept into another, logically equivalent concept. Reformulation rules are also deductive, truth-preserving transformations.*

Some of the most-used generalization rules, which are further described in the following, are:

- Turning constants into variables;
- Turning occurrences of a variable into different variables;
- Climbing the generalization hierarchies;
- Dropping conditions;
- Adding alternatives;
- Using theorems.

3.2.2.1 Turning constants into variables
This rule consists in generalizing an expression by replacing a constant with a variable. For instance, the expression

$$E_1 = \text{?X} \quad \text{IS} \qquad\qquad \text{CLEANER}$$
$$\phantom{E_1 = \text{?X}} \quad \text{REMOVES} \qquad \text{CONTACT-ADHESIVE1}$$

may be generalized to the following by turning the instance **CONTACT-ADHESIVE1** into the variable **?Y**:

$$E_2 = \text{?X} \quad \text{IS} \qquad\qquad \text{CLEANER}$$
$$\phantom{E_2 = \text{?X}} \quad \text{REMOVES} \qquad \text{?Y}$$
$$ \text{?Y} \quad \text{IS} \qquad\qquad \text{SOMETHING}$$

The first expression represents the set of cleaners that remove **CONTACT-ADHESIVE1** while the second expression represents the set of cleaners that remove **SOMETHING**. Since the second set includes the first one, it is more general.

To show that E_2 is a generalization of E_1 one simply needs to rewrite E_1 in the equivalent form:

$$E_1 = E'_1 = \text{?X} \quad \text{IS} \qquad\qquad \text{CLEANER}$$
$$\phantom{E_1 = E'_1 = \text{?X}} \quad \text{REMOVES} \qquad \text{?Y}$$
$$ \text{?Y} \quad \text{IS} \qquad\qquad \text{CONTACT-ADHESIVE1}$$

and to notice that **CONTACT-ADHESIVE1** is less general than **SOMETHING**.

3.2.2.2   Turning occurrences of a variable into different variables

According to this rule, the expression

$E_1 = \text{?T1}$	IS	TIN
	OBJECTS	?T
	WITH	?X
?T2	IS	TIN
	OBJECTS	?C
	WITH	?X
?T	IS	TERMINAL-WIRES
?X	IS	TINNING-BATH
?C	IS	COIL-ENDS

may be generalized to the following by turning two occurrences of the variable **?X** in E_1 into two variables, **?U** and **?V**:

$E_2 = \text{?T1}$	IS	TIN
	OBJECTS	?T
	WITH	?U
?T2	IS	TIN
	OBJECTS	?C
	WITH	?V
?T	IS	TERMINAL-WIRES
?U	IS	TINNING-BATH
?V	IS	TINNING-BATH
?C	IS	COIL-ENDS

The first expression represents the set of the sequences consisting of two tinning tasks, both using the same tinning-bath named ?**X**.

The second expression represents the set of the sequences consisting of two tinning tasks, using the tinning-baths named ?**U** and ?**V**, respectively. In particular, ?**U** and ?**V** may represent the same tinning-bath. Therefore, the second set includes the first one, and the second expression is more general than the first one.

In terms of substitutions, E_2 is a generalization of E_1 because $E_1 = \sigma E_2$, where $\sigma = ($?**U** ← ?**X**, ?**V** ← ?**X**$)$.

3.2.2.3   Climbing the generalization hierarchies

Using the concept hierarchy from the semantic network S of the language $\mathcal{L}$, one may generalize an expression by replacing a concept with a more general one. For instance, the expression

$$E_1 = \text{?\textbf{X}} \qquad \textbf{IS} \qquad \textbf{SOLVENT}$$

$$\textbf{COLOR} \qquad \text{?\textbf{Y}}$$

$$\text{?\textbf{Y}} \qquad \textbf{IS} \qquad \textbf{SOMETHING}$$

may be generalized to

$$E_2 = \text{?\textbf{X}} \qquad \textbf{IS} \qquad \textbf{SOFT-CLEANER}$$

$$\textbf{COLOR} \qquad \text{?\textbf{Y}}$$

$$\text{?\textbf{Y}} \qquad \textbf{IS} \qquad \textbf{SOMETHING}$$

by replacing the concept **SOLVENT** with the more general concept **SOFT-CLEANER** (see the generalization hierarchy in Figure 3.4).

Figure 3.11 presents an example of an ordered generalization hierarchy, where the nodes are ordered from left to right. This hierarchy may be used to generalize numbers representing human ages.

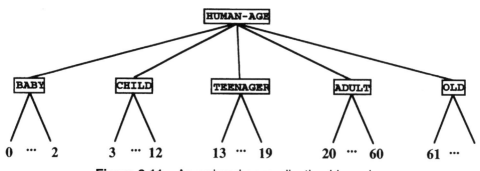

Figure 3.11   An ordered generalization hierarchy.

3.2.2.4 Dropping conditions

This rule consists of generalizing an expression by removing a constraint from its description. For instance, the expression

$$E_1 = \text{?X} \qquad \textbf{IS} \qquad\qquad \textbf{ADHESIVE}$$

$$\textbf{STATE} \qquad \textbf{fluid}$$

may be generalized to

$$E_2 = \text{?X} \qquad \textbf{IS} \qquad\qquad \textbf{ADHESIVE}$$

by removing a constraint on the **ADHESIVE** to be **fluid**.

3.2.2.5 Adding alternatives

A may be generalized to $A \lor B$, where A and B are either clauses or feature values in a clause. For instance

$$E_1 = \text{?X} \qquad \textbf{IS} \qquad\qquad \textbf{OBJECT}$$

$$\textbf{COLOR} \qquad \textbf{blue}$$

may be generalized to

$$E_2 = \text{?X} \qquad \textbf{IS} \qquad\qquad \textbf{OBJECT}$$

$$\textbf{COLOR} \qquad \textbf{\{blue red\}}$$

Thus E_1 represents the set of blue objects while E_2 represents the set of objects that are either blue or red.

3.2.2.6 Generalizing numbers to intervals

A number may be generalized to an interval containing it. For instance, 5 may be generalized to [4; 6]. Also, an interval may be generalized to another interval containing it. For instance, [4; 6] may be generalized to [0; 10].

3.2.2.7 Using theorems

If $B \to C$ is a theorem, then one may generalize B to C. Also, one may generalize $A \land B$ to $A \land C$. For instance, using the theorem

$$\forall\ \text{?X}\ \ \forall\ \text{?Y}\ [(\text{?X \textbf{GLUED-TO} ?Y}) \to (\text{?X \textbf{ATTACHED-TO} ?Y})]$$

one may generalize the expression

$$E_1 = \text{?R} \qquad \textbf{IS} \qquad\qquad \textbf{RING}$$

$$\textbf{GLUED-TO} \qquad \textbf{?C}$$

$$\text{?C} \quad \textbf{IS} \qquad\qquad \textbf{CHASSIS-MEMBRANE-ASSEMBLY}$$

to the expression

$$E_2 = \text{?R} \qquad \textbf{IS} \qquad\qquad \textbf{RING}$$

$$\textbf{ATTACHED-TO} \qquad \textbf{?C}$$

$$\text{?C} \quad \textbf{IS} \qquad\qquad \textbf{CHASSIS-MEMBRANE-ASSEMBLY}$$

Thus by applying the above theorem one may transform E_1 into the equivalent expression

$$E'_1 = \textbf{?R} \quad \textbf{IS} \qquad\qquad \textbf{RING}$$

	GLUED–TO	**?C**
	ATTACHED–TO	**?C**
?C	**IS**	**CHASSIS–MEMBRANE–ASSEMBLY**

Then, by dropping the relation **GLUED–TO**, one generalizes E'_1 to E_2.

3.2.3   Other definitions of generalizations

The generalization rules defined in the previous section allow us to give the following operational definition of generalization.

- *The concept P is more general than the concept Q if the description of Q may be transformed into the description of P by applying generalization rules.*

For instance, by applying generalization rules one can transform the concept

Q:	**?X**	**IS**	**MEMBRANE**
		MADE–OF	**?M**
	?M	**IS**	**PAPER**
	?Z	**IS**	**CONTACT–ADHESIVE**
		GLUES	**?M**
		STATE	**fluid**

into the more general concept

P:	**?X**	**IS**	**LOUDSPEAKER–COMPONENT**
		MADE–OF	**?M**
	?M	**IS**	**MATERIAL**
	?Z	**IS**	**ADHESIVE**
		GLUES	**?M**

Thus by applying the climbing generalization hierarchy rule three times to expression Q, one can replace the concepts **MEMBRANE**, **PAPER**, and **CONTACT–ADHESIVE** with the concepts **LOUDSPEAKER–COMPONENT**, **MATERIAL**, and **ADHESIVE**, which are more general according to the generalization hierarchies from Figure 3.12. Then, by applying the dropping condition rule to Q, one drops the property **STATE fluid** and obtains expression P. Therefore P is more general than Q.

Up to this point we have only defined when a concept is more general than another concept. Learning agents, however, would need to determine generalizations of sets of examples and concepts. In the following we define some of these generalizations.

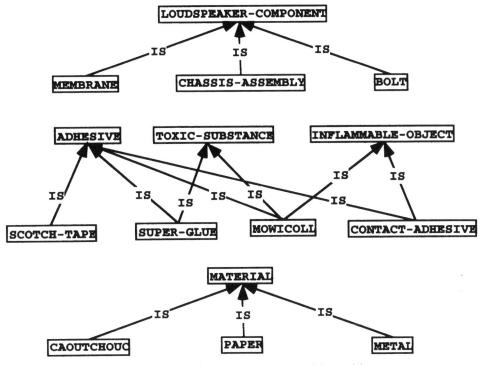

Figure 3.12   Three generalization hierarchies.

- *A **generalization** of two concepts A and B is a concept C that is more general than both A and B.*

Let us consider two clauses C_1 and C_2 corresponding to a same variable **?X** (see Figure 3.13). Any clause that consists of features belonging to both C_1 and C_2 and values that are generalizations of the corresponding values in C_1 and C_2 is a generalization of C_1 and C_2. Figure 3.13 shows three generalizations of C_1 and C_2.

To build a generalization of two clauses one first applies the dropping condition rule to remove the unmatched features (and possibly even matched features). Then one applies other generalization rules to determine the generalizations of the matched feature values. As one can see from Figure 3.13, there is more than one generalization of two expressions.

In a similar way one can determine a generalization G of two expressions E_1 and E_2, each consisting of a conjunction of clauses corresponding to the same set of variables. G consists of the conjunction of the generalizations of some of the corresponding clauses in the two expressions E_1 and E_2. Figure 3.14 shows several generalizations of E_1 and E_2.

The generalization mGG_1 in Figure 3.14 was obtained by dropping the clause of E_1 corresponding to **?N** and then by determining the generalization of each of the corresponding clauses. The generalizations of the corresponding features have been obtained by using the generalization hierarchies in Figure 3.12. For instance, **CONTACT–ADHESIVE** and **MOWICOLL** have been generalized to **ADHESIVE**. Also, the feature **STATE fluid**, has been dropped. mGG_2 was obtained in a similar fashion, except that **CONTACT–ADHESIVE** and **MOWICOLL** have been generalized to **INFLAMMABLE–OBJECT**. Neither mGG_1 nor mGG_2

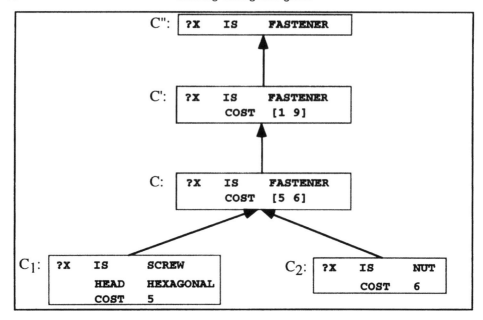

Figure 3.13   Different generalizations of C_1 and C_2.

is more general than the other. However, G is more general than each of them. In G, **?Z** could be either an **ADHESIVE** or an **INFLAMMABLE-OBJECT**.

Notice also that in all the above definitions and illustrations, we have assumed that the clauses to be generalized correspond to the same variables. If this assumption is not satisfied, then one would need first to match the variables, and then to compute the generalizations. In general, this process is computationally expensive because one may need to try different matchings. In the Disciple approach however, the above assumption is always satisfied, as will be shown in Chapter 4. This is an important factor that contributes to the efficiency of the learning processes of the Disciple agents.

As has been shown above, there are usually more than one generalization of two concepts, and further distinctions between these generalizations may be important. In the following we define some of these generalizations.

- *The concept G is a **minimally general generalization (mGG)** of A and B if and only if G is a generalization of A and B, and G is not more general than any other generalization of A and B.*

To determine a minimally general generalization of two clauses, one has to keep *all* the common features of the clauses and to determine a minimally general generalization of each of the matched feature values. In a similar way one determines the mGG of two conjunctions of clauses. One keeps *all* the matched clauses and determines the mGG of each pair of matched clauses.

Notice, however, that there may be more than one mGG of two expressions. For instance, according to the generalization hierarchy from the middle of Figure 3.12, there are two

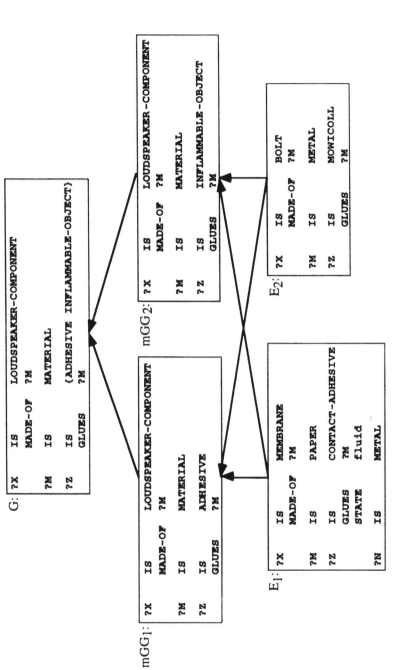

Figure 3.14 Several generalizations of the concepts E_1 and E_2.

mGGs of **MOWICOLL** and **CONTACT-ADHESIVE**. They are **ADHESIVE** and **INFLAM-MABLE-OBJECT**. Consequently, there are two mGGs of E_1 and E_2 in Figure 3.14. They are mGG_1 and mGG_2.

- *If there is only one minimally general generalization of two concepts A and B, then this generalization is called the **least general generalization (LGG)** of A and B.*

Most of the time, Disciple agents employ minimally-general generalizations, also called maximally specific generalizations (Plotkin, 1970; Kodratoff and Ganascia, 1986). However, in some cases, they also employ over-generalizations (Tecuci and Kodratoff, 1990; Tecuci, 1992b).

As has been mentioned in the previous section, the reverse of each generalization rule is a specialization rule. Therefore, one can determine a specialization of a concept by applying the reverse of the generalization rules.

For each of the above definitions of generalization there is a corresponding definition of a specialization. The following are some of these definitions.

- *A **specialization** of two concepts A and B is a concept C that is less general than each of them.*
- *The concept C is **a maximally general specialization (MGS)** of two concepts A and B if and only if C is a specialization of A and B and no other specialization of A and B is more general than C.*

The MGS of two clauses consists of the MGS of the matched feature-value pairs and all the unmatched feature-value pairs.

The MGS of two conjunctions of clauses C_1 and C_2 consists of the conjunction of the MGS of each of the matched clauses of C_1 and C_2 and all the unmatched clauses from C_1 and C_2.

Figure 3.15 shows several specializations of the concepts G_1 and G_2. MGS_1 and MGS_2 are two maximally general specializations of G_1 and G_2 because **CONTACT-ADHESIVE** and **MOWICOLL** are two maximally general specializations of **ADHESIVE** and **INFLAMMABLE-OBJECT**.

3.2.4   Rules as generalizations of examples of problem-solving episodes

The Disciple agents are trained to accomplish their tasks through examples of problem-solving episodes that are generalized into problem-solving rules.

The general structure of an example, shown in Figure 3.16, is similar to the structure of a rule (see Figure 3.7) because an example is just an instance of the rule.

The example shown in Figure 3.17 states that if the task to accomplish is to attach **MEMBRANE1** to **CHASSIS-ASSEMBLY1** then one should decompose this task into two simpler tasks, the task to apply **CONTACT-ADHESIVE1** to **MEMBRANE1**, and the task to press **MEMBRANE1** on **CHASSIS-ASSEMBLY1**.

A generalization of this example is the rule R shown in Figure 3.8. Using the knowledge about **MEMBRANE1**, **CHASSIS-ASSEMBLY1**, and **CONTACT-ADHESIVE1**, from the semantic network in Figure 3.18, one can infer that **MEMBRANE1** is made of **PAPER**, that **CHASSIS-ASSEMBLY1** is made of **METAL**, and that **CONTACT-ADHESIVE1** glues **PAPER** and **METAL** and is **fluid**.

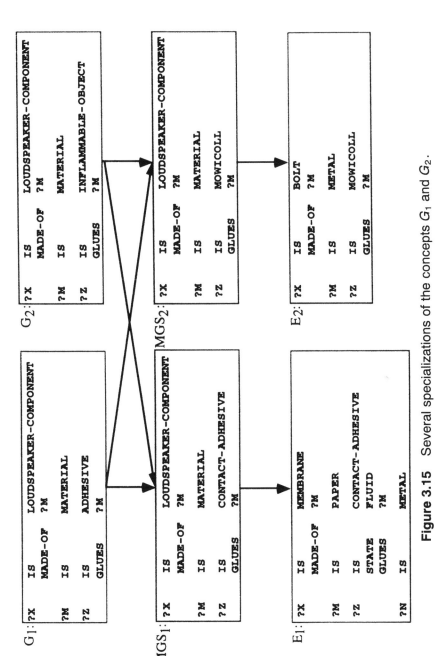

Figure 3.15 Several specializations of the concepts G_1 and G_2.

IF		; IF
	T	; T is a specific task to be accomplished
	K	; and K is the list of the descriptions of the object concepts from T and O
THEN		; THEN
	O	; apply the operation O

Figure 3.16   The general structure of an example of a problem-solving episode.

Therefore, example E can be rewritten in the equivalent form E', shown in the left-hand side of Table 3.5. Notice that, in order to simplify the presentation, we have introduced variable names in the description of the example that correspond to the variable names in the description of the rule.

Now one can show that each description from R is more general than a corresponding description from E', by using the knowledge from the semantic network in Figure 3.18. In particular **SOMETHING** is more general than both **MEMBRANE1** and **CHASSIS-ASSEM-BLY1**, **ADHESIVE** is more general than **CONTACT-ADHESIVE1**, and **MATERIAL** is more general than **PAPER** and **METAL**.

Table 3.5 shows also that one can generate an example of a rule by simply replacing the classes of the object concepts **?X, ?Y, ?Z, ?M**, and **?N** with the corresponding classes from the example: **MEMBRANE1, CHASSIS-ASSEMBLY1, CONTACT-ADHESIVE1, PAPER** and **METAL**, respectively.

Therefore, one can compactly represent the example from Figure 3.17 (which is an example of the rule from Figure 3.8) by the expression:

> **(?X IS MEMBRANE1, ?Y IS CHASSIS-ASSEMBLY1,**
>
> **?Z IS CONTACT-ADHESIVE1, ?M IS PAPER, ?N IS METAL)**

Figure 3.19 is a compact representation of the rule in Figure 3.8 and of its positive example in Figure 3.17.

```
IF
    ?A            IS          ATTACH
                  OBJECT      MEMBRANE1
                  TO          CHASSIS-ASSEMBLY1
THEN
    DECOMPOSE     TASK        ?A
                  INTO        ?AP    ?P

    ?AP           IS          APPLY
                  OBJECT      CONTACT-ADHESIVE1
                  ON          MEMBRANE1

    ?P            IS          PRESS
                  OBJECT      MEMBRANE1
                  ON          CHASSIS-ASSEMBLY1
```

Figure 3.17   An example of a problem-solving episode.

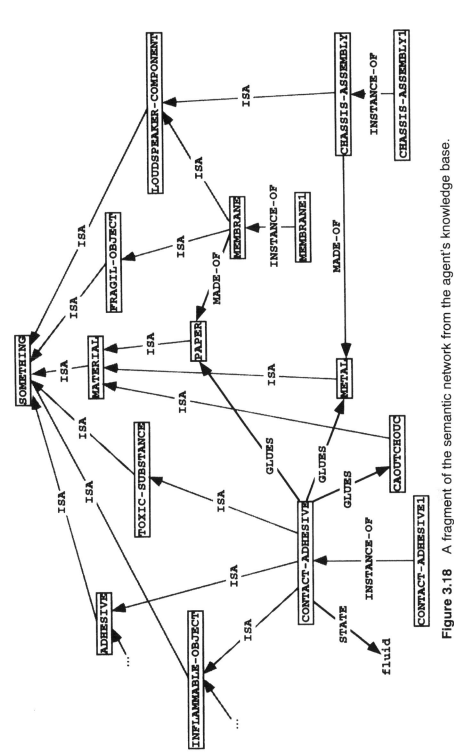

Figure 3.18 A fragment of the semantic network from the agent's knowledge base.

Table 3.5   The example E' and the rule R

E'			R		
IF			IF		
?A	IS	ATTACH	?A	IS	ATTACH
	OBJECT	?X		OBJECT	?X
	TO	?Y		TO	?Y
?X	IS	MEMBRANE1	?X	IS	SOMETHING
	MADE-OF	?M		MADE-OF	?M
?Y	IS	CHASSIS-ASSEMBLY1	?Y	IS	SOMETHING
	MADE-OF	?N		MADE-OF	?N
?Z	IS	CONTACT-ADHESIVE1	?Z	IS	ADHESIVE
	GLUES	?M		GLUES	?M
	GLUES	?N		GLUES	?N
	STATE	fluid		STATE	fluid
?M	IS	PAPER	?M	IS	MATERIAL
?N	IS	METAL	?N	IS	MATERIAL
THEN			THEN		
DECOMPOSE	TASK	?A	DECOMPOSE	TASK	?A
	INTO	?AP ?P		INTO	?AP ?P
AP	IS	APPLY	AP	IS	APPLY
	OBJECT	?Z		OBJECT	?Z
	ON	?X		ON	?X
?P	IS	PRESS	?P	IS	PRESS
	OBJECT	?X		OBJECT	?X
	ON	?Y		ON	?Y

```
    IF
         ?A    ATTACH,      OBJECT    ?X,    TO      ?Y
         ?X    SOMETHING,   MADE-OF   ?M
         ?Y    SOMETHING,   MADE-OF   ?N
         ?Z    ADHESIVE,    GLUES     ?M,    GLUES  ?N,    STATE   fluid
         ?M    MATERIAL
         ?N    MATERIAL
    THEN
         DECOMPOSE    TASK     ?A,    INTO ?AP  ?P
         ?AP   APPLY,   OBJECT   ?Z,    ON     ?X
         ?P    PRESS,   OBJECT   ?X,    ON     ?Y
    with  the  positive  example
         (?X   IS   MEMBRANE1,   ?Y  IS   CHASSIS-ASSEMBLY1,
          ?Z  IS  CONTACT-ADHESIVE1,   ?M  IS  PAPER,   ?N  IS  METAL)
```

Figure 3.19   A rule with a positive example.

3.3   Elementary problem-solving methods

In this section we will present several elementary, but general, problem-solving methods employed by the Disciple agents. The developer of a Disciple agent will have to build the domain-specific problem-solving component of the agent based upon these elementary problem-solving methods. Examples of such domain-specific problem solvers are given in the case studies from Chapters 6, 7 and 8.

3.3.1 Use of transitivity and inheritance

There are several important features of a Disciple agent's semantic network that are very useful in problem-solving. They are presented in the following.

3.3.1.1 Properties of ISA, INSTANCE-OF and IS
The **ISA** relation is transitive:

$$\forall ?X \ \forall ?Y \ \forall ?Z \ [(?X \ \mathbf{ISA} \ ?Y) \land (?Y \ \mathbf{ISA} \ ?Z) \rightarrow (?X \ \mathbf{ISA} \ ?Z)]$$

That is, if **?X** is a subconcept of **?Y** and **?Y** is a subconcept of **?Z**, then **?X** is a subconcept of **?Z**. As an example, consider the semantic network in Figure 3.18, and the concepts **CONTACT-ADHESIVE, ADHESIVE** and **SOMETHING**:

(CONTACT-ADHESIVE ISA ADHESIVE) $\land$ **(ADHESIVE ISA SOMETHING)**

$\rightarrow$ **(CONTACT-ADHESIVE ISA SOMETHING)**

The **IS** relation, being a generalization of **ISA**, is also transitive:

$$\forall ?X \ \forall ?Y \ \forall ?Z \ [(?X \ \mathbf{IS} \ ?Y) \land (?Y \ \mathbf{IS} \ ?Z) \rightarrow (?X \ \mathbf{IS} \ ?Z)]$$

The **INSTANCE-OF** relation has a property similar to transitivity:

$$\forall ?X \ \forall ?Y \ \forall ?Z \ [(?X \ \mathbf{INSTANCE\text{-}OF} \ ?Y) \land (?Y \ \mathbf{ISA} \ ?Z) \rightarrow (?X \ \mathbf{INSTANCE\text{-}OF} \ ?Z)]$$

That is, if **?X** is an instance of **?Y** and **?Y** is a subconcept of **?Z**, then **?X** is an instance of **?Z**. For instance, from the semantic network in Figure 3.18 one could make the following inferences:

(CONTACT-ADHESIVE1 INSTANCE-OF CONTACT-ADHESIVE)

$\land$ **(CONTACT-ADHESIVE ISA ADHESIVE)**

$\rightarrow$ **(CONTACT-ADHESIVE1 INSTANCE-OF ADHESIVE)**

(CONTACT-ADHESIVE1 INSTANCE-OF ADHESIVE) $\land$ **(ADHESIVE ISA SOMETHING)**

$\rightarrow$ **(CONTACT-ADHESIVE1 INSTANCE-OF SOMETHING)**

3.3.1.2 Inheritance of features
A theorem which is implicitly represented in a semantic network is the inheritance of features from a more general concept to a less general concept or instance. As indicated in the definition of instances, an instance inherits the properties of the concepts to which it belongs. That is:

$$\forall ?X \ \forall ?Y \ \forall ?Z \ [(?X \ \mathbf{INSTANCE\text{-}OF} \ ?Y) \land (?Y \ \mathbf{FEATURE} \ ?Z) \rightarrow (?X \ \mathbf{FEATURE} \ ?Z)]$$

Similarly, a concept inherits the properties of its superconcepts:

$$\forall ?X \ \forall ?Y \ \forall ?Z \ [(?X \ \mathbf{ISA} \ ?Y) \land (?Y \ \mathbf{FEATURE} \ ?Z) \rightarrow (?X \ \mathbf{FEATURE} \ ?Z)]$$

In general, because of the transitivity property of ISA, IS and INSTANCE-OF, an instance or a concept inherits all the features from all of its superconcepts:

$$\forall \ ?X \ \forall \ ?Y \ \forall \ ?Z \ [(?X \ IS \ ?Y) \wedge (?Y \ FEATURE \ ?Z) \rightarrow (?X \ FEATURE \ ?Z)]$$

For instance, from the semantic network in Figure 3.18 one can make the following inference:

$$(\textbf{CONTACT--ADHESIVE1 IS CONTACT--ADHESIVE})$$

$$\wedge \ (\textbf{CONTACT--ADHESIVE GLUES METAL})$$

$$\rightarrow (\textbf{CONTACT--ADHESIVE1 GLUES METAL})$$

The inheritance of properties is one of the important strengths of the semantic network representation, allowing a compact and economical representation of knowledge. If all the instances of a concept have the same property P with the same value, then it is enough to associate the property with the concept because it will be inherited by each of the concept's instances. There are, however, two special cases of inheritance that one should pay special attention to. They are presented in the following.

3.3.1.3 Default inheritance

There are some domains of knowledge in which exceptions to general rules exist. For example, it is generally useful to assume that all birds can fly. Certain birds such as the ostrich and the kiwi, however, cannot fly. In such a case, it is reasonable to use a representation scheme in which properties associated with concepts in a hierarchy are assumed to be true for all subconcepts and instances, unless specifically overridden by a denial or modification associated with the subconcept or instance.

Let us consider the example in Figure 3.20. The fact that a woodpecker flies is inherited from 'bird'. On the other hand, the ostrich does not inherit this property from 'bird' because it is explicitly represented that the ostrich does not fly. This overrides the default which is further up in the tree. Therefore, to find a feature of some object, the agent will first check whether the feature is explicitly associated with the object and take the corresponding value.

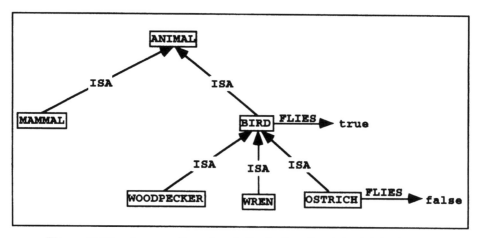

Figure 3.20 A semantic network illustrating the default inheritance.

Only if the feature is not explicitly associated with the object will the agent try to inherit it from the superconcepts of the object, by climbing the **ISA** hierarchy.

3.3.1.4 Multiple inheritance

It is possible for a concept to have more than one superconcept. For instance, in the semantic network from Figure 3.18, **CONTACT–ADHESIVE** is not only an **ADHESIVE** but also an **INFLAMMABLE–OBJECT** and a **TOXIC–SUBSTANCE**. Therefore, **CONTACT–ADHESIVE** will inherit features from all three of them and there is a potential for inheriting conflicting values. In such a case, the agent should use some strategy in selecting one of the values. A better solution, however, is to detect such conflicts when the semantic network is built or updated, and to directly associate the correct feature value with each node that would otherwise inherit conflicting values.

3.3.2 Network matching

Network matching is the operation of retrieving information from the semantic network of the agent's knowledge base. It consists of matching the description of the looked for information with a fragment of the semantic network. This operation can be used for answering questions about the objects from the agent's knowledge base, as illustrated in Figure 3.21.

The network matcher can make inferences during the matching process to match a network structure that is only implicitly represented in the agent's semantic network. Let us consider the problem of finding a network fragment in the agent's semantic network that matches the following object concept description:

$$C: \quad \textbf{?X} \quad \textbf{IS} \quad \textbf{ADHESIVE}$$

$$\textbf{STATE} \quad \textbf{fluid}$$

This is the description of the concept 'fluid adhesive' and the looked-for semantic network fragment should represent either this concept or a subconcept or an instance of it. To find it, the agent starts at the concept **ADHESIVE** (see Figure 3.18), and checks whether it or any of its subconcepts or instances is fluid. In the semantic network fragment from Figure 3.18, the agent will find the concept **CONTACT–ADHESIVE** and the instance **CONTACT–ADHESIVE1**. **CONTACT–ADHESIVE** has the property **STATE fluid** directly associated with it, while **CONTACT–ADHESIVE1** inherits this property from **CONTACT–ADHESIVE**. The top part of Figure 3.22 shows the general network corresponding to the concept C. The bottom part of Figure 3.22 shows two network fragments from the agent's semantic network that it has matched.

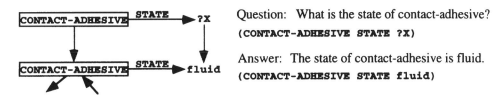

Question: What is the state of contact-adhesive?
(**CONTACT–ADHESIVE STATE ?X**)

Answer: The state of contact-adhesive is fluid.
(**CONTACT–ADHESIVE STATE fluid**)

Figure 3.21 Answering questions through network matching.

Notice that, in a problem-solving context, the expression C may be interpreted as the question:

'Is there a fluid adhesive?'

In such a case, **CONTACT-ADHESIVE** and **CONTACT-ADHESIVE1** would be two answers to this question.

Let us now consider the following concept (**?Z ?X ?M**) consisting of three clauses:

?Z	**IS**	**SOMETHING**
	GLUES	**?M**
	STATE	**fluid**
?X	**IS**	**SOMETHING**
	MADE-OF	**?M**
?M	**IS**	**MATERIAL**

In such a case the agent will look in the semantic network for a tuple of three object concepts or instances, each satisfying the descriptions of **?Z**, **?X** and **?M**, respectively. Let us start with **?Z**. In principle, because **?Z** is a subconcept of something, the agent would need to inspect every single object concept or instance from the semantic network to check whether it glues anything and is fluid. A much more efficient way is to look at the definition of the relations **GLUES** and **STATE**. Because **?z** has these features, it should be in the intersection of the domains of **GLUES** and **STATE**. As can be seen from Table 3.2, this is **ADHESIVE ∩ SOMETHING = ADHESIVE** because **ADHESIVE** is a subconcept of **SOME-THING**. Therefore, the agent should only check the concept **ADHESIVE** and its subconcepts and instances, to identify one that is fluid. It will find, for instance, **CONTACT-ADHESIVE** and **CONTACT-ADHESIVE1**. Then the agent will try to find values for **?X** and **?M**, starting with one of these values for **?Z**, for instance **CONTACT-ADHESIVE**. If no values for **?X** and **?M** are found that correspond to the value **CONTACT-ADHESIVE** of **?Z**, then the agent will try again with **CONTACT-ADHESIVE1**. So let us first consider the value **CONTACT-ADHESIVE** for **?Z**. **CONTACT-ADHESIVE** glues **METAL**, **PAPER** and **CAOUTCHOUC** (see Figure 3.18). Therefore **?M** should be one of these and the agent will check whether any of these values satisfies the description of **?M**. That is, it will check whether any of them is a **MATERIAL**. This is true for all of them. Therefore **?M** could be either **METAL** or **PAPER** or **CAOUTCHOUC**. Let **?M** be **METAL**. Now the agent will need to find an object **?X** that is made of some metal:

?X	**IS**	**SOMETHING**
	MADE-OF	**?M**
?M	**IS**	**METAL**

Again the agent will look at the definition of **MADE-OF**, but this does not help because both its domain and its range are **SOMETHING**. This means that the agent would need to inspect every object from the semantic network until it finds one made of metal. This may be very expensive computationally.

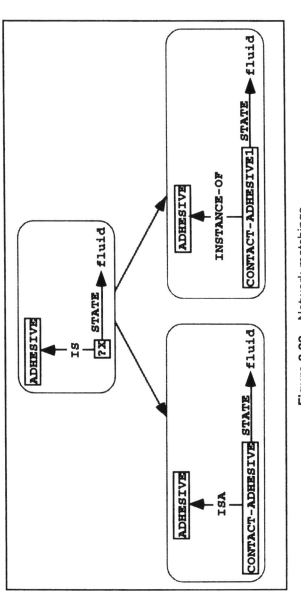

Figure 3.22 Network matchings.

However, as will be presented in detail in Section 4.1, the agent maintains some meta-knowledge about each object and relation from the semantic network. More precisely, associated with each object or relation are all the objects that use that object or relation in their definitions. That is, associated with **METAL** is the list of all the objects from the semantic network that use metal in their definition. Also, associated with **MADE-OF** are all the objects from the semantic network that use **MADE-OF** in their definition. Consequently, to find a value for **?X** one needs to only look for an object that is associated with both **METAL** and **MADE-OF**. This significantly reduces the list of objects the agent needs to check. In the presented example, such an object is **CHASSIS-ASSEMBLY**. Therefore, the following object concepts have been found that satisfy the descriptions of (**?Z, ?X, ?M**): (**CONTACT-ADHESIVE, CHASSIS-ASSEMBLY, METAL**).

The next section presents an example generation method that is based on network matching.

3.3.3   Example generation

Example generation is a method for building positive examples of rules. Let us consider, for instance, the rule in Figure 3.8. The agent will look in the semantic network for objects that satisfy the descriptions of the object concepts **?X, ?Y, ?Z, ?M** and **?N** from the rule's condition. Following the procedure described in the previous section, it will find the following objects:

(?X IS MEMBRANE, ?Y IS CHASSIS-ASSEMBLY, ?Z IS CONTACT-ADHESIVE,

?M IS PAPER, ?N IS METAL)

Replacing the class of the variables from the rule in Figure 3.8 with these values leads to the rule example from Figure 3.23. Notice that we have omitted the descriptions of **?X, ?Y**, and **?Z** because it is known that **MEMBRANE, CHASSIS-ASSEMBLY**, and **CONTACT-ADHESIVE** satisfy these descriptions. Notice also that the example in Figure 3.23 is in fact a general example because the objects from its description are not instances. Following the same procedure, the agent could also generate the example from Figure 3.17, starting from the rule from Figure 3.8.

Example generation is at the basis of the problem solver of the assessment agent described in Chapter 6.

3.3.4   Rule matching

Rule matching is the process of checking whether a rule R can be applied to perform a task T. The agent will have to check whether the task T_g from the condition of the rule is more general than the task T to be performed. If this is true then the agent will conclude that it can apply the operation from the THEN part of the rule to perform the task T.

```
IF
    ?A              IS          ATTACH
                    OBJECT      ?X
                    TO          ?Y

    ?X              IS          MEMBRANE

    ?Y              IS          CHASSIS-ASSEMBLY

    ?Z              IS          CONTACT-ADHESIVE
THEN
    DECOMPOSE   TASK        ?A
                INTO        ?AP    ?P

    ?AP             IS          APPLY
                    OBJECT      ?Z
                    ON          ?X

    ?P              IS          PRESS
                    OBJECT      ?X
                    ON          ?Y
```

Figure 3.23   A general example of the rule in Figure 3.8.

Let us consider the rule in Figure 3.8 and the following task description $T_g \wedge K_g$ from its condition:

$$T_g: \quad \text{?A} \quad \text{IS} \quad \text{ATTACH}$$

$$\text{OBJECT} \quad \text{?X}$$

$$\text{TO} \quad \text{?Y}$$

$$K_g: \quad \text{?X} \quad \text{IS} \quad \text{SOMETHING}$$

$$\text{MADE-OF} \quad \text{?M}$$

$$\text{?Y} \quad \text{IS} \quad \text{SOMETHING}$$

$$\text{MADE-OF} \quad \text{?N}$$

$$\text{?Z} \quad \text{IS} \quad \text{ADHESIVE}$$

$$\text{GLUES} \quad \text{?M}$$

$$\text{GLUES} \quad \text{?N}$$

$$\text{STATE} \quad \text{fluid}$$

$$\text{?M} \quad \text{IS} \quad \text{MATERIAL}$$

$$\text{?N} \quad \text{IS} \quad \text{MATERIAL}$$

Let us also consider that agent has the following task $T \land K$ to perform:

T:	?A1	IS	ATTACH
		OBJECT	?X1
		TO	?Y1
K:	?X1	IS	MEMBRANE
		MADE-OF	?P1
	?Y1	IS	CHASSIS-ASSEMBLY
	?P1	IS	PAPER

First, notice that the task description $T_g \land K_g$ from the rule is a conjunction of clauses as is the task $T \land K$ to perform. In order to determine whether the rule in Figure 3.8 can be applied to the task $T \land K$, the agent will need to show that $T_g \land K_g$ is more general than $T \land K$, in the context of the representation language $\mathcal{L}$ that contains the semantic network from Figure 3.18. This is done using conjunctive formula generalization. Each clause from $T_g \land K_g$ must be transformed by using the appropriate transformation rules such that it is more general than a corresponding clause from $T \land K$.

First the agent determines the substitution σ such that $\sigma(T_g) = T$:

σ(?A	IS	ATTACH	
	OBJECT	?X	
	TO	?Y) =	
(?A1	IS	ATTACH	
	OBJECT	?X1	
	TO	?Y1).	

This substitution is $\sigma = (\text{?A} \leftarrow \text{?A1}, \text{?X} \leftarrow \text{?X1}, \text{?Y} \leftarrow \text{?Y1})$.

Next the agent has to compute $\sigma(K_g)$:

$\sigma(K_g)$:	?X1	IS	SOMETHING
		MADE-OF	?M
	?Y1	IS	SOMETHING
		MADE-OF	?N
	?Z	IS	ADHESIVE
		GLUES	?M
		GLUES	?N
		STATE	fluid
	?M	IS	MATERIAL
	?N	IS	MATERIAL

The agent has to show that each clause of $\sigma(K_g)$ is more general than a corresponding clause of K. This may require extending the substitution σ and transforming the clauses of K_g and

K into equivalent clauses. Let us step through this process clause by clause. The first clause of K_g,

?X1	IS	SOMETHING
	MADE-OF	?M

is more general than the following clause from K:

?X1	IS	MEMBRANE
	MADE-OF	?P1

because **SOMETHING** is more general than **MEMBRANE**. Moreover, the substitution σ is extended with $\sigma_1 = ($?M $\leftarrow$?P1$)$ to $\sigma' = (\sigma \cup \sigma_1) = ($?A $\leftarrow$?A1, ?X $\leftarrow$?X1, ?Y $\leftarrow$?Y1, ?M $\leftarrow$?P1$)$. This leads to

$\sigma'(K_g)$:	?X1	IS	SOMETHING
		MADE-OF	?P1
	?Y1	IS	SOMETHING
		MADE-OF	?N
	?Z	IS	ADHESIVE
		GLUES	?P1
		GLUES	?N
		STATE	fluid
	?P1	IS	MATERIAL
	?N	IS	MATERIAL

Next the agent has to show that the second clause from $\sigma'(K_g)$

?Y1	IS	SOMETHING
	MADE-OF	?N

is more general than the ?Y1 clause from K:

?Y1	IS	CHASSIS-ASSEMBLY

Obviously **SOMETHING** is more general than **CHASSIS-ASSEMBLY**. Therefore, the agent needs only to check that **MADE-OF ?N** matches a relation of **CHASSIS-ASSEMBLY** from the semantic network of $\mathcal{L}$. It finds **CHASSIS-ASSEMBLY MADE-OF METAL** and extends the clause from K to:

?Y1	IS	CHASSIS-ASSEMBLY
	MADE-OF	?N
?N	IS	METAL

Therefore K is transformed into the equivalent form K'.

K':	?X1	IS	MEMBRANE
		MADE-OF	?P1
	?Y1	IS	CHASSIS-ASSEMBLY
		MADE-OF	?N
	?P1	IS	PAPER
	?N	IS	METAL

Because the clause ?z of $\sigma'(K_g)$ does not correspond to any of the clauses in K', the agent will need to show that there exists some object in the semantic network of $\mathcal{L}$ that satisfies the description of ?z. Using network matching, it will find **CONTACT-ADHESIVE** (as well as **PAPER** for ?P1 and **METAL** for ?N). Consequently, the following clause is added to K' (that becomes K''):

	?Z	IS	CONTACT-ADHESIVE
		GLUES	?P1
		GLUES	?N
		STATE	fluid

In $\mathcal{L}$, K'' is equivalent to K and K':

K'':	?X1	IS	MEMBRANE
		MADE-OF	?P1
	?Y1	IS	CHASSIS-ASSEMBLY
		MADE-OF	?N
	?Z	IS	CONTACT-ADHESIVE
		GLUES	?P1
		GLUES	?N
		STATE	fluid
	?P1	IS	PAPER
	?N	IS	METAL

Now it is obvious that $\sigma'(K_g)$ is more general than K''. Indeed, they have the same structure and each element from $\sigma'(K_g)$ is equal or more general than the corresponding element from K''. Therefore the agent has demonstrated that the task description $T_g \wedge K_g$ from the rule condition is more general than the task $T \wedge K$ to be performed, in the context of the representation language $\mathcal{L}$.

The agent can then apply the **DECOMPOSE** operation from the **THEN** part of the rule in Figure 3.8 to decompose the task:

?A1	IS	ATTACH
	OBJECT	?X1
	TO	?Y1
?X1	IS	MEMBRANE
	MADE-OF	?P1
?Y1	IS	CHASSIS-ASSEMBLY
?P1	IS	PAPER

into the following two subtasks:

?AP	IS	APPLY
	OBJECT	?Z
	ON	?X1
?P	IS	PRESS
	OBJECT	?X1
	ON	?Y1
?Z	IS	CONTACT-ADHESIVE

The problem solver of the loudspeaker manufacturing assistant presented in Chapter 2 is based on rule matching.

3.3.5   Reasoning with plausible version space rules

When trying to solve a problem, the agent will always look for a rule with a single applicability condition. However, if no such rule is applicable, the agent will try to use a plausible version space (PVS) rule. As has been presented in Section 3.1.3.4, such a rule has two conditions, a plausible upper bound condition and a plausible lower bound condition, that approximate the hypothetical exact condition of the rule (see Figure 3.10). Because the plausible lower bound condition is, as an approximation, less general than the hypothetical exact condition, any situation in which the plausible lower-bound condition is satisfied is a situation where the hypothetical exact condition of the rule is also very likely to be satisfied. Therefore, in such a situation, the conclusion of the rule is very likely to be true, as indicated in Figure 3.24. On the other hand, a situation where the plausible lower bound condition is not satisfied, but the plausible upper bound condition is satisfied, is a situation where the hypothetical exact condition may or may not be satisfied. Therefore, in such a case, the conclusion of the rule may be considered only plausible. Finally, because the plausible upper bound condition is, as an approximation, more general than the hypothetical exact condition,

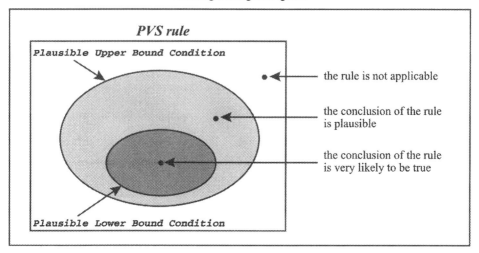

Figure 3.24   Reasoning with a PVS rule.

a situation where the plausible upper bound condition is not satisfied is a situation where the rule should not be applied.

For some types of agents (such as the design assistant presented in Chapter 8), the use of this kind of plausible reasoning is very beneficial, allowing the agent to cope with situations for which it has not been trained, and to learn from its experiences. For other types of agents (especially those used by non-experts) it is not recommended to use plausible reasoning because this may lead to wrong solutions. For instance, this is the case with the Assessment Agent presented in Chapters 2 and 6. This agent generates tests for students and these tests have to be correct. Therefore, in such situations the agent will only use the plausible lower bound conditions of the PVS rules as if the PVS rule were a rule with only that condition.

4
Knowledge Acquisition and Learning

In this chapter we will present the knowledge acquisition and learning methods of the Disciple agents that are used to build the knowledge base. As shown in Figure 4.1, the main phases of this knowledge base development process are:

- *Knowledge elicitation* – the expert defines knowledge that he/she could easily express. Some of the initial knowledge could also be imported from an existing knowledge base.
- *Rule learning* – the expert shows the agent how to solve typical domain-specific problems and helps it to understand their solutions. The agent uses learning from explanations and by analogy to learn a general plausible version space rule that will allow it to solve similar problems.
- *Rule refinement* – the agent employs learning by experimentation and analogy, inductive learning from examples and learning from explanations, to refine the rules in the knowledge base. These could be either rules learned during the rule learning process, rules directly defined by the expert during knowledge elicitation, or rules that have other origins (for instance, rules transferred from another knowledge base). Rule refinement will also cause a refinement of the concepts from the semantic network.
- *Exception handling* – the agent refines the knowledge base to remove the exceptions from the rule. A negative exception is a negative example that is covered by the rule and a

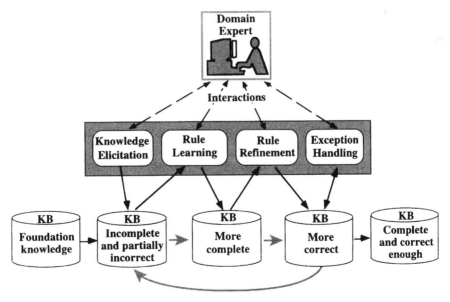

Figure 4.1 The main phases of the knowledge base development process.

positive exception is a positive example that is not covered by the rule. One common cause of the exceptions is the incompleteness of the knowledge base that does not contain the terms to distinguish between the rule's examples and exceptions. During exception handling, the agent hypothesizes additional knowledge and/or guides the expert to provide this knowledge that will extend the representation space for learning such that, in the new space, the rules could be modified to remove the exceptions.

It should be noticed that exceptions to general rules and concepts will also be accumulated during the normal problem-solving activity of the agent. Also, various changes in the application domain will require corresponding updates of the knowledge base. In such cases the same processes of knowledge elicitation, rule learning, rule refinement and exception handling will be invoked in a *retraining process*. In this approach, knowledge maintenance over the lifecycle of the knowledge base is no different from knowledge acquisition. Indeed, because the whole process of developing the knowledge base is one of creating and adapting knowledge pieces, this creation and adaptation may also occur in response to changes in the environment or goals of the system.

The following sections will present the methods employed in each of these four knowledge base development phases.

4.1   Knowledge elicitation

4.1.1   Knowledge elicitation goals and principles

The main goal of knowledge elicitation in Disciple is the manual construction of an initial knowledge base to be further extended and improved through apprenticeship and multistrategy learning and through further guided elicitation.

During the initial knowledge elicitation phase, the domain expert, possibly assisted by a knowledge engineer, has to manually define whatever knowledge pieces he/she could easily express. As described in Chapter 3, the knowledge base of a Disciple agent consists of object instances and concepts characterized by properties and relations, and of rules characterized by tasks, operations, and examples. Because it is much easier to define object instances and concepts than to define general rules, one could assume that the main result of the initial knowledge elicitation phase is a semantic network of object concepts and instances. Moreover, this semantic network is most likely to be incomplete and possibly even partially incorrect.

Although we assume that the initial knowledge base will be manually defined by the human expert from scratch, it is also possible to initialize the knowledge base from a library of reusable ontologies, common domain theories, and generic problem-solving strategies (Neches *et al.*, 1991; Genesereth and Ketchpel, 1994). Alternatively, the knowledge base could be initialized from an existing knowledge base, such as the CYC knowledge base (Guha and Lenat, 1994).

Knowledge elicitation is an interactive transfer of knowledge from the expert to the agent through definitions, refinements, and deletions of knowledge elements via the representation language defined in Chapter 3. In order to facilitate the building of an initial knowledge base that is as correct as possible, the knowledge elicitation methods of the Disciple agents are based on three important principles:

- *The dependency principle* – new knowledge elements should be defined only in terms of existing knowledge elements.

  This principle determines the order in which the knowledge elements should be defined in the knowledge base, as will be described in more detail in Section 4.1.3 and in Chapter 5. For instance, a new concept can be defined only after defining its superconcept, its properties and relations, and the concepts to which it is related. Similarly, a rule can be defined only after defining the task, operations and object concepts and instances used in its description.

- *The integrity principle* – new or updated knowledge elements should satisfy the constraints of the representation language.

  According to this principle, the agent will introduce into the knowledge base only those knowledge elements that satisfy the constraints imposed on them by the representation language. For instance, a knowledge element cannot have the same name as another knowledge element. Also, the value of a feature (property or relation) should be from the feature's range.

  When a plausible version space rule is manually defined or updated, several constraints are verified and the expert is warned of any attempt to override them. In particular, the plausible upper bound condition must be more general than the plausible lower bound condition, and both bounds should be consistent with the examples that are explicitly associated with the rule. For instance, they have to cover the positive examples that are explicitly associated with the rule. Of course, the expert could change the status of an example, for instance, from a positive exception to a positive example or from a negative example to a negative exception. In the end, however, the updated rule should satisfy the above constraints.

- *The consistency principle* – a knowledge element can be deleted only if it is not used in the description of any other knowledge element.

  The consistency principle assures that the knowledge base is consistent even if it is incomplete or partially incorrect. It also imposes an order on the updating operations of the knowledge elements. For instance, the descriptions of the knowledge elements using the element to be deleted must be modified first, and then the deletion request will be accepted by the agent.

4.1.2   Implicit associations between the knowledge elements

In order to maintain the consistency of the knowledge base and to optimize the learning and knowledge base maintenance operations, the agent automatically creates and maintains metaknowledge in the form of the following relations:

- 'Uses in its description' associates a knowledge element with all the knowledge elements from its description. This metaknowledge is created when the knowledge element is defined.

- 'Is used by the description of' associates a knowledge element with all the other knowledge elements that have this knowledge element in their descriptions. This metaknowledge is updated each time a new knowledge element is defined or an existing knowledge element is modified or deleted.

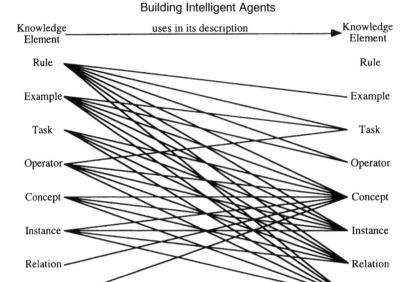

Figure 4.2 Knowledge element associations.

Figure 4.2 shows the different kinds of associations that could be created between the knowledge base elements, in accordance with the agent's representation language defined in Section 3.1.

Following the links from a knowledge element in the left-hand side of Figure 4.2 one can find all the knowledge elements that could be used in the definition of that element. For instance, a concept could use properties, relations, instances and other concepts in its definition. Similarly, by following the links from a knowledge element in the right-hand side of Figure 4.2 one can find all the knowledge elements that could use that element in their definitions. For instance, a concept could be used in the definitions of properties, relations, instances, concepts, tasks, operations, rules and examples.

Each association represented in Figure 4.2 has a specific name. For instance, the associations of a concept are called **USES-PROPERTIES**, **USES-RELATIONS**, **USES-INSTANCES**, **USES-CONCEPTS**, **USED-BY-PROPERTIES**, **USED-BY-RELATIONS**, **USED-BY-INSTANCES**, **USED-BY-CONCEPTS**, **USED-BY-TASKS**, **USED-BY-OPERATIONS**, **USED-BY-EXAMPLES** and **USED-BY-RULES**. Figure 4.3 shows the associations of the concept **SOLVENT** (see Section 3.1.2.4 for the definition of this concept).

Maintaining this kind of metaknowledge has several significant benefits that compensate for the additional memory and processing overhead. Firstly, it helps assure that the agent knowledge is consistent even if it is incomplete and partially incorrect. For instance, if the expert wants to delete a concept, the agent can immediately find all the rules that refer to it and could ask the expert to update these rules first. Secondly, it gives good insight into the content of the knowledge base and therefore allows for easy inspection and manipulation of the knowledge base by the agent's developer. Finally, it significantly speeds up the agent's learning and problem-solving processes.

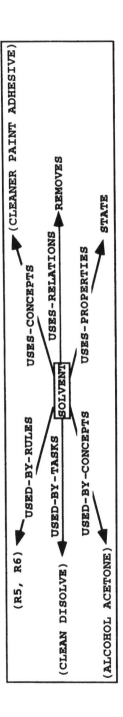

Figure 4.3 Sample associations of a concept.

4.1.3   Knowledge elicitation processes

A knowledge elicitation process may be initiated by the expert or by the agent. The expert may initiate an elicitation process by invoking any of the elicitation modules of the agent. Then the agent elicits all the necessary knowledge pieces, helping the expert to specify them. The agent may also initiate an elicitation process whenever an unknown term is used by the expert while communicating with the agent. In such a situation the agent will ask the expert to define the new term.

Each knowledge elicitation process follows the principles defined in Section 4.1.1. These principles apply to all the knowledge element types. However, in order to follow them, each knowledge element type requires a different course of action.

Common to all the knowledge elements is the naming requirement which enforces the uniqueness of names. Also common to all the elicitation processes is the specification of the natural language phrase describing the element. The remaining part of the elicitation process depends on the type of the knowledge element. Below is a brief summary of knowledge elicitation processes for all the knowledge elements:

- *Property elicitation* – elicit the property's domain, which should be an already defined abstract concept. If it is not defined, then elicit this concept first and resume the property elicitation process. Define the property's range. It does not depend on any knowledge elements.
- *Relation elicitation* – elicit the relation's domain, which should be a previously defined abstract concept. If it is not defined, then elicit this concept first and then resume the relation elicitation process. Elicit the relation's range, which should be an already defined abstract concept. If it is not, then elicit this concept first and resume the relation elicitation process.
- *Concept elicitation* – define the concept as a subconcept of existing concepts, link it with other concepts through the existing relations, and add additional characteristics by using properties and their values. If some of the concepts do not exist, then initiate the concept elicitation process to define them and resume the elicitation of the current concept. If some of the properties or relations do not exist, then initiate the property and relation elicitation processes to elicit them first.
- *Task elicitation* – elicit the list of relations and properties characterizing the task. If some of them are not defined, then initiate elicitation processes corresponding to these relations and properties.
- *Operation elicitation* – elicit the list of relations and properties characterizing the operation. If some of these features are not defined, then initiate the corresponding property and relation elicitation processes to elicit them.
- *Single condition rule elicitation* – elicit the components of the rule description – the condition and the conclusion. As part of condition elicitation, elicit an example of an already defined task. If the task is not defined, continue with the elicitation of the definition of this task and then resume the elicitation of the task's example. The elicitation of the task's example may trigger the elicitation of new object concepts or instances if they have to be part of the task example but are not defined in the knowledge base. As part of conclusion elicitation, elicit one or a sequence of already defined operations. If any of the operations is not defined, then continue with the elicitation of its definition and resume the conclusion elicitation. The elicitation of an operation may trigger the elicitation of new knowledge elements if they have to be part of the description but are not defined in the knowledge base.

- *Plausible version space (PVS) rule elicitation* – elicit the components of the rule description – the plausible upper bound condition, the plausible lower bound condition, and the conclusion, as in the single condition rule elicitation process. Check that the plausible upper bound condition is more general than the plausible lower bound condition and ask the expert to update the rule if this constraint is not satisfied.

The above elicitation processes are executed either during the construction of the initial knowledge base, or during the subsequent phases of knowledge base refinement. The following knowledge elicitation processes are usually executed only during knowledge base refinement:

- *Example elicitation* – elicit the following components of the example description – a task example and an operation (or a sequence of operations), as indicated in the description of the elicitation of single condition rules. The only difference is that the task and the operations will generally be specific descriptions.
- *Explanation elicitation* – elicit the explanation type (see Section 4.2.2.1 for the different types of explanations) and then ask the expert to select its components from the existing object instances and concepts and their properties and relations. If new knowledge elements need to be included in the explanation, then they need to be defined first.

Similar processes are executed for updating a knowledge element except that, in some cases, more checks could be performed. For instance, a manually updated rule should still be consistent with the examples from which it was learned, unless the status of these examples is explicitly changed by the expert.

With respect to the deletion of a knowledge element KE_i, this will be performed only if KE_i is not used in the definition of any other knowledge element KE_j from the current knowledge base. Otherwise, the expert will be asked to first modify the definition of KE_j to no longer refer to KE_i. When these operations are completed, the element KE_i can be deleted.

This initial knowledge base elicited from the expert will be further extended and refined as indicated in the following sections.

4.2 Rule learning

Assuming that the knowledge base is now complete enough to formulate examples of problem-solving episodes, the expert can begin teaching the agent through such examples. In this section we will present the method used by the Disciple agents to learn PVS rules from these examples and their explanations.

4.2.1 The rule-learning problem

The representation language has been introduced in Section 3.1 as a tuple $\mathcal{L} = (V, C, F, S, H, T, O, L)$, where S is an incomplete and possibly partially incorrect semantic network of object concepts and instances. A sample of such a semantic network for the manufacturing assistant is presented in Figure 4.4.

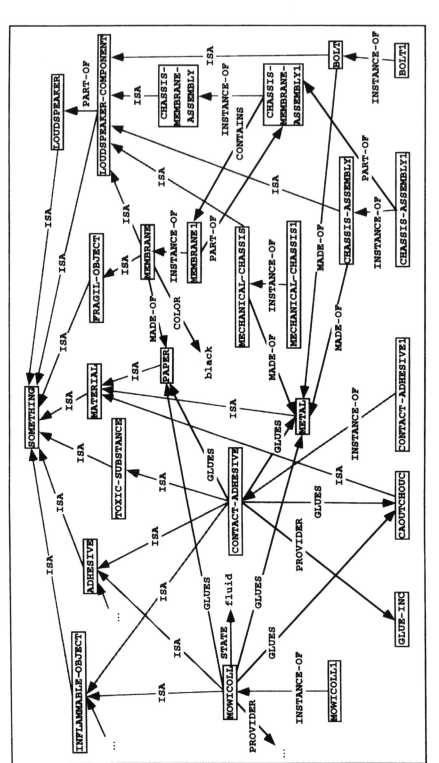

Figure 4.4 A sample semantic network.

Table 4.1 The rule learning problem

Given:
- A representation language $\mathcal{L}$ for concepts and rules that includes an incomplete and possibly partially incorrect semantic network of object concepts and instances;
- An example (problem-solving episode);
- An expert that understands why the given example is correct and may answer agent's questions.

Determine:
- A PVS rule that is an analogy-based generalization of the example;
- An improved representation language.

The rule learning problem is defined in Table 4.1. The agent receives an example and learns a rule which is an analogy-based generalization of the example. There is no restriction with respect to what the example actually represents. However, the example has to be described as a task and an operation that should be applied by the agent to perform the task. Therefore, this example may also be referred to as a problem-solving episode. The learned rule may represent a decomposition rule, an inference rule, or any other type of IF–THEN rule.

Figure 4.5 contains an example of a problem-solving episode for the manufacturing assistant. The expert training the agent should understand why the problem-solving episode is correct and should be able to answer the agent's questions. For instance, the expert should be able to distinguish between an explanation of the problem-solving episode and an irrelevant sentence, to recognize false sentences, to confirm whether an object has a certain feature, and/or to provide explanations.

The main result of the rule learning process is a general PVS rule that will allow the agent to solve problems by analogy with the example from which the rule was learned. The PVS rule learned from the example in Figure 4.5 is shown in Figure 4.6.

During rule learning, the agent might also improve the representation language in two ways. First, the semantic network might be extended with new object features or some of the existing feature representations might be corrected. For instance, the following property

```
IF  the  task  is                                  ; IF
        ?A          IS        ATTACH               ; the task is to attach
                    OBJECT    MEMBRANE1            ; membrane1
                    TO        CHASSIS-ASSEMBLY1    ; to chassis-assembly1
THEN                                               ; THEN
        DECOMPOSE TASK        ?A                   ; decompose this task
                  INTO        ?AP   ?P             ; into ?ap and ?p

        ?AP         IS        APPLY                ; ?ap is the task to apply
                    OBJECT    CONTACT-ADHESIVE1    ; contact-adhesive1
                    ON        MEMBRANE1            ; on membrane1

        ?P          IS        PRESS                ; ?p is the task to press
                    OBJECT    MEMBRANE1            ; membrane1
                    ON        CHASSIS-ASSEMBLY1    ; on chassis-assembly1
```

Figure 4.5 An example of a problem-solving episode.

```
Plausible  Upper  Bound  IF            ; IF (Plausible Upper Bound)
   ?A           IS       ATTACH        ; the task is to attach
                OBJECT   ?X            ; object ?x
                TO       ?Y            ; to ?y and

   ?X           IS       SOMETHING     ; ?x is some object
                MADE-OF  ?M            ; made of ?m and

   ?Y           IS       SOMETHING     ; ?y is some object
                MADE-OF  ?N            ; made of ?n and

   ?Z           IS       ADHESIVE      ; ?z is an adhesive that
                GLUES    ?M            ; glues ?m and
                GLUES    ?N            ; glues ?n and
                STATE    {solid fluid  gas}  ; is solid, fluid, or gas and

   ?M           IS       MATERIAL      ; ?m is a material and

   ?N           IS       MATERIAL      ; ?n is a material
Plausible  Lower  Bound  IF            ; IF (Plausible Lower Bound)
   ?A           IS       ATTACH        ; the task is to attach
                OBJECT   ?X            ; object ?x
                TO       ?Y            ; to ?y and

   ?X           IS       MEMBRANE1     ; ?x is membrane1
                MADE-OF  ?M            ; made of ?m and

   ?Y           IS       CHASSIS-ASSEMBLY1 ; ?y is chassis-assembly1
                MADE-OF  ?N            ; made of ?n and

   ?Z           IS       CONTACT-ADHESIVE1 ; ?z is contact-adhesive1 that
                GLUES    ?M            ; glues ?m and
                GLUES    ?N            ; glues ?n and
                STATE    fluid         ; is fluid and

   ?M           IS       PAPER         ; ?m is a paper and

   ?N           IS       METAL         ; ?n is a metal
THEN                                   ; THEN
   DECOMPOSE TASK        ?A            ; decompose the task ?a
                INTO     ?AP ?P        ; into ?ap and ?p

   ?AP          IS       APPLY         ; ?ap is the task to apply
                OBJECT   ?Z            ; object ?z
                ON       ?X            ; on ?x

   ?P           IS       PRESS         ; ?p is the task to press
                OBJECT   ?X            ; object ?x
                ON       ?Y            ; on ?y
with  the  positive  example
   (?X  IS MEMBRANE1,  ?Y  IS CHASSIS-ASSEMBLY1,
    ?Z  IS CONTACT-ADHESIVE1, ?M IS PAPER, ?N IS METAL)
```

Figure 4.6   A learned PVS rule.

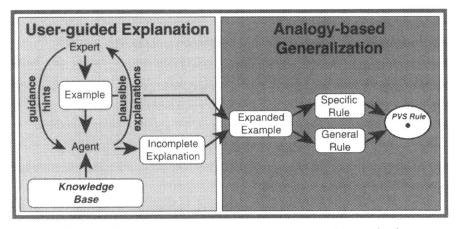

Figure 4.7   The main phases of the rule learning method.

might be added to the semantic network, if it was not already there:

CONTACT-ADHESIVE1 STATE fluid

Secondly, if **STATE** was not among the elements of the set F of the features from the representation language $\mathcal{L}$, then the expert will need to define this feature and accordingly extend F.

4.2.2   The rule-learning method

The rule learning method is presented schematically in Figure 4.7. As in explanation-based learning (DeJong and Mooney, 1986; Mitchell *et al.*, 1986), it consists of two phases, explanation and generalization. However, in the explanation phase the agent is not building a proof tree, and the generalization is not a deductive one.

In the explanation phase the agent, using the knowledge from the knowledge base and being guided by the expert, generates an explanation of why the example from Figure 4.5 is correct. In the generalization phase, the agent generalizes the example and the found explanation into the PVS rule from Figure 4.6 by using analogical reasoning.

In the following we will describe and illustrate this learning method. First we will define more precisely what an explanation is. Then we will present various strategies by which an expert could guide the agent in generating the explanations of the problem-solving episode. Finally we will present and justify the generalization method which is based on analogical reasoning.

4.2.2.1   *What is an explanation of an example?*
Let us consider the example (problem-solving episode) in Figure 4.5 written in the following simplified form:

```
IF the task is
    ATTACH OBJECT MEMBRANE1 TO CHASSIS-ASSEMBLY1
THEN decompose this task into the subtasks
    APPLY OBJECT CONTACT-ADHESIVE1 ON MEMBRANE1
    PRESS OBJECT MEMBRANE1 ON CHASSIS-ASSEMBLY1
```

| CONTACT-ADHESIVE1 STATE fluid |
| CONTACT-ADHESIVE1 GLUES PAPER ∧ MEMBRANE1 MADE-OF PAPER |
| CONTACT-ADHESIVE1 GLUES METAL ∧ CHASSIS-ASSEMBLY1 MADE-OF METAL |

Figure 4.8   The explanation of why the example in Figure 4.5 is correct.

If an expert would be asked why this problem-solving episode is correct then it would be natural to expect an explanation such as: '**CONTACT–ADHESIVE1** is fluid, it glues paper and **MEMBRANE1** is made of paper, and it also glues metal and **CHASSIS–ASSEMBLY1** is made of metal'.

As can be seen, this explanation consists of several explanation pieces, and each piece expresses a feature of some object, or a relation between two objects from the problem-solving episode, as indicated in Figure 4.8.

Each explanation piece is a path in the agent's semantic network of object concepts and instances, as shown in Figure 4.9. Notice that the path **CONTACT–ADHESIVE1 GLUES PAPER ∧ MEMBRANE1 MADE–OF PAPER** is not explicitly represented in the semantic network in Figure 4.4, but is inferred from the explicitly represented paths. In particular, **CONTACT–ADHESIVE1 GLUES PAPER** is inherited from **CONTACT–ADHESIVE GLUES PAPER** and **MEMBRANE1 MADE–OF PAPER** is inherited from **MEMBRANE MADE–OF PAPER**.

One can distinguish five types of explanation pieces that correspond to various types of path in the agent's semantic network:

- *Association* – a relation between two objects in the example (see Figure 4.10).
- *Correlation* – a common feature of two objects in the example (see Figure 4.11).
- *Property* – a property of an object in the example (see Figure 4.12).

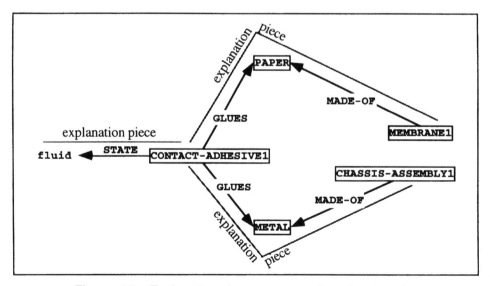

Figure 4.9   Explanation pieces as semantic network paths.

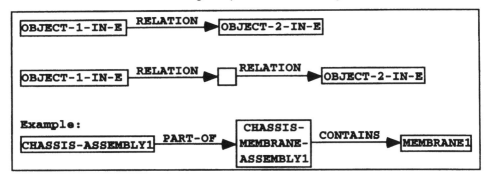

Figure 4.10   Explanations of type association.

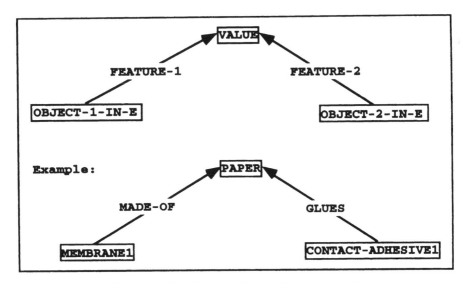

Figure 4.11   Explanations of type correlation.

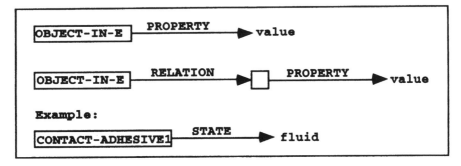

Figure 4.12   Explanations of type property.

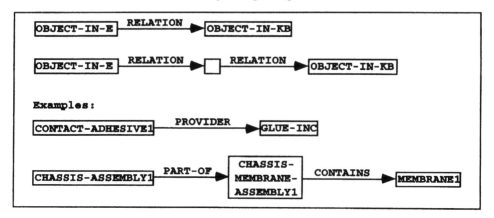

Figure 4.13   Explanations of type relation.

- *Relation* – a relation between an object in the example and an object in the knowledge base (see Figure 4.13).
- *Generalization* – a generalization of an object in the example (see Figure 4.14).

In conclusion, an explanation of an example (problem-solving episode) consists of several explanation pieces. Each explanation piece corresponds to a path in the semantic network between an object in the example (problem-solving episode) and another object. In principle, the path could have any length. In practice, however, one has to set upper limits to the maximum length of the path for a certain type of explanation, in order to reduce the number of plausible explanations generated by the agent.

The next section presents the explanation generation method.

4.2.2.2   The explanation generation method

In the baseline explanation generation method, the agent searches the semantic network for all the explanation types presented in the previous section and proposes them to the expert.

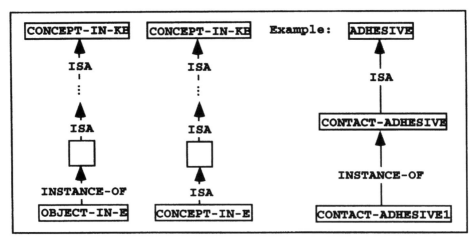

Figure 4.14   Explanations of type generalization.

The expert has to choose the correct explanations and can also define additional explanations. Choosing the correct explanations from an agent generated list has two advantages:

- it is easier for the expert to choose an explanation from a set than to define it;
- the generated explanation is already expressed in the representation language of the agent.

There are, however, two problems with the baseline approach. First of all, for a complex application domain, too many explanations will be generated. The process of analyzing each explanation and choosing the relevant ones will require a significant amount of time.

The other problem is when the expert has to define a new explanation by him/herself. This situation may appear quite often because the semantic network of the agent may be incomplete and this may not contain some relevant explanation pieces. By providing these explanations the expert not only helps the agent learn a rule, but also adds to the agent's knowledge. This, however, is a difficult communication problem. Indeed, we have noticed that it is not easy for the expert to define new explanations in terms of new concepts because he/she has difficulties in relating them appropriately to the concepts already defined. Both of these problems are addressed in the refined explanation generation method by using the limited-length heuristic and the attention-focusing method described below.

The limited-length heuristic. This heuristic consists of limiting the length of the semantic network path proposed by the agent as a plausible explanation of an example. It does not limit, however, the length of the explanations given by the expert. One justification of this heuristic is the combinatorial explosion resulting from the generation paths of any length. Another is that the longer the path in the semantic network the less likely the path represents an explanation of an example. This observation is confirmed by all the domains to which we have applied the Disciple approach.

Focusing agent's attention in explanation generation. The expert may ask the agent to generate only explanations related to a certain object. For instance, with respect to the problem-solving episode in Figure 4.5, the expert may ask the agent to generate only explanations involving **MEMBRANE1**. In such a case only the following explanations would be generated:

MEMBRANE1 HAS-COLOR black

MEMBRANE1 MADE-OF PAPER ∧ CONTACT-ADHESIVE1 GLUES PAPER

MEMBRANE1 PART-OF CHASSIS-MEMBRANE-ASSEMBLY1

    ∧ CHASSIS-ASSEMBLY1 PART-OF CHASSIS-MEMBRANE-ASSEMBLY1

The expert may further guide the agent by asking it to generate only certain types of explanations, such as, only explanations of type association or property.

Defining additional explanations. The expert may also define additional explanations. For instance, if the description of **CONTACT-ADHESIVE1** does not specify its state, then the explanation

        CONTACT-ADHESIVE1 STATE fluid

might be provided by the expert. This also results in the improvement of the description of **CONTACT-ADHESIVE1** and in the definition of the **STATE** property if this property has not been previously defined.

Table 4.2   The first part of the rule learning method: user-guided explanation

---

Let E be an example.

Repeat

- The expert focuses the agent's attention by pointing to the most relevant object(s) from the input example and/or by specifying the types of plausible explanations to be generated;
- The agent proposes all the plausible explanation pieces satisfying expert's constraints and consisting of paths composed of no more than n features (limited length heuristic);
- The expert chooses the relevant explanation pieces;
- The expert may define additional explanation pieces;

until the expert is satisfied with the explanation of the example.

The explanation of why the problem-solving episode in Figure 4.5 is correct is presented in Figure 4.8. The explanation generation method is presented in Table 4.2. The explanation of the example is most likely to be incomplete for at least three reasons:

- The semantic network of the agent may be incomplete, and therefore the agent may not be able to propose all the partial explanations of the example simply because they are not represented in the semantic network.
- The agent uses the limited-length heuristic to propose partial explanations and if an explanation is a longer path in the current semantic network, it will not be proposed to the expert.
- It is often the case that the human experts forget to provide explanations that correspond to commonsense knowledge which would be obvious to any human apprentice.

In the following sections we will show how the explanation and the example are generalized by using analogy-based generalization.

4.2.2.3 Analogical reasoning

The central intuition supporting the learning by analogy paradigm is that if two entities are similar in some respects then they could be similar in other respects as well. An important result of the learning by analogy research (Burstein, 1986; Carbonell, 1983, 1986; Gentner, 1983; Kedar-Cabelli, 1987; Prieditis, 1988; Russell, 1989; Winston, 1980) is that analogy involves mapping some underlying causal network of relations between analogous situations. By causal network of relations it is generally meant a set of relations related by higher-order relations such as 'physical-cause(r_i, r_j)', 'logically-implies(r_i, r_j)', 'enables(r_i, r_j)', 'explains(r_i, r_j)', 'justifies(r_i, r_j)', 'determines(r_i, r_j)' etc. The idea is that such similar 'causes' are expected to have similar effects (see Figure 4.15). That is, because A' is similar to A that causes B, one would expect that A' would cause B' that would be similar to B.

In Disciple, the relation between the explanation and the example is such a causal relation. The statements

CONTACT-ADHESIVE1 STATE fluid

CONTACT-ADHESIVE1 GLUES PAPER ∧ MEMBRANE1 MADE-OF PAPER

CONTACT-ADHESIVE1 GLUES METAL ∧ CHASSIS-ASSEMBLY1 MADE-OF METAL

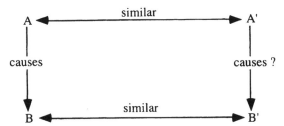

Figure 4.15 Similar causes are expected to have similar effects.

explain why the problem-solving episode

    IF the task is

        ATTACH OBJECT MEMBRANE1 TO CHASSIS-ASSEMBLY1

    THEN decompose this task into the subtasks

        APPLY OBJECT CONTACT-ADHESIVE1 ON MEMBRANE1

        PRESS OBJECT MEMBRANE1 ON CHASSIS-ASSEMBLY1

is correct. This is schematically represented in Figure 4.16.

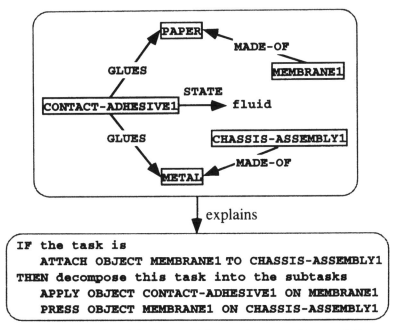

Figure 4.16 'Explains' as a higher-order causal-type relation

Let us now consider the following true statements that are similar to the explanation of the input problem-solving episode:

MOWICOLL1 STATE fluid

MOWICOLL1 GLUES METAL ∧ BOLT1 MADE-OF METAL

MOWICOLL1 GLUES METAL ∧ MECHANICAL-CHASSIS1 MADE-OF METAL

In this case we could conclude that the following episode,

 IF the task is

  ATTACH OBJECT BOLT1 TO MECHANICAL-CHASSIS1

 THEN decompose this task into the subtasks

  APPLY OBJECT MOWICOLL1 ON BOLT1

  PRESS OBJECT BOLT1 ON MECHANICAL-CHASSIS1

is likely to be correct.

Now let us point out that an agent making the above analogical inference must equally infer that the following problem-solving episode,

 IF the task is

  ATTACH OBJECT ?X TO ?Y

 THEN decompose this task into the subtasks

  APPLY OBJECT ?Z ON ?X

  PRESS OBJECT ?X ON ?Y

is likely to be correct for any objects **?X**, **?Y**, and **?Z** for which the expression

   ?Z STATE fluid

   ?Z GLUES ?M ∧ ?X MADE-OF ?M

   ?Z GLUES ?N ∧ ?Y MADE-OF ?N

is true.

Notice that, while two situations are usually considered to be analogous if they match within a certain predefined threshold, in Disciple, two situations are considered to be analogous if they generalize within a predefined threshold (the analogy criterion). This is not at all surprising since generalization may be reduced to structural matching (Kodratoff and Ganascia, 1986).

This type of analogical reasoning is schematically represented in Figure 4.17. The explanation from the left-hand side of Figure 4.17 explains why the input example is correct. The expression from the right-hand side of Figure 4.17 is similar with this explanation because both of them are less general than the analogy criterion from the top of Figure 4.17. Therefore, one may infer by analogy that the expression from the right-hand side of Figure 4.17 explains an example that is similar to the initial example.

4.2.2.4 The analogy-based generalization method

Based on the observations from the previous section one can generalize the initial problem-solving episode indicated by the expert into a plausible version space rule that will allow the agent to perform similar problem-solving episodes.

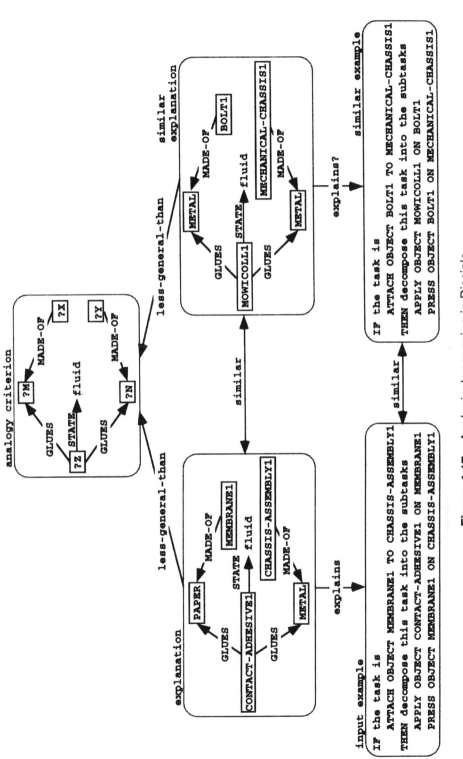

Figure 4.17 Analogical reasoning in Disciple.

Plausible	Lower	Bound	IF		; IF (Plausible Lower Bound)
? A		IS	ATTACH		; the task is to attach
		OBJECT	? X		; object ?x
		TO	? Y		; to ?y and
? X		IS	MEMBRANE1		; ?x is membrane1
		MADE-OF	? M		; made of ?m and
? Y		IS	CHASSIS-ASSEMBLY1		; ?y is chassis-assembly1
		MADE-OF	? N		; made of ?n and
? Z		IS	CONTACT-ADHESIVE1		; ?z is contact-adhesive1 that
		GLUES	? M		; glues ?m and
		GLUES	? N		; glues ?n and
		STATE	fluid		; is fluid and
? M		IS	PAPER		; ?m is a paper and
? N		IS	METAL		; ?n is a metal
THEN					; THEN
	DECOMPOSE	TASK	? A		; decompose the task ?a
		INTO	?AP ? P		; into ?ap and ?p
? AP		IS	APPLY		; ?ap is the task to apply
		OBJECT	? Z		; object ?z
		ON	? X		; on ?x
? P		IS	PRESS		; ?p is the task to press
		OBJECT	? X		; object ?x
		ON	? Y		; on ?y

Figure 4.18   The plausible lower bound rule *LR* that covers only the initial example.

First, let us note that the problem-solving episode from Figure 4.5 and its explanation from Figure 4.8 could be joined into the expanded example description shown in Figure 4.18. This expanded problem-solving episode could be interpreted as a very specific rule that covers only this episode (i.e. its variables may only be instantiated to values from the problem-solving episode). We will further refer to this rule as the plausible lower bound rule *LR* because it corresponds to the plausible lower bound and the conclusion of the rule to be learned (see Figure 4.6).

As justified in the previous section, we would expect that such a problem-solving episode will also be correct when the variables take as values objects that have the indicated features.

We have shown in Section 3.1.2.3 that each feature f is defined in terms of two sets: its domain (the set D_f of objects that could have this feature) and its range (the set R_f of the possible values for the feature). Consequently, if a variable $?x$ has the feature f_1 and is the value of the feature f_2, then the set of possible values of $?x$ is restricted to $D_{f1} \cap R_{f2}$.

Obviously this extends to the case where $?x$ has the features $f_{11}, ..., f_{1m}$, and appears as value of the features $f_{21}, ..., f_{2n}$. In such a case, the set of possible values of $?x$ is restricted to:

$$D_{f11} \cap D_{f12} \cap \cdots \cap D_{f1m} \cap R_{f21} \cap R_{f22} \cap \cdots \cap R_{f2n}$$

For instance, the features from the rule in Figure 4.18 are defined as indicated in Table 4.3.

Table 4.3   Definitions of some features

Name	Description	Domain	Range
IS	is	SOMETHING	SOMETHING
OBJECT	object	TASK	SOMETHING
TO	to	TASK	SOMETHING
MADE-OF	made of	SOMETHING	MATERIAL
GLUES	glues	ADHESIVE	MATERIAL
STATE	state	SOMETHING	{solid fluid gas}
TASK	task	OPERATION	TASK
INTO	into	OPERATION	TASK
ON	on	TASK	SOMETHING
PART-OF	part of	SOMETHING	SOMETHING

These feature definitions restrict the set of possible values of the variables from the rule in Figure 4.18. Let us consider, for instance, the variable **?M**. It has the relation **IS** and is a value of the relations **MADE-OF** and **GLUES**. Therefore, it can only take values from the following set:

$$D_{\mathbf{IS}} \cap R_{\mathbf{MADE\text{-}OF}} \cap R_{\mathbf{GLUES}} = \mathbf{SOMETHING} \cap \mathbf{MATERIAL} \cap \mathbf{MATERIAL} = \mathbf{MATERIAL}$$

Consequently, one could generalize the rule in Figure 4.18 to the rule in Figure 4.19. We will further refer to this rule as the plausible upper bound rule *UR* because it corresponds to the plausible upper bound and the conclusion of the rule to be learned (see Figure 4.6).

The plausible upper bound rule *UR* and the plausible lower bound rule *LR* have different conditions but the same conclusion. Moreover, the condition of *UR* is a generalization of the condition of *LR*. Therefore, the two rules could be compactly represented into a single rule with two conditions and one conclusion. This rule, called the PVS rule is the learned rule (see Figure 4.6). The method for generating such a rule from an example and its explanation is summarized in Table 4.4.

4.2.3   Characterization of the learned PVS rule

As shown in the previous section, the plausible upper bound condition of the learned rule (see Figure 4.6) is an analogy criterion that allows the agent to solve problems by analogy with the example from which the rule was learned (see Figure 4.5). Because analogy is only a plausible reasoning process, some of the examples covered by the rule may be wrong. The plausible upper bound of the rule is therefore only an approximation of a hypothetical exact condition that will cover only positive examples of the rule. That is why it is called plausible upper bound. The plausible lower bound condition of the rule in Figure 4.6 covers only the input example that is known to be correct. Therefore, this bound is a true lower bound for the hypothetical exact condition. However, we call it plausible lower bound because it may be over-generalized during the process of refining the rule (see Section 4.3) and may become a plausible condition as well. Figure 4.20 shows the most likely relation between the plausible lower bound, the plausible upper bound and the hypothetical exact condition of the rule. Notice that there are instances of the plausible upper bound that are not instances of the hypothetical exact condition of the rule. This means that the learned rule

Plausible	Upper	Bound	IF		; IF (Plausible Upper Bound)
	?A	IS	ATTACH		; the task is to attach
		OBJECT	?X		; object ?x
		TO	?Y		; to ?y and
	?X	IS	SOMETHING		; ?x is some object
		MADE-OF	?M		; made of ?m and
	?Y	IS	SOMETHING		; ?y is some object
		MADE-OF	?N		; made of ?n and
	?Z	IS	ADHESIVE		; ?z is an adhesive that
		GLUES	?M		; glues ?m and
		GLUES	?N		; glues ?n and
		STATE	{solid fluid gas}		; is solid, fluid, or gas and
	?M	IS	MATERIAL		; ?m is a material and
	?N	IS	MATERIAL		; ?n is a material
THEN					; THEN
	DECOMPOSE	TASK	?A		; decompose the task ?a
		INTO	?AP ?P		; into ?ap and ?p
	?AP	IS	APPLY		; ?ap is the task to apply
		OBJECT	?Z		; object ?z
		ON	?X		; on ?x
	?P	IS	PRESS		; ?p is the task to press
		OBJECT	?X		; object ?x
		ON	?Y		; on ?y

Figure 4.19   Generalization of the rule in Figure 4.18.

could also generate wrong solutions to some problems, as already mentioned. Also, there are instances of the hypothetical exact condition that are not instances of the plausible upper bound. This means that the plausible upper bound does not cover all the cases in which the solution provided by the rule would be correct. Both of these situations are a consequence of the fact that the explanation of the initial example might be incomplete, and are consistent with what one would expect from an agent performing analogical reasoning.

Table 4.4 The second part of the rule learning method: analogy-based generalization

Let E be an example and EX its explanation.

- Create an expanded example that includes the explanation.
- Transform the expanded example into a very specific rule that covers only this example by introducing a variable for each instance and restricting the value of the variable to the corresponding instance. Call this rule the plausible lower bound rule LR.
- Build the plausible upper bound rule UR as the generalization of LR where each variable may take any value allowed by the definitions of the features from LR that relate to that variable.
- Merge the two rules UR and LR into a plausible version space rule $PVSR$. The plausible upper bound condition of $PVSR$ is the condition of UR, the plausible lower bound condition is the condition of LR, and the THEN part is the common THEN part of UR and LR.

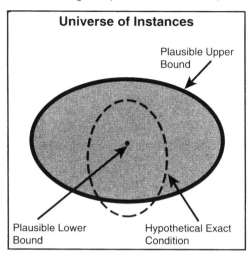

Figure 4.20   The relations between the conditions of the learned rule.

The next section presents several methods for refining PVS rules such that both its bounds approximate better the hypothetical exact condition and possibly become identical with it.

4.3   Rule refinement

4.3.1   The rule refinement problem

As a result of the knowledge elicitation and rule learning processes, the knowledge base of the agent will consist of a semantic network and a set of PVS rules.

The semantic network was created by the expert who defined all the concepts that he/she could easily express. It may have also been imported from an external knowledge base. This semantic network has been further extended with new terms as these terms were used by the expert to define problem-solving episodes to train the agent or to provide explanations of these episodes. It is natural to assume that this semantic network is incomplete and possibly partially incorrect.

The rules learned by the agent are PVS rules. The lower bound of each PVS rule covers only one instance and the upper bound covers instances that are only expected but not guaranteed to be correct, because they are based on analogical reasoning. Therefore these rules are incomplete and possibly partially incorrect.

The knowledge base may also contain rules manually defined by the expert or imported from an external knowledge base. These rules are most likely to be rules with a single condition and could also be incomplete and partially incorrect.

The goal of rule refinement is to improve the rules and to correspondingly extend and correct the semantic network until all the instances of the rules in the semantic network are correct. Because the rule refinement methods apply to PVS rules, the imported rules and the rules manually defined by the expert are first transformed by the agent into PVS rules. The

Table 4.5   The rule refinement problem

Given:
- A PVS rule;
- A positive or a negative example of the rule;
- An incomplete and partially incorrect semantic network;
- An expert that may answer agent's questions.

Determine:
- An improved rule that covers the example if it is positive, or does not cover the example if it is negative;
- An improved semantic network.

---

basic idea is to interpret the rule's condition as a plausible lower bound condition, and to generate the plausible upper bound condition as a generalization of the plausible lower bound condition, by using the generalization method described in Section 4.2.2.4.

The rule refinement problem is defined in Table 4.5. The triggering event for rule refinement is the discovery of a new example of the rule. There are various possible origins for these examples:

- The examples are generated by the agent itself through active experimentation (see Section 4.3.3) or rule verification (see Section 4.3.4).
- The examples are provided by the expert.
- The examples are obtained by the agent during its normal problem-solving operation.
- The examples are obtained from an external source (such as a repository of examples).

Regardless of the origin of a rule's example, the goal of the agent is to refine the rule such that it is consistent with the example or, if this is not possible, to learn a new rule starting from this example. A likely effect of rule refinement is the improvement of the semantic network of the agent by the addition of new knowledge pieces or by the correction of existing ones.

The next section presents an overview of the rule refinement method.

4.3.2   The rule refinement method

A high-level description of the rule refinement method is given in Table 4.6. More details are given in the following sections.

There are several strategies employed by the agent to perform rule refinement:

- active experimentation with the plausible upper bound (this covers case 2 of the method in Table 4.6);
- rule verification through active experimentation with the plausible lower bound (this covers case 1 of the method in Table 4.6);
- rule refinement with external examples (this covers all the three cases of the method in Table 4.6).

We will present each of these strategies in turn.

Table 4.6 The rule refinement method

Let R be a PVS rule, U its plausible upper bound condition, L its plausible lower bound condition, and E a new example of the rule.

1. *If E is covered by L* **then**
   - If E is a positive example **then** R need not to be refined.
   - If E is a negative example **then** both U and L need to be specialized as little as possible to no longer cover this example while still covering the known positive examples of the rule. If this is not possible, then E represents a negative exception to the rule. The rule would need to be refined as part of the exception-handling process.

   This first case is discussed in Section 4.3.4.

2. *If E is covered by U but it is not covered by L* **then**
   - If E is a positive example **then** L needs to be generalized as little as possible to cover it while remaining less general or at most as general as U.
   - If E is a negative example **then** U needs to be specialized as little as possible to no longer cover it while remaining more general than or at least as general as L. Alternatively, both bounds are specialized.

   This second case is discussed in Section 4.3.3.

3. *If E is not covered by U* **then**
   - If E is a positive example **then** it represents a positive exception to the rule. The rule would need to be refined as part of the exception-handling process described in Section 4.4.
   - If E is a negative example **then** no refinement is necessary.

   This third case is discussed in Section 4.4.

4.3.3 Rule refinement through active experimentation with its plausible upper bound

Let us consider that the rule to be refined is the one in Figure 4.21. The only difference between this rule and the rule in Figure 4.6 is the absence of the property **STATE** of the variable **?Z**, in both bounds of the rule. This is the rule that would have been learned from the example in Figure 4.5 if the explanation **CONTACT-ADHESIVE1 STATE fluid** were not found during the rule learning process described in Section 4.2.

One purpose for choosing this form of the rule is to show that it is not necessary that all the explanations of an example are found during the rule-learning process. Indeed, additional explanations could also be found during the rule refinement process.

4.3.3.1 The active experimentation process

The method for rule refinement through active experimentation with its plausible upper bound is illustrated in Figure 4.22. The agent generates rule examples that are covered by the plausible upper bound without being covered by the plausible lower bound. The rule examples are generated by applying the example generation method described in Section 3.3.3. For instance, if the rule is the one from Figure 4.21, then the agent looks in the semantic network for a tuple of objects that are examples of the **?X, ?Y, ?Z, ?M,** and **?N** concepts from the plausible upper bound condition of the rule, without being examples of the corresponding concepts from the plausible lower bound condition of the rule. Then it

```
Plausible Upper Bound IF
        ?A              IS              ATTACH
                        OBJECT          ?X
                        TO              ?Y

        ?X              IS              SOMETHING
                        MADE-OF         ?M

        ?Y              IS              SOMETHING
                        MADE-OF         ?N

        ?Z              IS              ADHESIVE
                        GLUES           ?M
                        GLUES           ?N

        ?M              IS              MATERIAL

        ?N              IS              MATERIAL

Plausible Lower Bound IF
        ?A              IS              ATTACH
                        OBJECT          ?X
                        TO              ?Y

        ?X              IS              MEMBRANE1
                        MADE-OF         ?M

        ?Y              IS              CHASSIS-ASSEMBLY1
                        MADE-OF         ?N

        ?Z              IS              CONTACT-ADHESIVE1
                        GLUES           ?M
                        GLUES           ?N

        ?M              IS              PAPER

        ?N              IS              METAL
THEN
        DECOMPOSE TASK                  ?A
                        INTO            ?AP ?P

        ?AP             IS              APPLY
                        OBJECT          ?Z
                        ON              ?X

        ?P              IS              PRESS
                        OBJECT          ?X
                        ON              ?Y

with the positive example
    ( ?X IS MEMBRANE1, ?Y IS CHASSIS-ASSEMBLY1,
      ?Z IS CONTACT-ADHESIVE1, ?M IS PAPER, ?N IS METAL)
```

Figure 4.21   A PVS rule to be refined.

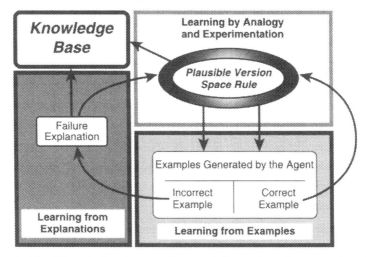

Figure 4.22   Knowledge base refinement through active experimentation.

builds the rule example corresponding to this tuple of objects (we will use the phrase 'example of the plausible upper bound' sometimes to refer to such a tuple of objects and sometimes to refer to the rule example corresponding to the tuple). The rule example generated in this way is similar with the example from which the rule was initially learned, as indicated in Section 4.2.2.3. This example is then shown to the expert who is asked to accept it as correct or to reject it, thus characterizing it as a positive or a negative example of the PVS rule. The positive example is used to generalize the plausible lower bound of the rule's condition through empirical induction. The negative example is used to elicit additional explanations from the expert and to specialize both bounds or only the plausible upper bound.

Table 4.7 presents the main steps of the method from Figure 4.22. Each of the steps is described in detail in the following sections.

4.3.3.2   Experimentation and verification

As indicated in the previous section, to generate an example of the rule R from Figure 4.21, the agent looks in the semantic network for objects satisfying the plausible upper bound condition of R. The top part of Figure 4.23 shows the semantic network representation of the object concepts from the plausible upper-bound condition of R. The middle part of Figure 4.23 shows three examples of these concepts. All these examples are fragments from the semantic network of the agent. The network example in the left-hand side of Figure 4.23 corresponds to the tuple (**BOLT1**, **MECHANICAL-CHASSIS1**, **MOWICOLL1**, **METAL**, **METAL**). By replacing each of the variables **?X**, **?Y**, **?Z**, **?M**, and **?N** in the task and operation of the rule R with the corresponding value from this tuple, the agent builds an example of the rule that satisfies the plausible upper bound without satisfying the plausible lower bound. This example is shown in the bottom left of Figure 4.23, in a simplified representation. The agent shows the generated example to the expert, asking him/her to characterize it as correct or incorrect, as shown in Figure 4.24.

Table 4.7   The method for rule refinement through active experimentation with the plausible upper bound

---

Let R be a PVS rule, U its plausible upper bound condition, L its plausible lower bound condition, and P the initial positive example from which the rule was learned.

1. *Experimentation and verification*
   Look for an example E_u of U in the semantic network that is not an example of L.
   If no such example exists
       then Return the current PVS rule and stop.
       else Build the example E of R based on E_u.
               Ask the expert to analyze the generated example E.
               If E is accepted by the expert as correct
                   then Continue with Step 2.
                   else Continue with Step 3.
               end
   end

2. *Refining the PVS rule with a positive example*
   Generalize L as little as possible to cover the positive example E while remaining less general than or at most as general as U by following the method presented in Table 4.9.
   If the new L is identical with U
       then An exact rule has been learned. Return it and stop.
       else Continue with Step 1.
   end

3. *Refining the PVS rule with a negative example*
   3.1 *Failure explanation*
        Compare the negative example E with the positive example P and look for an explanation of why P is a positive example of R while E is not.
        If such an explanation is found
            then Continue with Step 3.2.
            else Continue with Step 3.3.
        end
   3.2 *Explanation-based specialization*
        Specialize both bounds of the rule R by following the method presented in Table 4.10.
        Continue with Step 1.
   3.3 *Example-based specialization*
        Specialize the plausible upper bound of the rule R to uncover the negative example E while remaining more general than or as general as L by following the method presented in Table 4.12.
        If the new U is identical with L
            then An exact rule has been learned. Return it and stop.
            else Continue with Step 1.
        end

   The expert can control the experimentation process in different ways. For instance, the expert could ask the agent to generate examples that are as general as possible, as is the case in the middle of Figure 4.23. While such a general example may be more difficult to analyze, it speeds up significantly the rule refinement process because it is equivalent to the set of all of its instances. Indeed, accepting (rejecting) it as a correct (incorrect) problem-solving episode is equivalent to accepting (rejecting) all the instances covered by it.

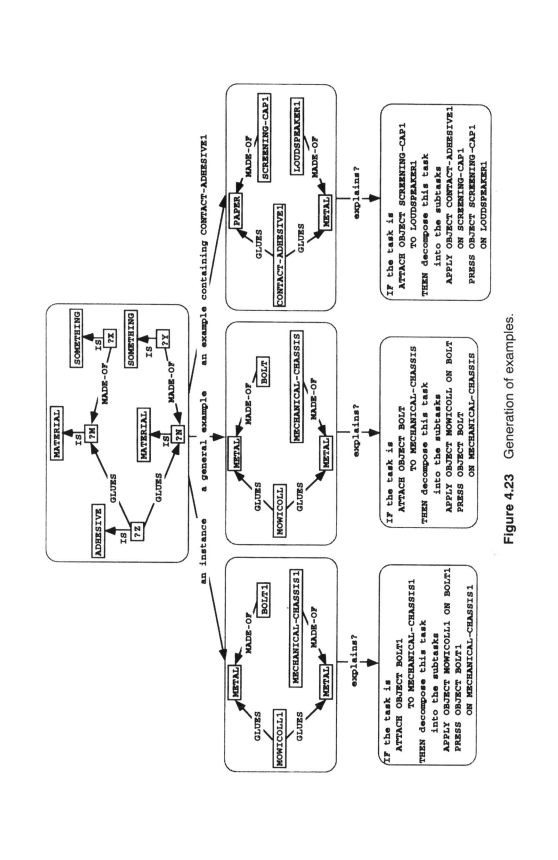

Figure 4.23 Generation of examples.

Analyze the following problem solving episode.
Accept it if it is correct, and reject it if it is incorrect.

```
IF
    ?A              IS          ATTACH
                    OBJECT      BOLT1
                    TO          MECHANICAL-CHASSIS1

THEN
    DECOMPOSE  TASK        ?A
               INTO        ?AP ?P

    ?AP             IS          APPLY
                    OBJECT      MOWICOLL1
                    ON          MECHANICAL-CHASSIS1

    ?P              IS          PRESS
                    OBJECT      BOLT1
                    ON          MECHANICAL-CHASSIS1
```

Figure 4.24   A generated problem-solving episode.

The expert may also direct the agent to generate examples that are as specific as possible. These are clearly easier to analyze but more of them may be required to refine the rule.

Another option is to direct the agent to generate examples that have some objects in common with the example from which the rule was initially learned. The expert could do this by 'fixing' the corresponding variables in the rule. Such an instance is the one from the bottom right part of Figure 4.23. It shares the object **CONTACT-ADHESIVE1** with the initial example shown in Figure 4.5. One justification of this option is to simplify the process of analyzing a generated example, because it makes it more similar to the initial example. If the generated example is not correct, then it is a near miss (Winston, 1975). Another justification is to reduce the combinatorial explosion of example generation that could take place in large domains.

Yet another option is to direct the agent to generate examples in which only different concepts are used. While these examples may still be very specific, a smaller number of them will be required to refine the rule because the use of different concepts results in more general generalizations.

The experimentation and verification method is summarized in Table 4.8.

4.3.3.3   Refining the PVS rule with a positive example

Let us suppose that the expert accepts the problem-solving episode from Figure 4.24. Therefore this episode represents a new positive example of the rule that needs to be generalized to cover it. The right-hand side of Figure 4.25 represents the instance of the plausible upper-bound condition of the rule corresponding to the newly generated positive example. The left-hand side of Figure 4.25 represents the plausible lower-bound condition of the rule. The new plausible lower bound of the rule is the minimal generalization of these two expressions that is less general than or at most as general as the plausible upper bound. The minimal generalization in Figure 4.25 has been made by making the minimal generalizations of

Table 4.8   The experimentation and verification method

Experimentation and verification

1. The expert constrains the example to be generated by the agent:
   - to be as specific as possible, or as general as possible, or of any degree of generality,
   - to contain specific concepts, or
   - to contain only distinct concepts.

2. The agent looks into the semantic network for a tuple of objects satisfying all the constraints from the plausible upper bound and from the expert, without satisfying all the constraints from the plausible lower bound.
   If a tuple of objects is found
       then Continue with step 3.
       else If there have been expert-defined constraints
              then Go to step 1 asking the expert to relax the constraints.
             else Stop because no new example can be generated.
          end
   end

3. The agent uses the found tuple of objects to generate an example of the rule and asks the expert to confirm or to reject it.

4. The expert accepts or rejects the example constructed by the agent, characterizing it as a positive or as a negative example of the rule.

---

the concepts from the plausible lower bound and the example, corresponding to each of the rule's variables, as indicated in Figure 4.26.

Therefore, the problem of finding the minimal generalization of the plausible lower-bound condition and the example's condition reduces to a set of problems, each consisting of finding the minimal generalization of two concepts (one from the plausible lower bound and the other from the new positive example) that is less general or as general as another concept (the corresponding concept from the plausible upper bound). To solve each of these problems one uses the climbing generalization hierarchy rule described in Section 3.2.2.

Before we describe the generalization method in more detail, let us consider some of the cases from Figure 4.26. For instance, with respect to variable **?X**, one has to generalize **MEMBRANE1** as little as possible to cover **BOLT1** and, at the same time, to be less general than or at most as general as **SOMETHING**, as indicated in Figure 4.27. As could be seen from Figure 4.4, **LOUDSPEAKER-COMPONENT** is the least general generalization of **MEM-BRANE1** and **BOLT1** and it is less general than **SOMETHING**. Therefore, **MEMBRANE1** is generalized to **LOUDSPEAKER-COMPONENT** which becomes the new *L* element, as indicated in Figure 4.27(b) and (c).

Figure 4.28 shows a case where there is more than one minimal generalization of the plausible lower bound. As shown in Figure 4.4 and in Figure 4.28(b), **CONTACT-ADHE-SIVE1** and **MOWICOLL1** are both instances of **ADHESIVE** and **INFLAMMABLE-OBJECT**, while **CONTACT-ADHESIVE1** is also an instance of a **TOXIC-SUBSTANCE**. In this case, minimal generalizations of **CONTACT-ADHESIVE1** and **MOWICOLL1** are **ADHESIVE** and **INFLAMMABLE-OBJECT**. However, **INFLAMMABLE-OBJECT** is not less general than the plausible upper-bound element which is **ADHESIVE**. Therefore, the new plausible lower-bound element is **ADHESIVE**, as indicated in Figure 4.28(c).

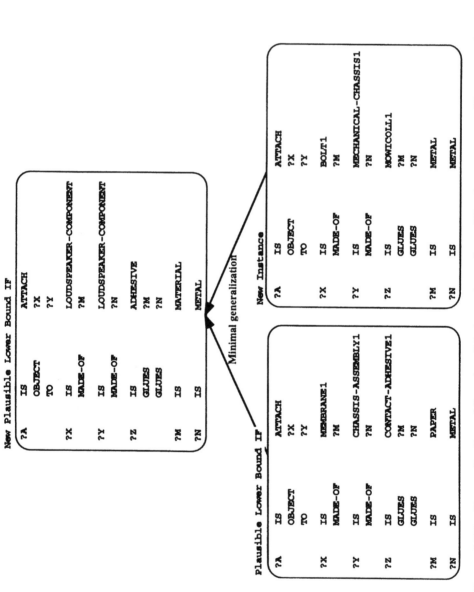

Figure 4.25 Generalization of the plausible lower-bound condition to cover a positive example.

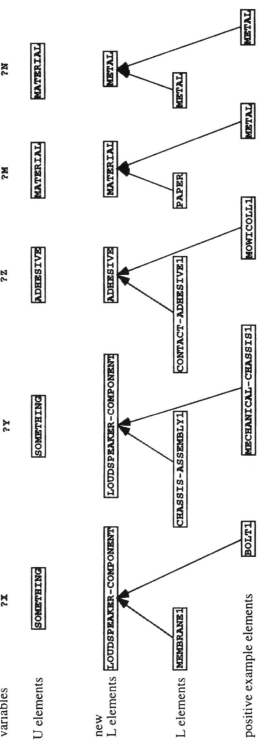

Figure 4.26 Minimal generalizations of the concepts from the plausible lower-bound condition.

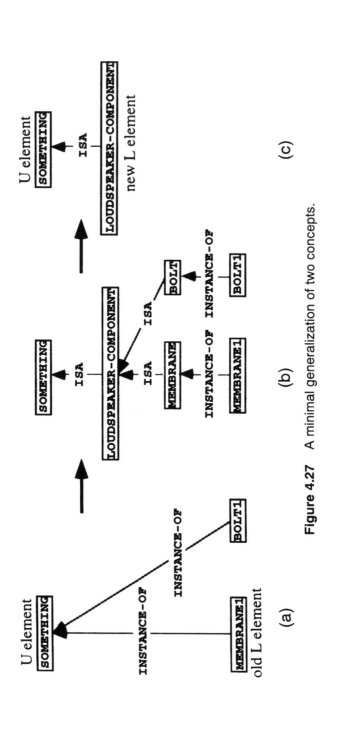

Figure 4.27 A minimal generalization of two concepts.

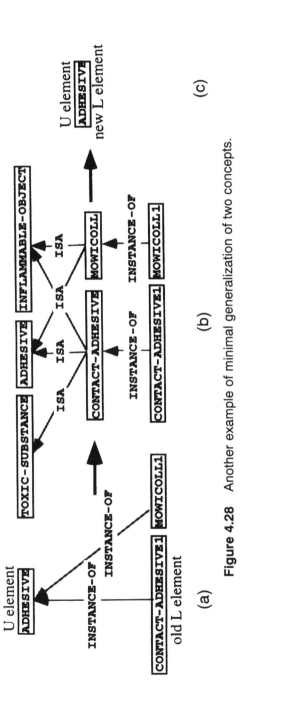

Figure 4.28 Another example of minimal generalization of two concepts.

```
Plausible Upper Bound IF
        ?A          IS          ATTACH
                    OBJECT      ?X
                    TO          ?Y

        ?X          IS          SOMETHING
                    MADE-OF     ?M

        ?Y          IS          SOMETHING
                    MADE-OF     ?N

        ?Z          IS          ADHESIVE
                    GLUES       ?M
                    GLUES       ?N

        ?M          IS          MATERIAL

        ?N          IS          MATERIAL
Plausible Lower Bound IF
        ?A          IS          ATTACH
                    OBJECT      ?X
                    TO          ?Y

        ?X          IS          LOUDSPEAKER-COMPONENT
                    MADE-OF     ?M

        ?Y          IS          LOUDSPEAKER-COMPONENT
                    MADE-OF     ?N

        ?Z          IS          ADHESIVE
                    GLUES       ?M
                    GLUES       ?N

        ?M          IS          MATERIAL

        ?N          IS          METAL
THEN
        DECOMPOSE   TASK        ?A
                    INTO        ?AP ?P

        ?AP         IS          APPLY
                    OBJECT      ?Z
                    ON          ?X

        ?P          IS          PRESS
                    OBJECT      ?X
                    ON          ?Y

with the positive examples
    ( ?X IS MEMBRANE1, ?Y IS CHASSIS-ASSEMBLY1,
      ?Z IS CONTACT-ADHESIVE1, ?M IS PAPER, ?N IS METAL)

    ( ?X IS BOLT1, ?Y IS MECHANICAL-CHASSIS1,
      ?Z IS MOWICOLL1, ?M IS METAL, ?N IS METAL)
```

Figure 4.29 The PVS rule after refinement with a new positive example.

As a consequence of these generalizations the new PVS rule is the one in Figure 4.29.

Because the generalization/specialization language for learning (as represented by the agent's semantic network) is incomplete, the generalization of the plausible lower bound indicated in Figure 4.25 may be in fact an over-generalization and the lower bound may need to be specialized, as indicated in the next section.

In the case of the illustrated rule, all the generalized elements were single concepts. In general, however, each could be a list of concepts. The generalization method corresponding to this general case is presented in Table 4.9. It is the same with the part of the candidate elimination method (Mitchell, 1997) corresponding to the treatment of a positive example. This method also extends to the cases where the concepts to be generalized are represented as sets of constants or as number intervals.

One of the most difficult problems in computing a good generalization of some expressions is to establish the objects to be matched (Kodratoff and Ganascia, 1986). This problem, however, is trivial in Disciple because both the plausible lower bound and the condition of the example have exactly the same structure, and the corresponding variables have the same names, as shown in the bottom part of Figure 4.25. This is a direct consequence of the fact that the example is generated from the plausible upper bound condition of the rule.

4.3.3.4   Refining the PVS rule with a negative example

Let us consider another example generated by the agent. This example is shown in the bottom right part of Figure 4.30 and is rejected by the expert. It is thus characterized as a

Table 4.9   The example-based generalization method

Let R be a PVS rule, U its plausible upper bound condition, L its plausible lower bound condition, and P a positive example of R covered by U and not covered by L.

Repeat for each variable **?X** from U representing an object concept
- Let the class of **?X** from U be $U_x = \{u_1 \ldots u_m\}$.
- Let the class of **?X** from L be $L_x = \{l_1 \ldots l_n\}$.
  Each concept u_i from U_x is a maximal generalization of all the known positive examples of **?X** that does not cover any of the known negative examples of **?X** and is more general than (or as general as) at least one concept l_j from L_x.
  Each concept l_i from L_x is a minimal generalization of the known positive examples of **?X** that does not cover any of the known negative examples of **?X** and is less general than (or as general as) at least one concept u_j from U_x, and is not covered by any other element l_k of L_x.
- Let C_x be the new positive example of **?X** from the rule example P. C_x is covered by at least one element u_i of U_x.
- Remove from U_X any element that does not cover C_x.
- **Repeat** for each l_i from L_x that does not cover C_x
  - Remove l_i from L_x.
  - Add to L_x all minimal generalizations of l_i and C_x that are less general or at most as general as an element of U_x.
  - Remove from L_x all the elements which are more general than or as general as other elements from L_x.

  end
end

Return the generalized rule R with the updated conditions U and L, and P in the list of positive examples of R.

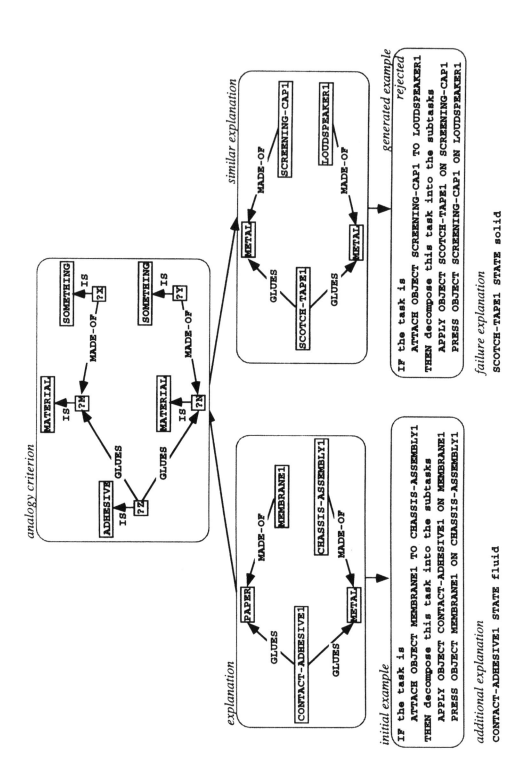

Figure 4.30 A negative example generated by analogy with the initial example.

negative example of the PVS rule. The fact that an example generated by analogy with the initial example from which the rule was learned is incorrect shows that the explanation of the initial example is incomplete. The agent first tries to find an explanation of why the generated example is incorrect. Then it uses this failure explanation and the similarity between the two examples to find or elicit from the expert the corresponding explanation piece of why the initial example is correct.

Failure explanation.   The problem-solving episode from the bottom right side of Figure 4.30 is incorrect because **SCOTCH-TAPE1 STATE solid** prevents **SCOTCH-TAPE1** from being properly applied on the surface of **SCREENING-CAP1**. By comparing this incorrect problem-solving episode and its failure explanation with the initial problem-solving episode shown in the bottom left side of Figure 4.30, one could easily find the corresponding explanation of why this episode is correct: **CONTACT-ADHESIVE1 STATE fluid**. This is the additional explanation piece of why the initial example is correct. The process of finding the failure explanation follows the same steps as that of finding the explanations of the initial example (see Table 4.2).

Explanation-based specialization.   As shown in the previous section, the following is an additional piece of explanation for the initial example from which the rule was learned:

$$EX: \quad \textbf{CONTACT-ADHESIVE1 STATE fluid}$$

This explanation piece is first generalized to:

$$EX_{ug}: \quad \textbf{?Z} \quad \textbf{IS} \qquad \textbf{SOMETHING}$$

$$\textbf{STATE} \quad \textbf{\{solid fluid gas\}}$$

by applying the method described in Section 4.2.2.4 and Table 4.4.

Next, one has to replace the plausible upper bound U of R with the maximally general specialization of U and EX_{ug} that does not cover the negative example and is more general than (or as general as) the plausible lower bound L, as indicated in Figure 4.31.

The process of updating the plausible lower bound condition of the rule is more complicated. First, one needs to find the pieces of explanations EX_i of the positive examples of the rule that correspond to the new piece of explanation EX: **CONTACT-ADHESIVE1 STATE fluid**, of the example from which the rule was initially learned. Some of these examples may not satisfy this explanation and will become positive exceptions to the updated rule.

In the illustrated case there is only one additional positive example of the current PVS rule, characterized by:

(?X IS BOLT1, ?Y IS MECHANICAL-CHASSIS1,

?Z IS MOWICOLL1, ?M IS METAL, ?N IS METAL)

The additional piece of explanation corresponding to this example would need to have the following form:

MOWICOLL1 STATE ?V

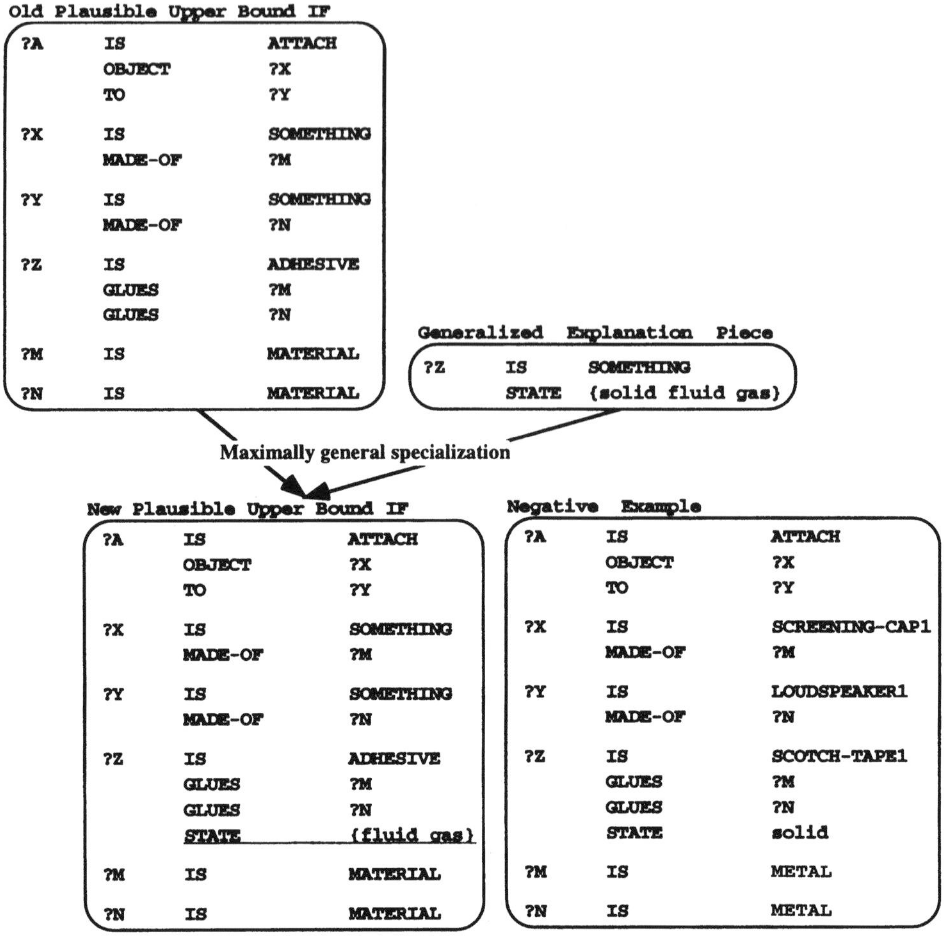

Old Plausible Upper Bound IF

?A	IS	ATTACH
	OBJECT	?X
	TO	?Y
?X	IS	SOMETHING
	MADE-OF	?M
?Y	IS	SOMETHING
	MADE-OF	?N
?Z	IS	ADHESIVE
	GLUES	?M
	GLUES	?N
?M	IS	MATERIAL
?N	IS	MATERIAL

Generalized Explanation Piece

?Z	IS	SOMETHING
	STATE	{solid fluid gas}

Maximally general specialization

New Plausible Upper Bound IF

?A	IS	ATTACH
	OBJECT	?X
	TO	?Y
?X	IS	SOMETHING
	MADE-OF	?M
?Y	IS	SOMETHING
	MADE-OF	?N
?Z	IS	ADHESIVE
	GLUES	?M
	GLUES	?N
	STATE	{fluid gas}
?M	IS	MATERIAL
?N	IS	MATERIAL

Negative Example

?A	IS	ATTACH
	OBJECT	?X
	TO	?Y
?X	IS	SCREENING-CAP1
	MADE-OF	?M
?Y	IS	LOUDSPEAKER1
	MADE-OF	?N
?Z	IS	SCOTCH-TAPE1
	GLUES	?M
	GLUES	?N
	STATE	solid
?M	IS	METAL
?N	IS	METAL

Figure 4.31   The specialization of the plausible upper bound.

The agent will look in the knowledge base (see Figure 4.4) for a value of **?V** and will find **fluid**. Therefore, the additional explanation piece is:

MOWICOLL1 STATE fluid

It could have been the case that the current knowledge base does not contain this piece of knowledge. In this case the agent will attempt to elicit it from the expert by asking the question:

What is the **STATE** of **MOWICOLL1**?

Each positive example of the current PVS rule has now an additional piece of explanation:

CONTACT–ADHESIVE1 STATE fluid

MOWICOLL1 STATE fluid

The agent has to determine the minimal generalization EX_{lg} of these pieces that are less general than (or as general as) EX_{ug}. By using the method described in Table 4.9 one finds:

EX_{lg}:   **?Z   IS   {ADHESIVE INFLAMMABLE-OBJECT}**

        STATE   fluid

Next the agent has to determine the new plausible lower bound L as the maximally general specialization of the current plausible lower bound L and EX_{lg}, a specialization that should also be less general than (or as general as) the plausible lower bound, as indicated in Figure 4.32.

The updated PVS rule is presented in Figure 4.33.

Let us now illustrate the case of a more complex failure explanation that is not true for all the positive examples of the rule. Let us assume that the following is an additional piece of explanation of the first positive example of the rule:

MEMBRANE1 PART-OF CHASSIS-MEMBRANE-ASSEMBLY1

  ∧ CHASSIS-ASSEMBLY1 PART-OF CHASSIS-MEMBRANE-ASSEMBLY1

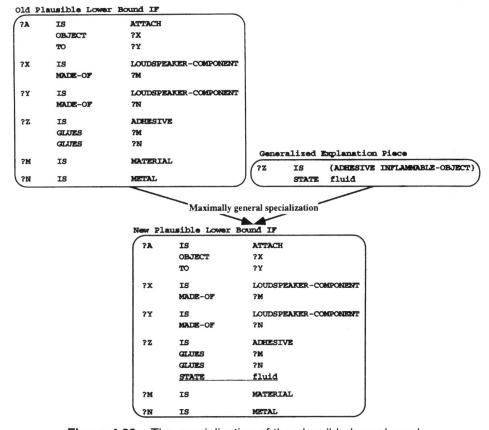

Figure 4.32   The specialization of the plausible lower bound.

```
Plausible Upper Bound IF
        ?A          IS              ATTACH
                    OBJECT          ?X
                    TO              ?Y

        ?X          IS              SOMETHING
                    MADE-OF         ?M

        ?Y          IS              SOMETHING
                    MADE-OF         ?N

        ?Z          IS              ADHESIVE
                    GLUES           ?M
                    GLUES           ?N
                    STATE           (fluid gas)

        ?M          IS              MATERIAL

        ?N          IS              MATERIAL
Plausible Lower Bound IF
        ?A          IS              ATTACH
                    OBJECT          ?X
                    TO              ?Y

        ?X          IS              LOUDSPEAKER-COMPONENT
                    MADE-OF         ?M

        ?Y          IS              LOUDSPEAKER-COMPONENT
                    MADE-OF         ?N

        ?Z          IS              ADHESIVE
                    GLUES           ?M
                    GLUES           ?N
                    STATE           fluid

        ?M          IS              MATERIAL

        ?N          IS              METAL
THEN
        DECOMPOSE   TASK            ?A
                    INTO            ?AP ?P

        ?AP         IS              APPLY
                    OBJECT          ?Z
                    ON              ?X

        ?P          IS              PRESS
                    OBJECT          ?X
                    ON              ?Y
with the positive examples
    ( ?X IS MEMBRANE1, ?Y IS CHASSIS-ASSEMBLY1,
      ?Z IS CONTACT-ADHESIVE1, ?M IS PAPER, ?N IS METAL)

    ( ?X IS BOLT1, ?Y IS MECHANICAL-CHASSIS1,
      ?Z IS MOWICOLL1, ?M IS METAL, ?N IS METAL)
with the negative example
    ( ?X IS SCREENING-CAP1, ?Y IS LOUDSPEAKER1,
      ?Z IS SCOTCH-TAPE1, ?M IS METAL, ?N IS METAL)
```

Figure 4.33   The PVS rule after refinement with a negative example.

This explanation piece is generalized to:

$$EX_{ug}: \quad \begin{array}{lll}
\textbf{?X} & \textbf{IS} & \textbf{SOMETHING} \\
& \textbf{PART-OF} & \textbf{?U} \\
\textbf{?Y} & \textbf{IS} & \textbf{SOMETHING} \\
& \textbf{PART-OF} & \textbf{?U} \\
\textbf{?U} & \textbf{IS} & \textbf{SOMETHING}
\end{array}$$

by applying the method from Table 4.4. Next the agent has to replace the plausible upper bound condition U with the maximally general specialization of U and EX_{ug}, as indicated above.

Let us now turn to the updating of the plausible lower bound condition L. As before, let us consider that there is only one additional positive example of the current PVS rule, characterized by:

(?X IS BOLT1, ?Y IS MECHANICAL-CHASSIS1,

?Z IS MOWICOLL1, ?M IS METAL, ?N IS METAL)

The additional piece of explanation corresponding to this example would need to have the following form:

BOLT1 PART-OF ?V ∧ MECHANICAL-CHASSIS1 PART-OF ?V

The agent will try to find a value of **?V** for which the above expression is true, either from the semantic network or from the expert. Let us assume, however, that no such value exists. Therefore, this positive example of the rule does not satisfy the additional piece of explanation and will have to be considered as a positive exception of the rule.

The only positive example of the rule remains the initial one. The instance of L corresponding to this example is the one from Figure 4.34. In such a case, this is precisely the new

```
Plausible Lower Bound IF
        ?A     IS        ATTACH
               OBJECT    ?X
               TO        ?Y

        ?X     IS        MEMBRANE1
               MADE-OF   ?M
               PART-OF   ?U

        ?Y     IS        CHASSIS-ASSEMBLY1
               MADE-OF   ?N
               PART-OF   ?U

        ?Z     IS        CONTACT-ADHESIVE1
               GLUES     ?M
               GLUES     ?N

        ?M     IS        PAPER

        ?N     IS        METAL

        ?U     IS        CHASSIS-MEMBRANE-ASSEMBLY1
```

Figure 4.34   Instance of the plausible lower bound condition.

```
Plausible Upper Bound IF
        ?A              IS              ATTACH
                        OBJECT          ?X
                        TO              ?Y

        ?X              IS              SOMETHING
                        MADE-OF         ?M
                        PART-OF         ?U

        ?Y              IS              SOMETHING
                        MADE-OF         ?N
                        PART-OF         ?U

        ?Z              IS              ADHESIVE
                        GLUES           ?M
                        GLUES           ?N

        ?M              IS              MATERIAL

        ?N              IS              MATERIAL

        ?U              IS              SOMETHING

Plausible Lower Bound IF
        ?A              IS              ATTACH
                        OBJECT          ?X
                        TO              ?Y

        ?X              IS              MEMBRANE1
                        MADE-OF         ?M
                        PART-OF         ?U

        ?Y              IS              CHASSIS-ASSEMBLY1
                        MADE-OF         ?N
                        PART-OF         ?U

        ?Z              IS              CONTACT-ADHESIVE1
                        GLUES           ?M
                        GLUES           ?N

        ?M              IS              PAPER

        ?N              IS              METAL

        ?U              IS              CHASSIS-MEMBRANE-ASSEMBLY1
THEN
        DECOMPOSE        TASK           ?A
                         INTO          ?AP ?P

        ?AP              IS              APPLY
                         OBJECT         ?Z
                         ON             ?X

        ?P               IS              PRESS
                         OBJECT         ?X
                         ON             ?Y

with the positive example
        ( ?X IS MEMBRANE1, ?Y IS CHASSIS-ASSEMBLY1,
          ?Z IS CONTACT-ADHESIVE1, ?M IS PAPER, ?N IS METAL,
          ?U IS CHASSIS-MEMBRANE-ASSEMBLY1)

with the positive exception
        ( ?X IS BOLT1, ?Y IS MECHANICAL-CHASSIS1,
          ?Z IS MOWICOLL1, ?M IS METAL, ?N IS METAL)
```

Figure 4.35   The updated PVS rule in a hypothetical case.

plausible lower bound condition of the rule, as shown in Figure 4.35. However, if there would have been more such examples then the new plausible lower bound condition would have been a minimal generalization of them (which would also need to be less general than, or as general as, the plausible upper bound condition U).

A summary of the explanation-based specialization method is given in Table 4.10.

Example-based specialization.   It may well be the case that no failure explanation is found for the current negative example. If this happens then one has to specialize the plausible upper bound U as little as possible to uncover the negative example while still remaining more general than (or as general as) the plausible lower bound L.

Because several specializations may be possible, the agent will first try to elicit some guidance from the expert. For instance, a natural question to ask the expert is which object concept from the negative example should be blamed for the failure. If the expert points to such an object then the agent will try to specialize the plausible upper bound condition so as to no longer cover this object and to remain more general than the plausible lower bound. If the expert does not blame any particular object, then the agent will randomly pick one to perform the specialization.

Table 4.10 · The explanation-based specialization method

Let R be a PVS rule, U its plausible upper bound condition, L its plausible lower bound condition, $P_1, ..., P_k$ the set of covered positive examples, and EX an additional explanation piece for the positive example from which the rule R was learned. EX may have been generated in response to a negative example N.

1. Generalize EX to EX_{ug} by applying the method in Table 4.4.

2. **If** EX has been generated in response to a negative example N
    then Replace U with the maximally general specialization of U and EX_{ug} that does not cover N.
    else Replace U with the maximally general specialization of U and EX_{ug}.
    end

3. **Repeat** for each positive example P_i of R
    Attempt to find an explanation piece EX_i similar to EX by applying the method in Table 4.11.
    end

4. Let LE be the list of found explanation pieces.

5. Let PE be the list of positive examples that do not have such explanations.

6. Update the rule R by removing the examples in PE from the list of positive examples and including them in the list of positive exceptions of the rule.

7. Determine the maximally specific generalization EX_{lg} of all the explanations from LE, generalization that is less general than (or, at most, as general as) E_{ug}.

8. Replace L with the maximally general specialization of L and EX_{lg} that is less general than (or, at most, as general as) U.

9. Return the updated rule R.

Table 4.11   The method for explanation-based elicitation of knowledge

---

Let R be a PVS rule, U its plausible upper bound condition, L its plausible lower bound condition, $P_1, ..., P_k$ the set of covered positive examples, and EX_g an additional generalized explanation piece for the positive examples of the rule R.

- Let LE and PE be two empty lists.
- **Repeat** for each positive example P_i of R
  - Let EX_i be the instance of EX_g corresponding to P_i.
  - Look for an instance EX_i' of EX_i in the KB (semantic network).
  - **If** EX_i' is found

       then   Insert EX_i' in the list LE.

       else   Ask the expert whether in some instance EX_i' of EX_i is true.

          If EX_i' is true

             then Represent it into the KB and insert EX_i' into LE.

             else P_i will be a positive exception of R.

                Insert P_i in the list PE.

       end

  end

  end
- Elicit additional instances of EX_g from the expert and introduce them into the KB.
- Return the lists LE and PE.

To illustrate this method let us consider again the negative example from the bottom right of Figure 4.30. Let us now assume that instead of providing a failure explanation, the expert has blamed the object **SCOTCH-TAPE1** for this failure. **SCOTCH-TAPE1** is therefore a negative example of the concept represented by the variable **?Z** in the rule R (see Figure 4.29). The class of **?Z** is **ADHESIVE** in both U and L. Therefore U cannot be specialized to uncover **SCOTCH-TAPE1** and to remain more general than L. If the expert believes that **SCOTCH-TAPE1** is the only object that should be blamed for this failure, in the sense that if another object were in its place then the example would be correct, then he/she may decide to leave the rule unchanged and to keep the current rule example as a negative exception.

Let us now assume that the expert does not blame any object and therefore provides no guidance to the agent on how to specialize the rule in response to the negative example from the bottom right side of Figure 4.30. In this situation the agent picks a class from U that could be specialized. Such a class is **SOMETHING** that corresponds to the variable **?Y** (see Figure 4.29). This has to be specialized to uncover **LOUDSPEAKER1** and to cover **LOUD-SPEAKER-COMPONENT**, as it is illustrated in Figure 4.36.

Notice that although this is not a justified specialization, the refined rule is correct and the only problem is that it may not be as general as it could be.

The method for inductive specialization of R to uncover a negative example is presented in Table 4.12. Notice that step 3 is the same as the part of the candidate elimination method (Mitchell, 1997) corresponding to the treatment of a negative example.

The learning process decreases the distance between the two bounds of the version space. This process should, in principle, continue until the lower bound becomes identical to the upper one. However, because the semantic network that is at the basis of the generalization

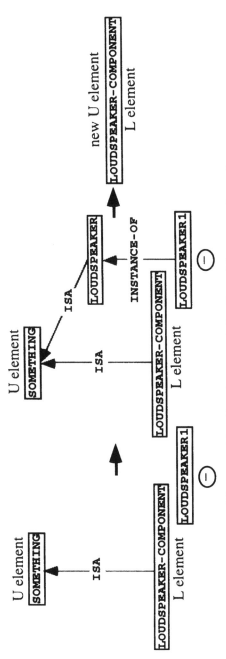

Figure 4.36 Random specialization of the upper bound.

Table 4.12   The example-based specialization method

Let R be a PVS rule, U its plausible upper bound condition, L its plausible lower bound condition, and N a negative example covered by U and not covered by L.

1. *Blaming an object concept for the failure*
    1.1 Ask the expert to point to an object from the negative example that could be considered responsible for the failure.
        If an object C_x is blamed
            then Continue with step 1.2 **else** continue with step 2.
    1.2 Let **?X** be the variable from the rule's conditions that corresponds to the blamed object.
        Let $(U_x L_x)$ be the classes of **?X** in the two bounds.
        If each concept from L_x covers C_x
            then Continue with step 1.3 **else** Continue with step 3.
    1.3 The rule cannot be specialized to uncover the current negative example.
        Ask the expert whether the current negative example N should be associated with the rule as a negative exception.
        If the expert agrees
            then Return the updated rule R and stop **else** continue with step 1.1.

2. *Random specialization of U*
    2.1 Choose a pair $(U_x L_x)$ of classes corresponding to a variable **?X** from U and L such that at least one element l_i from L_x does not cover the object C_x corresponding to **?X** in the negative example N. At least one such pair exists.

        Let $U_x = \{u_1 \dots u_m\}$ and $L_x = \{l_1 \dots l_n\}$.

        Each concept u_i is a maximal generalization of all the known positive examples of **?X** that does not cover any known negative example of **?X** and is more general than (or as general as) at least one element l_j from L_x. At least one concept u_i covers the object C_x from the current negative example N.
        Each concept l_i is a minimal generalization of the known positive examples of **?X** that does not cover any negative example of **?X**. At least one concept l_i does not cover C_x.
    2.2 Continue with step 3.

3. *Inductive specialization with an object*
    There are concepts in L_x that do not cover C_x. The rule can be specialized to uncover N by specializing both L_x and U_x (which is known to be more general than C_x).
    3.1 Remove from L_x any element that covers C_x.
    3.2 **Repeat** for each element u_i of U_x that covers C_x
        • Remove u_i from U_x.
        • Add to U_x all maximally general specializations of u_i that do not cover C_x and are more general than or at least as general as a concept from L_x.
        • Remove from U_x all the concepts which are less general than or as general as other concepts from U_x.
        end

4. Return the specialized rule R with N in its list of negative examples.

---

language is incomplete, we should expect that this will not always happen. Therefore, the agent will be forced to preserve two conditions (the plausible upper bound and the plausible lower bound), instead of a single applicability condition.

The remainder of this chapter summarizes the other methods used by the agent to refine the rules.

4.3.4 Rule verification through active experimentation with its plausible lower bound

Disciple agents are able to learn and refine rules very quickly by using only a small number of examples. While this is an important strength, it also presents a drawback in that the rules may not have been tested thoroughly and may have unknown exceptions. For this reason, the expert may decide to verify the rule with a set of examples covered by the plausible lower bound, generated by the agent itself. If any of the generated examples is not correct, then the rule would have to be updated by using the approach presented in Section 4.3.3.4. The rule verification method is presented in Table 4.13.

4.3.5 Rule refinement with external examples

We have presented the methods of refining a PVS rule with examples generated by the agent. There are, however, other sources of the examples as well. For instance, the expert may wish to give the agent additional examples. Alternatively the examples may be obtained by the agent during its normal problem-solving operation, or there may be an external repository of

Table 4.13 The method for rule verification through active experimentation with the plausible lower bound

---

Let R be a PVS rule, U its plausible upper bound condition, L its plausible lower bound condition, and P the example from which the rule was initially learned.

1. Generate a testing set TS of examples covered by L.

2. **Repeat** for each example E from TS

    2.1 *Verify the example*

        If E is accepted by the expert
            then Include E in the list of positive examples and continue with Step 2.
            else Continue with Step 2.2.
        end

    2.2 *Repair the rule conditions*

        2.2.1 *Failure explanation*

            Compare the negative example E with the positive example P and look for an explanation of why P is a positive example of R while E is not.
            If an explanation is found
                then Continue with Step 2.2.2.
                else Continue with Step 2.2.3.
            end

        2.2.2 *Explanation-based specialization*

            Specialize both bounds of the rule R by following the method presented in Table 4.10. Continue with Step 2.

        2.2.3 *Exception*

            Include this example in the list of negative exceptions of the rule. Continue with Step 2.

    end

Table 4.14   The method of refining the rule with external examples

---

Let E be a new example of the PVS rule R characterized by the plausible upper bound condition U and the plausible lower bound condition L.

1. **If** E is covered by L **then** Apply the method in Table 4.13.

2. **If** E is covered by U but it is not covered by L **then** Apply the method in Table 4.7.

3. **If** E is not covered by U **then**
    If E is a positive example
        then E represents a positive exception to the rule.
            Include E in the list of positive exceptions of R.
            The rule needs to be refined as part of the exception-handling process.
    end
    If E is a negative example
        then No refinement is necessary.
            Include E in the list of the negative examples of R.
    end
    end

---

examples. The method of refining a PVS rule with such external examples is presented in Table 4.14.

4.3.6   Characterization of the refined rule

The refinement process should, in principle, continue until the two conditions of the rule become identical. However, because the agent's knowledge is incomplete and possibly partially incorrect, we should expect that this will not always happen. Therefore, the agent will sometimes be forced to preserve two conditions instead of a single applicability condition.

Figure 4.37 shows the expected form of the refined PVS rule. It has the characteristics described in Section 3.1.3.4, except that the examples from which the rule has been learned and refined are explicitly associated with the rule. These examples are grouped into four categories: positive examples, negative examples, positive exceptions and negative exceptions. In the positive examples group are the positive examples covered by the plausible lower bound condition. In the negative examples group are the negative examples that are not covered by the plausible upper bound condition. In the negative exceptions group are the negative examples that are covered by both conditions. In the positive exceptions group are the positive examples that are not covered by the conditions.

- A ***negative exception*** *to the rule is a negative example that is covered by the plausible lower bound and any specialization of the plausible lower bound to uncover it results in uncovering some positive examples.*
- A ***positive exception*** *to the rule is a positive example that is not covered by the plausible upper bound and any generalization of the plausible bounds to cover it results in covering some negative examples.*

The next section presents several knowledge base refinement methods that the Disciple agents use to eliminate, or at least reduce, the number of rule exceptions.

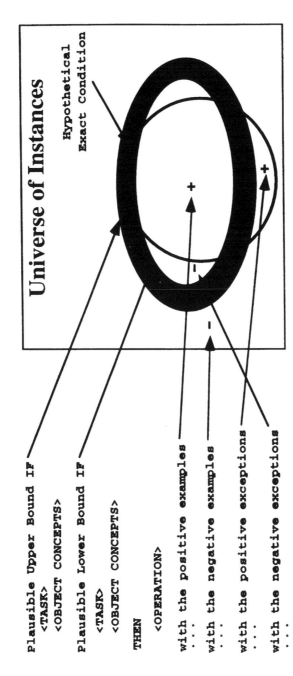

Figure 4.37   The expected form of a refined PVS rule.

4.4   Exception-driven knowledge base refinement

4.4.1   The exception-driven knowledge base refinement problem

As indicated in Figure 4.37, the result of rule refinement is most likely a PVS rule with exceptions. Exception-driven refinement of the knowledge base consists of refining the semantic network in such a way as to reduce the number of exceptions of a rule, or replacing the rule with several conjunctive rules that have fewer exceptions. The problem of exception-based refinement of the knowledge base is defined in Table 4.15.

There are two main possible causes of a rule's exception:

- The exception may be caused by the fact that the semantic network is incomplete and partially incorrect. Therefore, it does not contain the features or object concepts that discriminate between the examples and exceptions, or some of the existing features and concepts are wrong. In such a case the agent can remove the exceptions by improving its semantic network.
- The exception may also be caused by the fact that the agent is trying to learn one conjunctive rule for a given set of positive and negative examples when it should learn several conjunctive rules that together are equivalent with a disjunctive rule. In such a case one can remove the exceptions by simply learning several conjunctive rules.

There are several methods that a Disciple agent can use to extend and improve the semantic network in order to reduce the number of rule exceptions. These methods are presented in the following sections. As shown in Figure 4.38, the agent could attempt to discover or elicit from the expert new object features or even new concepts that discriminate between the rule's examples and exceptions. As a result of applying these methods, the rules will have fewer (if any) exceptions and the semantic network will be more complete and more correct.

If, after applying these discovery and elicitation methods, the rule R still has exceptions, then the agent can remove them by replacing the (conjunctive) rule R with several conjunctive rules. For instance, one can remove a negative exception NX by specializing both of its plausible bounds as little as possible to uncover NX and to cover as many positive examples as possible. The positive examples that are no longer covered by R become positive

---

Table 4.15   The problem of exception-driven refinement of the knowledge base

---

Given:
- An incomplete and possibly partially incorrect semantic network S;
- A PVS rule R;
- A set of positive examples P;
- A set of negative examples N;
- A set of positive exceptions PX;
- A set of negative exceptions NX.

Determine:
- An improved semantic network S obtained by adding, modifying, or deleting object concepts or features;
- An improved conjunctive rule R that has fewer exceptions (if any) or several conjunctive rules $R_1, ..., R_n$ that have fewer exceptions (if any).

---

exceptions. Such a process reduces the number of negative exceptions by increasing the number of the positive exceptions. Therefore one may prefer to keep some of the negative exceptions unchanged. Next the agent removes all the positive exceptions that are associated with the rule R and learns other PVS rules for which these positive exceptions are positive examples. For instance, the agent creates a new rule R' starting from one of the positive exceptions PX_i, by using the rule-learning method presented in Section 4.2. In such a case, the agent can use the explanations of the rule R as a starting point for determining the explanations of PX_i. Then the agent refines the rule R' by using the positive exceptions of R as positive examples of R' and the negative examples and exceptions of R as negative examples of R'. This process will lead to a refined rule R' that covers several of the positive exceptions of R. The rule R' may itself have positive and negative exceptions. Therefore, the process may be recursively applied. The agent and the expert may also decide to keep an exception explicitly associated with the rule when removing it would require too complex modification of the knowledge base.

In the following sections we will present in more detail the exception-handling methods from Figure 4.38. Although we will consider PVS rules with exceptions, the same methods apply to single-condition rules. Such a single-condition rule could be regarded as a PVS rule where both the plausible lower bound and the plausible upper bound are identical with the rule's condition. Therefore, in the methods to be presented, one could simply replace both the plausible lower bound condition and the plausible upper bound condition with the exact condition of the rule.

Finally, let us notice that a rule with negative exceptions is considered inconsistent. Eliciting new knowledge to remove negative exceptions is therefore a type of consistency-driven elicitation. Similarly, a rule with positive exceptions is considered incomplete. Eliciting new knowledge to remove positive exceptions is a type of completeness-driven elicitation.

In the following sections we will refer often to the objects, variables, properties and relations from a rule and its examples. We will use the following notation to denote an example or an exception of a rule R:

$$E_i = (..., ?j \text{ IS } O_{ij}, ..., ?p \text{ IS } O_{ip}, ...)$$

O_{ij} is the object from the example (exception) that corresponds to the variable $?j$ from the conditions of R:

    $?j$ IS Cu_j ; Cu_j is the class of $?j$ in the plausible upper bound condition of R

       ...   ; the other properties and relations of $?j$

    $?j$ IS Cl_j ; Cl_j is the class of $?j$ in the plausible lower bound condition of R

       ...   ; the other properties and relations of $?j$

4.4.2   The consistency-driven knowledge base refinement method

In order to remove the negative exceptions of a rule, the agent will analyze the objects from the set of positive examples and will attempt to elicit from the expert new object properties and relations and even new object concepts that distinguish between the positive examples and some or all of the negative exceptions. The overall method is presented in Table 4.16 and further details are given in the following sections.

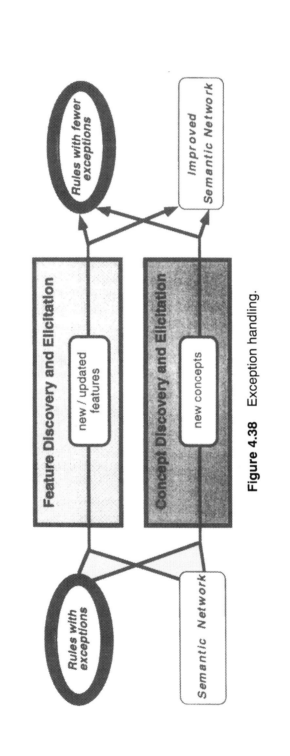

Figure 4.38 Exception handling.

Table 4.16   The consistency-driven elicitation method

Let R be an inconsistent PVS rule, U its plausible upper bound condition, L its plausible lower bound condition, $\{E_1, ..., E_k\}$ the set of its positive examples, and $\{N_n, ..., N_s\}$ the set of its negative exceptions.

1. *Property discovery and elicitation*

   Repeat for each negative exception N_i of the rule R

   - Use the method in Table 4.19 to attempt to transform the negative exception N_i into a negative example by discovering or eliciting from the expert property-value pairs $(P\ v)$ common only to the positive examples.
   - **If** R has been updated with the property P of the concept $?j$

     then Attempt to specialize the value of the property P in U to uncover the remaining negative exceptions of R while remaining more general than the corresponding value in L.

   - Transfer the uncovered negative exceptions from the list of negative exceptions to the list of negative examples.
   - Any other negative exception that is not characterized by the property-value pairs $(P\ v)$ has also become a negative example. Remove it from the list of negative exceptions of the rule and add it to the list of negative examples of the rule.

   end

2. *Relation discovery and elicitation*

   Repeat for each remaining negative exception N_i of the rule R

   - Use the method in Table 4.20 to attempt to transform the negative exception N_i into a negative example by discovering or eliciting from the expert a relation REL that holds only between objects of the positive examples.
   - **If** R has been updated with the relation REL between the concept $?j$ and the concept $?p$

     then Any negative exception N_q for which the relation REL does not hold between O_{qj} and O_{qp} has also become a negative example. Remove each such exception from the list of negative exceptions and add it to the list of negative examples.

   end

3. *Concept discovery and elicitation*

   Repeat for each remaining negative exception N_i of the rule R

   - Use the method in Table 4.22 to attempt to transform the negative exception N_i into a negative example by eliciting from the expert a concept that covers all the corresponding objects from the positive examples without covering the corresponding object from the negative exception N_i.
   - **If** R has been updated with the concept C_j' representing the class of $?j$

     then Any negative exception for which the object corresponding to $?j$ is not covered by C_j' has also become a negative example. Remove it from the list of negative exceptions of R and add it to the list of negative examples.

   end

To illustrate this method let us consider the rule in Figure 4.39 that has a negative exception, and the semantic network in Figure 4.4. The negative exception is similar with the negative example generated by the agent during the rule refinement process (see Section 4.3.3.4). The only difference is that the value of the variable **?Y** is **CHASSIS–MEMBRANE–ASSEMBLY1** instead of **LOUDSPEAKER1**. In such a case, it is not possible to specialize the plausible upper bound condition to uncover the negative example while still remaining more

```
Plausible Upper Bound IF
          ?A           IS           ATTACH
                       OBJECT       ?X
                       TO           ?Y

          ?X           IS           SOMETHING
                       MADE-OF      ?M

          ?Y           IS           SOMETHING
                       MADE-OF      ?N

          ?Z           IS           ADHESIVE
                       GLUES        ?M
                       GLUES        ?N

          ?M           IS           MATERIAL

          ?N           IS           MATERIAL

Plausible Lower Bound IF
          ?A           IS           ATTACH
                       OBJECT       ?X
                       TO           ?Y

          ?X           IS           LOUDSPEAKER-COMPONENT
                       MADE-OF      ?M

          ?Y           IS           LOUDSPEAKER-COMPONENT
                       MADE-OF      ?N

          ?Z           IS           ADHESIVE
                       GLUES        ?M
                       GLUES        ?N

          ?M           IS           MATERIAL

          ?N           IS           METAL
THEN
          DECOMPOSE TASK            ?A
                    INTO           ?AP ?P

          ?AP          IS           APPLY
                       OBJECT       ?Z
                       ON           ?X

          ?P           IS           PRESS
                       OBJECT       ?X
                       ON           ?Y

with the positive examples
  ( ?X IS MEMBRANE1, ?Y IS CHASSIS-ASSEMBLY1,
    ?Z IS CONTACT-ADHESIVE1, ?M IS PAPER, ?N IS METAL)

  ( ?X IS BOLT1, ?Y IS MECHANICAL-CHASSIS1,
    ?Z IS MOWICOLL1, ?M IS METAL, ?N IS METAL)

with the negative exception
  ( ?X IS SCREENING-CAP1, ?Y CHASSIS-MEMBRANE ASSEMBLY1,
    ?Z IS SCOTCH-TAPE1, ?M IS METAL, ?N IS METAL)
```

Figure 4.39   A rule with a negative exception.

general than or at least as general as the plausible lower bound condition. Therefore, in such a case, the agent cannot apply the rule refinement methods described in Section 4.3 and has to rely on the exception-handling methods described below.

4.4.3 Consistency-driven discovery and elicitation of features

A covered negative example of a rule may be uncovered by discovering or eliciting a new object feature that discriminates between all the covered positive examples and the covered negative example. For instance, to uncover the negative exception of the rule from Figure 4.39, the agent will first try to find in the semantic network, or elicit from the expert, a property of the positive instances of **?X** (**MEMBRANE1** and **BOLT1**) which is not a property of the negative instance of **?X** (**SCREENING-CAP1**). Alternatively, the agent may apply the same procedure to any other variable such as **?Z**. That is, it will try to find or elicit a property of the positive instances of **?Z** (**CONTACT-ADHESIVE1**, **MOWICOLL1**), which is not a property of the negative instance of **?Z** (**SCOTCH-TAPE1**). More specifically:

- the agent may look in the semantic network for a property of **CONTACT-ADHESIVE1** that may be a property of **MOWICOLL1**, without being a property of **SCOTCH-TAPE1**;
- or, it may look for a property of **MOWICOLL1** that may be a property of **CONTACT-ADHESIVE1**, without being a property of **SCOTCH-TAPE1**.

An illustration of this feature discovery and elicitation dialog is shown in Table 4.17. In this table and the subsequent examples the expert's answers are shown in italics. As a result of this dialog, the agent has found the property (**?Z STATE fluid**) which discriminates between the positive examples of R_1 and its negative exception. The answer 'Irrelevant' means that the property proposed by the agent does not make sense for the corresponding object concept.

By introducing the discriminating property into the rule's condition, the negative exception is transformed into a negative example.

Another result of the above dialog is that the descriptions of several object instances and concepts are refined by adding the elicited property. This is a case of goal-driven property transfer from one instance or concept (**MOWICOLL1**) to other instances or concepts (**CONTACT-ADHESIVE1**, **CONTACT-ADHESIVE**, **SCOTCH-TAPE**, etc.).

Table 4.17 Sample agent–expert dialog during property discovery and elicitation

---

I know that '**MOWICOLL1 STATE fluid**'.

Is it true that '**CONTACT-ADHESIVE1 STATE fluid**'? [Yes/No/Irrelevant]: *Yes*

This means that '**?Z STATE fluid**' is a property characterizing all the covered positive examples and may distinguish them from some of the covered negative examples.

Is it false that '**SCOTCH-TAPE1 STATE fluid**'? [Yes/No/Irrelevant]: *Yes*

Describe this property for each object that it applies to:

   CONTACT-ADHESIVE STATE fluid
   SCOTCH-TAPE STATE solid

It is not unusual for an expert to define a feature of a concept, but then to forget to specify it when describing another concept. With the method presented above, the agent may discover and repair such cases of incomplete knowledge.

It may also be the case that the agent cannot find a property to transfer. In such a case, it will try to elicit a new property by using a technique similar to the Triad method employed in the elicitation of the repertory grids (Boose and Bradshaw, 1988; Shaw and Gaines, 1987). If the semantic network in Figure 4.4 does not contain the property **MOWICOLL STATE fluid**, the dialog for eliciting a new property would be as shown in Table 4.18.

An outline of the consistency-driven property discovery and elicitation method is presented in Table 4.19.

Similarly, one could elicit relations between object concepts in order to eliminate the negative exceptions. Let us consider again the positive examples and the negative exception of the rule R in Figure 4.39. To uncover the negative exception the agent may look for a relation between the instance of **?Z** and the instance of **?X**, which may hold for all the positive examples but does not hold for some of the negative exceptions. More specifically, the agent may look for a relation between **CONTACT-ADHESIVE1** and **MEMBRANE1** that may also hold between **MOWICOLL1** and **BOLT1** but does not hold between **SCOTCH-TAPE1** and **SCREENING-CAP1**. Such a relation, for instance, could be **MAY-BE-APPLIED-ON**:

      CONTACT-ADHESIVE1 MAY-BE-APPLIED-ON MEMBRANE1

      MOWICOLL1 MAY-BE-APPLIED-ON BOLT1

      not(SCOTCH-TAPE1 MAY-BE-APPLIED-ON SCREENING-CAP1)

Table 4.18   Sample agent–expert dialog during property elicitation

---

Consider the following two groups of objects:

   group1: **CONTACT-ADHESIVE1, MOWICOLL1**
   group2: **SCOTCH-TAPE1**

in the context of the following problem-solving episode where **?Z** may take values from group1:

```
IF
      ?A ATTACH, OBJECT ?X, TO ?Y
THEN
      DECOMPOSE TASK ?A, INTO ?AP ?P
      ?AP APPLY, OBJECT ?Z, ON ?X
      ?P PRESS, OBJECT ?X, ON ?Y
```

Could you think of some property that discriminates between **CONTACT-ADHESIVE1, MOWICOLL1**, on one hand, and **SCOTCH-TAPE1**, on the other hand? [Yes/No]: *Yes*

Describe this property for each object that it applies to:

   CONTACT-ADHESIVE1 STATE fluid
   MOWICOLL1 STATE fluid
   SCOTCH-TAPE1 STATE solid
   ...

Table 4.19   The method for consistency-driven property discovery and elicitation

---

Let R be an inconsistent PVS rule, U its plausible upper bound condition, L its plausible lower bound condition, $\{E_1, ..., E_k\}$ the set of its positive examples, and N a negative exception.

Repeat for each object variable $?j$ from one of the conditions of the rule R

- Let $O_{1j}, ..., O_{kj}$ be the objects corresponding to $?j$ in the positive examples $E_1, ..., E_k$.
- Let O_{nj} be the object corresponding to $?j$ in the negative exception N.
- Look in the semantic network or elicit from the expert one or a set of property-value pairs $(P\ v_1), ..., (P\ v_q)$ such that:
  - each of the objects $O_{1j}, ..., O_{kj}$ has the property P with one of the values $v_1, ..., v_q$ (although this property may have not been represented in the semantic network and has only now been elicited from the expert);
  - the object O_{nj} from the negative exception N does not have the property P with any of the values $v_1, ..., v_q$.
- Refine the descriptions of the objects $O_{1j}, ..., O_{kj}$ in the semantic network by adding the corresponding property–value pair $(P\ v_i)$. Interact with the expert to determine whether these property-value pairs can be associated with ancestors of $O_{1j}, ..., O_{kj}$ in agent's semantic network. If this is possible then $O_{1j}, ..., O_{kj}$ will inherit the property P from these ancestors.
- Determine the minimally general generalization E_{lg} of all the properties $(O_{1j}\ P\ v_1), ..., (O_{kj}\ P\ v_q)$:

$$
\begin{array}{llll}
E_{lg}: & ?j & \text{IS} & mGG(O_{1j}, ..., O_{kj}) \\
        &   & P   & mGG(v_1, ..., v_q)
\end{array}
$$

- Replace L with the maximally general specialization of L and E_{lg}.
- Determine the following maximal generalization E_{ug}:

$$
\begin{array}{llll}
E_{ug}: & ?j & \text{IS} & \text{domain of } P \\
        &   & P   & \text{range of } P
\end{array}
$$

- Replace U with the maximally general specialization of U and E_{ug}.
- Remove N from the list of negative exceptions of R and add it to the list of negative examples.

until one or a set of property-value pairs $(P\ v)$ has been identified or all the object variables from the condition of the rule have been considered.

After such a relation is discovered or elicited from the expert, the rule is updated in a similar way as in the case of the discovered properties. The corresponding consistency-driven relation discovery and elicitation method is presented in Table 4.20.

4.4.4   Consistency driven discovery and elicitation of concepts

Another method of uncovering a negative exception of a rule consists of discovering or eliciting a new concept that discriminates between the positive examples and the negative example. Let us consider again the negative exception of the rule in Figure 4.39. The concept covering **CONTACT-ADHESIVE1**, **MOWICOLL1**, and **SCOTCH-TAPE1** in both conditions of the rule R is **ADHESIVE**. In order to uncover the negative exception, the expert may define the concept C'_j covering **CONTACT-ADHESIVE1** and **MOWICOLL1**, but not covering **SCOTCH-TAPE1**. Such a concept might be, for instance, **FLUID-ADHESIVE**, as indicated in Figure 4.40. The concept elicitation dialog is shown in Table 4.21. As a consequence, the agent replaces **ADHESIVE** with **FLUID-ADHESIVE** in both bounds of the rule R.

Table 4.20   The method for consistency-driven relation discovery and elicitation

Let R be an inconsistent PVS rule, U its plausible upper bound condition, L its plausible lower bound condition, $\{E_1, ..., E_k\}$ the set of its positive examples, and N a negative exception.

Repeat for each object variable $?j$ from one of the conditions of the rule R
- Let $O_{1j}, ..., O_{kj}$ be the objects corresponding to $?j$ in the positive examples $E_1, ..., E_k$.
- Let O_{nj} be the object corresponding to $?j$ in the negative exception N.
- Look in the semantic network or elicit from the expert a relation REL between two objects of a positive example such that:
  - $(O_{ij}$ REL $O_{ip})$ is true for any object O_{ij} from the list of objects $O_{1j}, ..., O_{kj}$ (this relation may have not been represented in the semantic network and has only now been discovered or elicited from the expert);
  - $(O_{nj}$ REL $O_{np})$ is not true, i.e. REL is not a relation between the corresponding objects from the negative exception N.
- Refine the descriptions of the objects $O_{1j}, ..., O_{kj}$ in the semantic network by adding the relations $(O_{1j}$ REL $O_{1p}), ..., (O_{kj}$ REL $O_{kp})$.
- Determine the minimally general generalization E_{lg} of all the relations $(O_{1j}$ REL $O_{1p}), ..., (O_{kj}$ REL $O_{kp})$:

$$
\begin{array}{llll}
E_{lg}: & ?j & \text{IS} & mGG(O_{1j}, ..., O_{kj}) \\
        &   & \text{REL} & ?p \\
        & ?p & \text{IS} & mGG(O_{1p}, ..., O_{kp})
\end{array}
$$

- Replace L with the maximally general specialization of L and E_{lg}.
- Determine the maximal generalization E_{ug} of all the relations $(O_{1j}$ REL $O_{1p}), ..., (O_{kj}$ REL $O_{kp})$:

$$
\begin{array}{llll}
E_{ug}: & ?j & \text{IS} & \text{domain of REL} \\
        &   & \text{REL} & ?p \\
        & ?p & \text{IS} & \text{range of REL}
\end{array}
$$

- Replace U with the maximally general specialization of U and E_{ug}.
- Remove N from the list of negative exceptions of R and add it to the list of negative examples.
until a relation REL has been identified or all the object variables from the condition of the rule have been considered.

One may notice that this method is also a type of goal-driven conceptual clustering in which known concepts are clustered under newly defined ones, to improve the consistency of the learned rules. An outline of the method is presented in Table 4.22.

4.4.5   The completeness-driven knowledge base refinement method

In general, the strategy to cover a positive exception is to generalize the rule's condition by either removing features from rule's clauses or by replacing the object concepts with more general concepts from the semantic network. However, in Disciple, these features correspond to the explanations from which that rule has been learned or are features that have been introduced to eliminate negative exceptions. Therefore the agent will not attempt to remove them. Also, the concepts in the clauses of the rule's conditions have been obtained by minimal specializations of the bounds to uncover negative examples. Therefore, a generalization of these concepts in the current semantic network would lead to the covering

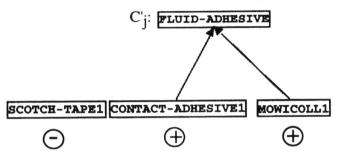

Figure 4.40   A concept that would allow the uncovering of a negative exception.

of negative examples. Based on these considerations one may conclude that a more likely cause of the positive exceptions is that the descriptions of the objects from these positive exceptions may be incomplete and possibly partially incorrect (Table 4.23).

First, the agent will attempt to cover the positive exceptions that are not covered because some of their objects do not have the features specified in the rule's conditions. For these objects the agent will attempt to elicit the corresponding features from the expert. As a result, some of the positive exceptions will become positive examples.

Second, the agent will attempt to cover the positive exceptions that are not covered because some of their objects are not covered by the corresponding concepts in the rule's conditions. In such cases the agent will attempt to discover and elicit new concepts from the expert that are more general than the concepts from the rule's conditions and discriminate between the objects from the positive examples and exceptions on one side, and the objects from the negative examples on the other side. This may further reduce the number of positive exceptions. For the remaining positive exceptions, the agent will assume that some of their features that are not covered by the rule's conditions may be wrong and will confirm them with the expert. This may further reduce the number of positive exceptions. The remaining

Table 4.21   Sample agent–expert dialog during concept discovery and elicitation

Consider the following problem-solving episode:

```
IF
    ?A ATTACH, OBJECT ?X, TO ?Y
THEN
    DECOMPOSE TASK ?A, INTO ?AP ?P
    ?AP APPLY, OBJECT ?Z, ON ?X
    ?P PRESS, OBJECT ?X, ON ?Y
```

Could you think of a concept corresponding to **?Z** that cover **CONTACT–ADHESIVE1** and **MOWICOLL1** without covering the object **SCOTCH–TAPE1** [Yes/No]: *Yes*

Give this concept a meaningful name: *FLUID–ADHESIVE*

Define the concept **FLUID–ADHESIVE** and its relations with other concepts:

```
FLUID-ADHESIVE ISA ADHESIVE
CONTACT-ADHESIVE ISA FLUID-ADHESIVE
MOWICOLL ISA FLUID-ADHESIVE
```

Table 4.22   The method for consistency-driven concept discovery and elicitation

---

Let R be an inconsistent PVS rule, U its plausible upper bound condition, L its plausible lower bound condition, $\{E_1, ..., E_k\}$ the set of its positive examples, and N a negative exception of R.

Repeat for each object variable $?j$ from the condition of the rule R
- Let $O_{1j}, ..., O_{kj}$ be the objects corresponding to $?j$ in the positive examples $E_1, ..., E_k$.
- Let O_{nj} be the object corresponding to $?j$ in the negative exception N.
- Let C_{uj} be the object concept in U that covers the objects $O_{1j}, ..., O_{kj}$, and O_{nj}.
- Let C_{lj} be the object concept in L that covers the objects $O_{1j}, ..., O_{kj}$, and O_{nj}.
- Attempt to elicit a new object concept C'_j such that
  - C'_j covers the objects $O_{1j}, ..., O_{kj}$ from ALL the positive examples;
  - C'_j does not cover the corresponding object O_{nj} from the negative exception N;
  - C'_j may be given a meaningful name by the expert.
- Ask the expert to define C'_j and introduce C'_j into the hierarchical semantic network.
- Modify U by replacing C_{uj} with C'_j.
- Modify L by replacing C_{lj} with C'_j.
- Remove N from the list of the negative exceptions and add it to the list of the negative examples.
until an object concept C'_j has been discovered or elicited or all the object variables from the condition of the rule have been considered.

Table 4.23   The method for completeness-driven knowledge base refinement

---

Let R be an incomplete rule, L its plausible lower bound condition, $\{PX_1, ..., PX_k\}$ the set of its positive exceptions, and $\{N_n, ..., N_s\}$ the set of its negative examples.

Repeat for each positive exception PX_i

    Repeat for each object concept $?j$ from L that does not cover the corresponding object O_{ij} of the positive exception PX_i
    If the object O_{ij} is not covered because of the property-value pair $(P\ v)$ from L
        then Continue with the property elicitation method described in Table 4.24.
        else if the object O_{ij} is not covered because of the relation $(?j\ REL\ ?p)$ from L
            then Continue with the relation elicitation method described in Table 4.25.
            else if the object O_{ij} is not covered by the class C_{lj} of the clause
                then Continue with the concept discovery and elicitation method described in
                    Table 4.26.
            end
        end
    end
    If the property/relation/concept elicitation succeeded
        then Continue checking the coverage of the positive exception PX_i.
        else Conclude that the rule cannot cover this positive example.
            Transfer it from the list of positive exceptions of this rule to the list ER of positive
            examples for which new rules have to be learned.
            This finishes the processing of this positive exception.
    end
    end
end

Transfer the list of the newly covered examples from the list of positive exceptions to the list of positive examples.

Table 4.24   The method for completeness-driven property elicitation

---

Let LC be the plausible lower bound clause of the rule R that contains the property–value pair $(P\,v)$ and therefore does not cover the corresponding object O_{ij} from the positive exception PX_i.

If the description of object O_{ij} does not contain the property P
   then Ask the expert whether the object O_{ij} is characterized by $(P\,v)$.
      If the expert confirms **then** Update the description of O_{ij} in the KB **end**
   else If O_{ij} is not covered because it has the property P with value v_1 that is not covered by v
         then Ask the expert whether the property-value pair of the object O_{ij} is $(P\,v)$.
            If the expert confirms
                  then Update the description of O_{ij} in the KB as well as all the knowledge pieces
                        from the KB that refer to the object O_{ij}.
            end
      end
end

---

positive exceptions cannot be covered by the current rule. For them the agent will attempt to learn new rules. An outline of the method is presented in Table 4.23.

4.4.6   Completeness-driven elicitation of features

A positive example that is not covered by the rule because the description of some of its objects do not have some of the properties from the rule's conditions may become covered by eliciting these properties from the expert. An outline of the method is presented in Table 4.24.

Similarly, a positive example that is not covered by the rule because the description of some of its objects are not in the relations specified in the rule's conditions may become covered by eliciting these relations from the expert. An outline of the method is presented in Table 4.25.

4.4.7   Completeness-driven discovery and elicitation of concepts

A positive example may not be covered by the rule because one or several of the component objects are not instances of the corresponding concepts (classes) from the plausible lower bound condition of the rule. The agent can ask the expert whether these objects should be

Table 4.25   The method for completeness-driven relation elicitation

---

Let LC be the plausible lower bound clause of the rule R that contains the relation $(?j\ \text{REL}\ ?p)$ and does not cover the positive exception PX_i because its objects do not satisfy the relation $(O_{ij}\ \text{REL}\ O_{ip})$.

- Ask the expert whether the relation $(O_{ij}\ \text{REL}\ O_{ip})$ is true.
- **If** the expert confirms
      then Update the description of the object O_{ij} in the KB as well as all the knowledge pieces from
            the KB that refer to the object O_{ij}.
   end

Table 4.26 The method for completeness-driven discovery and elicitation of concepts

---

Let R be an incomplete rule, PE_i a positive exception, and $\{E_1, ..., E_m\}$ the set of its positive examples.

- Let C_{lj} be the object concept in the lower bound condition of R that does not cover the corresponding object O_{ij} from the positive exception PE_i, and covers the corresponding objects $O_{1j}, ..., O_{mj}$ from the positive examples.
- Ask the expert whether C_{lj} should cover the object O_{ij}.
- **If** the expert confirms
    then Update accordingly the description of O_{ij}.
    else Attempt to elicit a new object concept C_j from the expert such that:
    - C_j covers both the objects $O_{1j}, ..., O_{mj}$ and the object O_{ij};
    - C_j may be given a meaningful name by the expert.
    If a concept C_j has been elicited
        then • Ask the expert to define C_j and introduce C_j into the hierarchical semantic network.
        - Modify the plausible lower bound condition of R by replacing C_{lj} with C_j.
        - Modify the plausible upper bound condition of R by replacing C_{uj} with C_j.
        - Check the status of the negative examples of the rule. Some of them might now be covered.
        end
    end

---

instances of those concepts or whether the expert could define new concepts that also cover these objects. An outline of the method is presented in Table 4.26.

4.5 An analysis of the expert–agent interactions

4.5.1 Types of interactions

The Disciple approach has been used in several applications, mainly by its developers but also by graduate students in machine learning and knowledge acquisition. Based on this, we could draw some general conclusions on how difficult the different types of learning interactions are for the expert that teaches and trains the agent. The general idea of this approach is to allow the expert to teach the agent in a variety of ways, and to intervene whenever the expert wishes during the teaching process. On the other hand, the agent has a very proactive strategy of soliciting explanations in a variety of ways in order to remedy its failures. At a conceptual level, there are five types of learning interactions. Each of them is briefly described in the following:

- *Knowledge specification* – the expert directly adds to the knowledge base of the agent. As opposed to the mainstream knowledge acquisition research (Buchanan and Wilkins, 1993), which relies significantly on this type of interaction to produce the *final* knowledge base of the agent, in the Disciple approach this is just the beginning of the knowledge base construction process. The main purpose of this type of interaction is to allow the expert to define whatever knowledge he/she could easily express, because this initial knowledge base will be progressively refined as a direct result of training. For instance, the expert is only required to provide descriptions of object instances and concepts, because the rules are learned during future interactions. Moreover, these descriptions need not be complete. In addition, as has been described in Section 4.1, for each type of knowledge element there is a specific knowledge elicitation process that guides the expert in describing that knowledge element and in composing its description from the already defined knowledge elements. These processes are based on the dependency, integrity and consistency principles described in Section 4.1.1. All of these significantly reduce the burden placed on the expert during this type of interaction. Although this interaction occurs primarily during the construction of the initial knowledge base, it is also available through the entire learning process, and is integrated with the other interactions.

- *Training example specification* – the expert gives the agent examples of problems and solutions identified as positive or negative. Most machine-learning systems rely upon this type of interaction as their only input from the expert. In addition, this interaction is extremely one-sided – in empirical inductive learning the expert prepares an input file in a predefined format, specifying *all* of the positive and negative examples. In the Disciple approach, as illustrated in the previous sections, the expert is only required to give the agent one positive example at the start of the learning process. Thereafter, the system analogically generates additional training examples for the expert to validate, thus avoiding further training example specification.

- *Oracle interaction* – the agent proposes solutions to problems and asks the expert to indicate if they are correct or not. This type of interaction is the simplest for the expert as it only requires a 'yes' or 'no' answer to a question that must be familiar to the expert. Very few machine-learning systems carry out the type of experimentation that a Disciple agent uses, with the associated validation from the expert. CLINT (De Raedt, 1993) and ILP (Muggleton, 1992) are some of the notable exceptions. Most learning systems typically require the expert to specify all the training examples, rather than generate them for the expert.

- *Explanation* – the expert explains to the agent why a problem and its solution are correct, or why they are incorrect. As presented in Section 4.2.2.1, there are several forms of explanations and a large variety of explanation interactions ranging from those which are easy for the expert to those interactions which are more challenging. An easier interaction is asking the expert to choose the correct explanations from a set generated by the agent. Other simple explanation interactions consist of focusing the agent's attention to some parts of the example to be explained. It is more difficult for the expert to explain him/herself why a solution is right or wrong, but the agent facilitates this process by guiding the expert in defining the explanation.

- *Guided elicitation* – the expert is guided to specify a feature or a concept name that discriminates between two groups of objects. This type of interaction occurs primarily during the exception-driven refinement of the knowledge base. It is used in knowledge elicitation systems but is very rare among the learning systems. Most of these types of interactions are quite simple, such as asking if an object has or does not have a certain feature. They are quite similar to the type of interactions that take place in repertory grid elicitation (Gaines and Shaw, 1992a). We found, however, that there are certain questions that are more difficult to answer, such as asking the expert to indicate the name of a concept covering a given set of instances.

There are several other approaches to building agents that are comparable to the Disciple approach, from the point of view of the employed interaction:

- Learning apprentices like LEAP (Mitchell *et al.*, 1985), Odysseus (Wilkins, 1990), and CLINT (De Raedt, 1993) learn from their users. A Disciple-based agent is also a learning apprentice. It differs, however, from these systems in that it employs more types of interaction. LEAP is based on example specification interactions. Odysseus learns by watching the user and not by interacting with the user. CLINT relies on training example specification and oracle interaction. The early work on the Instructable Production System (Rychener, 1983) has identified many of the issues later developed by the learning apprentices, such as supervising the system as it solves problems and providing context-dependent instruction. A new generation of learning apprentices are the learning personal assistants such as Maxims (Maes *et al.*, 1994) and CAP (Mitchell *et al.*, 1994). They mostly use the training example specification and oracle interaction to guide an empirical learning from examples approach, as well as relying significantly on learning by watching the user.
- Case-based reasoning agents like PROTOS (Porter *et al.*, 1990), CREEK (Aamodt, 1995), and CABINS (Sycara and Miyashita, 1995) rely heavily on various types of interactions with the user (primarily knowledge specification, training example specification, and oracle interaction, but also explanations in the case of Protos). In that respect they are very similar to the Disciple approach. However, their learning methods are significantly different, being case-based as opposed to rule and semantic network based in the Disciple approach.
- Advice-taking systems like FOO (Mostow, 1983) and the system of Gordon and Subramanian (1993) receive general knowledge from the trainer (a form of explanation interaction) that needs to be specialized (operationalized), as opposed to Disciple, which receives specific knowledge (examples) that need to be generalized. We believe that for many tasks it is easier for a user to give situated examples as opposed to such general advice.
- Knowledge elicitation systems like KSS0 (Gaines and Shaw, 1992a and b) or NANO-KLAUS (Hass and Hendrix, 1983) are primarily concerned with helping an expert to express his/her knowledge in a formal way. KSS0 uses the repertory grid representation and is based on two types of interactions: knowledge specification and guided elicitation. NANOKLAUS primarily deals with knowledge specification, its emphasis being on the use of natural language.

4.5.2   The utility of explanations

As will be shown in the case studies in the following chapters, a Disciple agent is able to learn complex rules from a small number of examples. This efficiency is achieved through the use of simple plausible version spaces and a human guided heuristic search of these spaces. Plausible version spaces have been inspired by the classical version space concept introduced by Mitchell (1978), developing it along several directions such as:

- the ability to learn from only a few examples since the expert's explanations identify the relevant features of the examples;

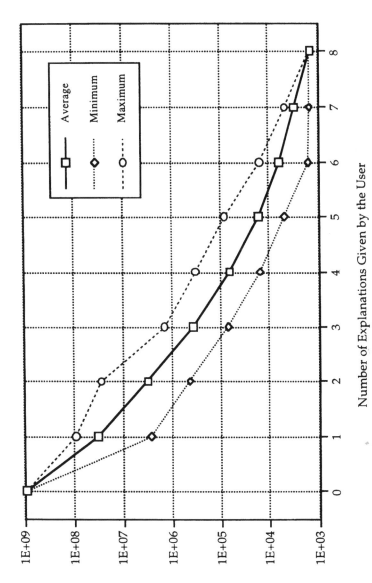

Figure 4.41 Reduction of search space by explanations.

- the ability to learn partially inconsistent rules, when the representation language is incomplete, as well as to guide the elicitation of additional knowledge from the expert, to reduce this incompleteness;
- the use of a heuristic search by limiting the upper and lower bounds to only one conjunctive expression. This avoids a combinatorial explosion of the version space bounds and is not a limitation because several conjunctive rules, equivalent with a disjunctive rule, could be learned.

In order to quantify the value of explanations we have run a series of experiments. These experiments were a variation of the leave-one-out experiments done with empirical inductive learners. First, an ideal rule was learned for a certain problem in the ModSAF domain (see Chapter 9) by giving eight explanations. Then, a set of rules were learned by withholding explanations from the agent in a progressive fashion (giving first one less, then two less and so on). We have also varied the withheld explanations to generate all possible combinations of the remaining explanations. Thus by leaving out one explanation, we had eight combinations, by leaving out two explanations we had 28 combinations, and so on. The initial search space (which in the Disciple approach is approximated by the set of instances of the plausible upper bound) was measured for each rule to determine what the effect of leaving out the explanations was. The graph in Figure 4.41 was obtained by averaging the search space obtained for each number of explanations given. We have found a wide variation of the search space since the individual utility of the explanations varied. Both the minimum and maximum values of the search space at each number of explanations are also presented on the graph to show this variation. Figure 4.41 shows that there is roughly an order of magnitude drop in the search space that the method uses for each given explanation. The maximum search space is 9.54E8. The initial explanations vastly reduce this search space. As an example, after two explanations are given, it averages 3.1E6 instances. This result explains why the expert can teach the agent the rule with only a small number of examples.

Other experimental results are presented in Section 6.6.

4.6   Final remarks

The Disciple learning and knowledge acquisition methods have been initially developed by Gheorghe Tecuci and have been further refined and developed in collaboration with his students. Tomasz Dybala has developed the knowledge elicitation approach described in Section 4.1 and has written most of that section. He also had the idea of splitting the rule learning and refining method into two independent methods. The experiments presented in Section 4.5.2 have been performed by Michael Hieb who contributed to the writing of Section 4.5. Support for the more recent development of the learning and knowledge acquisition approach described in this chapter has been provided by the AFOSR grant F49620-97-1-0188, as part of the DARPA's High Performance Knowledge Base Program.

5
The Disciple Shell and Methodology

This chapter presents the architecture of the Disciple learning agent shell. It also presents the methodology of developing it into a specialized intelligent agent, describing and illustrating the interactions between the expert and the agent during the development process.

5.1   Architecture of the Disciple shell

The Disciple shell has a modular architecture which facilitates the development of learning agents for a variety of domains. Its general architecture was briefly presented in Chapter 2. Figure 5.1 presents the same architecture but in somewhat greater detail.

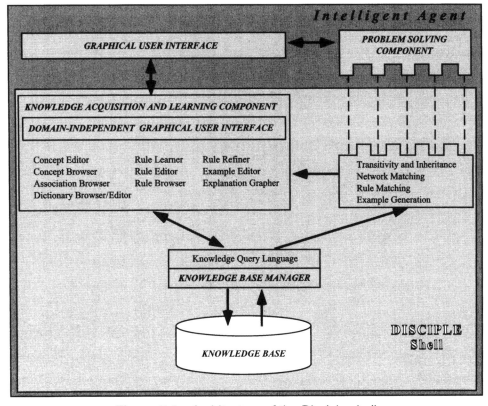

Figure 5.1   Architecture of the Disciple shell.

The Disciple shell consists of the four main components in the light gray area which are domain independent. They are:

- a *knowledge acquisition and learning component* for developing and improving the knowledge base, with a general graphical user interface to enable the expert to interact with the shell for the purpose of developing the knowledge base;
- a basic *problem-solving component* which provides various facilities used by the knowledge acquisition and learning component and provides basic agent operations;
- a *knowledge base manager* which controls access and updates to the knowledge base;
- an *empty knowledge base* to be developed for the specific application domain.

The two components in the dark gray area are the domain-dependent components that need to be developed and integrated with the Disciple shell to form a customized agent that performs specific tasks in an application domain. They are:

- a *graphical user interface* which supports specialized knowledge elicitation and agent operation, determined in part by the nature of the tasks to be performed and in part by the objects in the domain;
- a *problem-solving component* which provides the specific functionality of the agent.

The domain-specific problem-solving component is built on top of the domain-independent problem-solving operations of the Disciple shell, forming together the inference engine of a specific agent. The domain-specific interface is also built for the specific agent to allow the domain experts to communicate with the agent as close as possible to the way they communicate in their environment. Some specific agents, such as the assessment agent (see Chapter 6), need additional user interfaces to support unique domain requirements for knowledge representation.

5.1.1   Knowledge acquisition and learning component

The main purpose of the knowledge acquisition and learning component is to assist the expert during the development of the knowledge base. This component contains modules for knowledge elicitation, rule learning, and rule refinement. There are also modules for exception-driven knowledge base refinement, but they were not yet fully integrated within the Disciple shell when this book was written and are therefore not shown in Figure 5.1. These modules of the knowledge acquisition and learning component implement the methods described in Chapter 3 and Chapter 4.

The component uses the services of the knowledge base manager to create, access and modify/delete elements in the knowledge base. It also uses the services of the problem-solving component.

The expert interacts with the agent's knowledge acquisition and learning facilities via a domain-independent graphical user interface which provides overall control over the process of developing the knowledge base. A fundamental premise behind Disciple's interface is that it should incorporate an element-centered design where the expert can focus on the knowledge elements, browsing for any element and changing it directly. As the expert interacts with each element, the appropriate description and relations are displayed automatically. As illustrated in Figure 5.1, Disciple's interface consists of browsing and editing facilities, which support the knowledge elicitation requirements for building a knowledge base, and rule-learning and refinement facilities.

During knowledge elicitation, the expert interacts with Disciple's editing and browsing capabilities to define knowledge that can be easily expressed. The interfaces associated with these capabilities include the *dictionary browser/editor*, the *concept browser*, the *concept editor*, the *association browser*, the *rule browser* and the *rule editor*. The *concept editor*, the *rule editor*, and the *dictionary browser/editor* are the main interfaces which allow the expert to directly influence the content and the structure of the agent's knowledge.

When teaching the agent to perform its tasks, the expert uses the *rule learner* interface and the *rule refiner* interface, and is supported by the *example editor* for the creation of problem-solving episodes and the *explanation grapher* for the definition of explanations.

5.1.2   Problem-solving component

The main purpose of the problem-solving component of the shell is to provide basic capabilities which support the inferential processes necessary for rule learning and refinement and the basic operation of the agent. There are several basic elementary operations that can be used by a domain-specific problem solver: transitivity of certain relations, inheritance of features in a semantic network, network matching, rule matching and example generation.

Transitivity and inheritance are basic reasoning operations in a semantic network that are used by the other problem-solving operations of the agent. Transitivity is used to determine the relationship of an instance or concept with another which is elsewhere (whether a super-concept or a subconcept/instance) in the generalization hierarchy. Once such a relationship is established, the inheritance of features from a more general concept to a less general concept or instance can be used.

Network matching is the operation of retrieving information from the semantic network of the agent's knowledge base. It consists of matching the description of the looked-for information with a fragment of the semantic network. This operation can be used for answering questions about the objects from the agent's knowledge base.

Rule matching is the operation of checking whether a rule can be applied to perform a task. The agent checks whether the task from the condition of the rule is more general than the task to be performed; if this is true, then it concludes that it can apply the operation from the THEN part of the rule to perform the task. Rule matching can use either the plausible lower bound condition of the rule or the plausible upper bound condition of the rule.

Example generation is the operation of building a positive example of a rule by matching the clauses from the rule's condition with the object concepts from the agent's semantic network. The agent can also attempt to generate an example of the rule starting from a partially specified example. Moreover, it can generate an example of a rule that satisfies certain constraints. For instance, it can generate an instance of the rule or a generalized example.

5.1.3   Knowledge base manager

By the very nature of the process, the knowledge base under development will be incomplete and possibly partially incorrect. The basic function of Disciple's knowledge base manager then is to support this development process, providing a means to easily update the evolving knowledge base while ensuring consistency between knowledge elements.

The knowledge base manager serves as the primary interface for the entry of and access to the knowledge elements within the knowledge base. The main role of the knowledge base manager is threefold:

- To maintain a consistent knowledge base during its development. It does this by ensuring uniqueness among the names of the elements through the use of a *dictionary* data structure and operators, and by monitoring associations between knowledge elements.
- To support uniform interactions between the other modules of Disciple and the knowledge base over a diversity of knowledge elicitation and learning operations. Uniform interactions are achieved by a *knowledge query language* which is used to access the services of the knowledge base manager. Each element of the knowledge base may be directly accessed by its name or as a result of a query for an element with specific characteristics.
- To be responsible for the administrative operations associated with loading, initializing or saving the knowledge base, and to provide similar operations on associated elements, such as a set of examples.

5.1.3.1   Dictionary

Each knowledge element has a unique name maintained in a data structure called the dictionary. The knowledge base manager uses the dictionary to ensure that knowledge element names are unique when created and remain unique when modified. The dictionary is also used to confirm that an element with a given name exists. The four basic operations on the dictionary are:

- *Put into the dictionary*, which inserts a new element name into the dictionary and sorts the dictionary in alphabetical order. If an element with the same name already exists, then this operation does not introduce a duplicate name.
- *Delete from the dictionary*, which removes a specified element from the dictionary. If the name specified does not exist in the dictionary, an error is reported.
- *Get from the dictionary*, which returns a specified name if it exists in the dictionary. This operation is mainly used to determine whether an element with that name exists.
- *Scan the dictionary*, which maps a specified condition over the elements included in the dictionary and returns the element names which satisfy the condition.

5.1.3.2   Knowledge query language

All the capabilities of the knowledge base manager are externally accessed via the knowledge query language. This language defines four groups of operations – constructors, destructors, modifiers and queries – that implement functions to define, delete, modify and get knowledge elements, respectively. Each group of operations has a set of functions with a specific syntax and must meet specific requirements before the knowledge base is updated. The knowledge query language isolates the low-level knowledge base operations from the high-level knowledge elicitation, learning, and problem-solving operations. It also allows the learning and problem-solving methods to be designed and implemented in an abstract, more knowledge-representation-independent way. In the following, we briefly describe the operations from the knowledge query language.

Constructors.   These are used to define new knowledge elements. In every case, a unique name must be specified. Other constraints must also be satisfied, depending upon the element being defined. Before the knowledge base is extended with the new element, the element description is checked for consistency (ensuring the uniqueness of the new name and the presence of related concepts in the dictionary). If the input is correct, the dictionary structure is updated, the element description is introduced into the knowledge base and any associations with other elements are added to the element's descriptions. The following example shows how a new instance is created:

> **(define-instance**   `BOLT1`
>
>   `:description`   **"Bolt Number 1"**
>
>   `:instance-of`   `BOLT`**)**

Destructors.   These are used to delete knowledge elements or some parts of their descriptions. The deletion takes effect only when it does not violate descriptions of other knowledge elements. Thus the destructor operations heavily depend upon the associations among the elements to speed up the search process for possible links. For instance, a concept cannot be deleted if it is used by a relation to define its domain. When an element is deleted, all of its associations with other elements are removed.

The following example shows a function call to delete the relation **SUBSTANCE** from the task **APPLY**. The description of **SUBSTANCE** is also updated to remove its association with the task **APPLY**.

> **(delete-task**   `APPLY`
>
>   `:feature`   `SUBSTANCE`**)**

Modifiers.   These are used to extend element descriptions with new parts or to change existing knowledge elements. When an element is modified, the relevant associations are updated. For example, if a function call is like the example below, then the list containing the feature names characterizing the task **APPLY** is extended with the feature **OBJECT**. This happens only if the feature **OBJECT** is currently defined.

> **(modify-task**   `APPLY`
>
>   `:features`   `OBJECT`**)**

Queries.   These are used to search for knowledge elements satisfying some conditions, or to return an element description or some of its parts according to the specified request. The query functions use some internal properties of the knowledge representation, like the associations or the multiple inheritance of features, to answer a specified query. For example, the response to the query below will be the list of the inherited features of the concept **BOLT**.

> **(get-concept**   `BOLT`
>
>   `:request`   `:inherited-features`**)**

5.2   The methodology for building intelligent agents

The Disciple learning shell allows rapid development of a customized learning agent for a particular application domain. There are two main types of agent which can be developed:

- an agent which is used by and serves as an assistant to the expert; and
- an agent which is used by and provides a service to a non-expert user.

In the first case, the expert is both the developer and the user of the system. The expert initially teaches the agent basic knowledge about the application domain then the agent is put to use as an assistant. The agent continues to improve its knowledge during its interactions with the expert.

In the second case, the agent is initially developed by the expert then is delivered to the non-expert users. Any necessary changes to its knowledge or operation means the involvement of the expert.

5.2.1   An overview of the Disciple methodology

An overview of the Disciple methodology is given in Figure 5.2. Depending upon the complexity of the application domain, the expert may require the assistance of a separate developer (a software and/or knowledge engineer) to build the domain-dependent modules of the agent shown in the dark gray area of Figure 5.1. The expert and the developer should work closely together to determine the customization requirements and build the agent. The dark arrows in Figure 5.2 indicate the extent of the agent-building process if no customization is required but only the development of the knowledge base. The gray arrows highlight

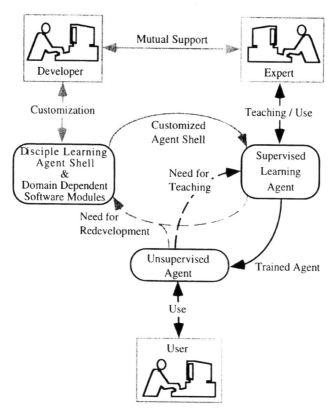

Figure 5.2   Disciple's agent-building methodology.

the customization activities required during the agent-building process. Broken arrows indicate where the agent development process may require some iterative development efforts, due to changes in the application domain or in the requirements for the agent, after the agent has been developed.

According to this portrayal, there are three stages and three different participants in the agent's lifetime:

1. The agent developer (software and knowledge engineer), cooperating with the domain expert, customizes the Disciple shell by developing a domain-specific interface on top of the domain-independent graphical user interface of the Disciple shell. This domain-specific interface allows the domain experts to express their knowledge as naturally as possible. The result of this effort is a customized agent with learning capabilities, a generic problem-solving component, and an empty knowledge base. This agent can interact with the expert during the knowledge base development process. The domain expert and the agent developer also decide on the nature and extent of a domain-specific problem solver, based upon the type and purpose of the agent to be developed. The agent developer continues the customization of the agent with the development of the domain-specific problem solver. This agent is fully customized and can interact with the expert for problem-solving and learning.
2. The domain expert interacts with the agent to develop its initial knowledge base and teach it domain-specific tasks. The expert both teaches and uses a customized learning agent which is to be his or her assistant. If the agent is to be used by another user, the expert teaches, verifies and validates the agent and then releases it to the user.
3. The agent is released to the user and performs the tasks it was taught. In some cases the agent may require retraining by the expert or further customization by the developer to update or refine its interfaces or problem-solving methods. The process of retraining and redevelopment does not differ from the agent's initial training and development.

The first two stages of the evolving agent consist of the following activities:

- customization of the Disciple shell to support specialized knowledge elicitation;
- development of the semantic network;
- customization of the Disciple shell with domain-dependent modules for agent operation (the problem solver and the user interface);
- training of the agent for its domain-specific tasks;
- verification and validation of the agent.

The next four chapters of the book present and illustrate each of them in various degrees of detail, in the context of building four specialized intelligent agents.

5.2.2   Issues in developing Disciple agents

The Disciple agent-building methodology addresses several critical issues that have been found to be limiting factors in building intelligent agents for complex real-world domains. First and most important, it is an attempt to overcome the knowledge acquisition bottleneck and the knowledge adaptation bottleneck. The knowledge acquisition bottleneck expresses the difficulty of encoding knowledge in the knowledge base of an intelligent agent. The main idea for overcoming it is to allow the expert to express his or her expertise in a natural way. Ideally, this would be identical with the way the expert would teach the human apprentice.

   The knowledge adaptation bottleneck expresses the difficulty of changing the knowledge in the knowledge base of the agent in response to changes in the application domain or in the requirements of the agent. While a natural approach to this type of bottleneck would be autonomous learning by the agent from its own experience, this approach alone is not powerful enough for complex application environments. Disciple supports an approach to the knowledge adaptation bottleneck based on a retraining process. This means that, when the mismatch between the agent's model of the world and the world itself is above a certain threshold, the agent enters a retraining phase in which it is again directly taught by the expert. By training and retraining the agent (rather than manually encoding and manually modifying its knowledge base later) one greatly facilitates the knowledge acquisition and adaptation processes.

   Another critical issue addressed by the Disciple approach is the scalability of the agent-building process. This is mainly achieved in two ways. The first is the use of an advanced model of interaction between the expert and the agent that allows the expert to guide the agent in building a large knowledge base (see the following sections). The second is the use of efficient multistrategy learning methods based on the plausible version space representation.

   Finally, there is the issue of finding a suitable balance between using general (and therefore reusable) modules and specific (and therefore powerful) modules in building the agent. Using general modules significantly speeds up the development process. However, the agent may not be well adapted to its specific application domain and may not be that useful. Contrary to this, building the agent from domain-specific modules leads to a well-adapted and useful agent, but the development process is very difficult. The Disciple shell provides a set of general and powerful modules for knowledge acquisition and learning. They are domain-independent and are incorporated as such in a developed agent. However, for the interface and the problem solver, which are domain dependent, the Disciple shell contains a generic graphical user interface and problem-solving modules that support only basic problem-solving operations. Therefore, for a given application domain, one has to develop additional, domain-specific interfaces and problem-solving modules, in order to create an easy-to-train and useful agent. Moreover, if the agent has to execute in, or communicate with, an existing application, such as the MMTS (see Chapter 6) or ModSAF (see Chapter 9), then one also has to develop the interface with the application.

   In the following sections we present the interactions between the expert and the agent during the main processes of knowledge base development: knowledge elicitation, rule learning and rule refinement.

5.3   Expert–agent interactions during the knowledge elicitation process

As presented in Chapter 4, the expert defines the initial knowledge elements of the knowledge base during the initial knowledge elicitation phase. These initial elements primarily consist of concepts and instances which together constitute the semantic network, and the specific knowledge elements from which examples, used for both rule learning and refinement, are constructed: the tasks which the agent is expected to perform and the types of operations which the agent is expected to execute. Both tasks and operations are also complex elements, described in terms of their features.

Although rules may also be manually specified by the expert during the initial knowledge elicitation phase, this is not generally an easy task. Thus, the expert may prefer to use Disciple's rule-learning facilities as much as possible and only manually specify rules when necessary. For this reason, the description of rule elicitation is presented in Section 5.4 on rule learning. This section presents the knowledge elicitation interactions between an expert and the agent by illustrating the definition of a concept in the loudspeaker manufacturing domain.

5.3.1   Overview of the interaction process

Figure 5.3 illustrates the process one goes through in order to build the initial knowledge base. The initial semantic network of concepts and instances is built using the concept editor to define the elements and the concept browser to display the hierarchy as it is being developed. Once the initial set of concepts is defined, the required descriptive elements

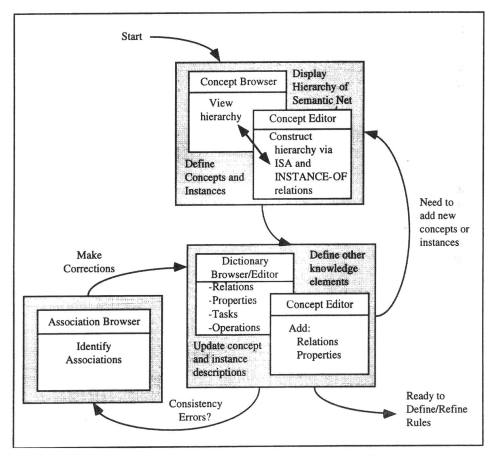

Figure 5.3   A process for knowledge elicitation.

(relations and properties) are defined using the dictionary browser/editor and the concept descriptions are updated using the concept editor. New concepts (and instances) are defined in terms of more general concepts, with additional descriptive features provided as needed. This process iterates while there are more concepts or instances to define.

The knowledge base is updated as each new feature, concept or instance is defined. However, before the knowledge base is updated, Disciple ensures that the expert has not introduced any inconsistencies as a result of modifications or deletions of knowledge elements. If so, the association browser indicates the various associations of the element with others in the knowledge base so that the expert can make the necessary corrections using the appropriate editor tool.

Although tasks and problem-solving operations are not part of the agent's semantic network, they are an intrinsic part of examples and rules, hence must be defined before rules can be learned from examples or defined manually. Similar to concepts and instances, before tasks and actions can be defined, the features associated with these elements must first be defined. The definition of tasks and problem-solving operations is done using the dictionary browser/editor.

Thus, the primary Disciple knowledge elicitation interfaces include the concept browser, the concept editor, the dictionary browser/editor, and the association browser. Let us describe the functionality of these tools while stepping through the creation of the concept **SUPER-GLUE**.

5.3.2   Interactions with the concept browser

The main goal of the concept browser is to give an insight into the agent's semantic network of concepts (see Figure 5.4). The main component of the concept browser window is the graphical display of the concept hierarchy. All the links displayed represent the **IS-A** or **INSTANCE-OF** relations between concepts. Below the concept hierarchy are panes which display information about the currently selected concept or instance: its name, its superconcepts, subconcepts, instances, features (direct, inherited or both) and feature values. By using the concept browser the expert can get a good understanding of the agent's current state of the concept hierarchy.

One superconcept of **SUPER-GLUE** is **TOXIC-SUBSTANCE**. In Figure 5.4, the selected concept is **TOXIC-SUBSTANCE**, indicated as such by its highlighting in the concept hierarchy and its name in the concept pane below the graphic. The panes below show that **TOXIC-SUBSTANCE** has **SOMETHING** as a superconcept, already has **CONTACT-ADHESIVE** as a subconcept and is described by the feature **TYPE-OF**. The feature **TYPE-OF** has value **SOMETHING**.

The expert can navigate up or down the hierarchy by clicking on the superconcept **SOMETHING** or the subconcept **CONTACT-ADHESIVE** shown in the left panes. The graphical display of the hierarchy is modified such that the selected concept is centered and highlighted, and the lower panes are updated accordingly. The expert can also specify a concept or instance of interest in the graphical display itself or directly type a concept or instance name into the selected concept field. This then becomes the new selected concept and the concept browser is updated accordingly.

The description of **TOXIC-SUBSTANCE** may be modified by pressing *Edit* in the bottom right corner. This launches the concept editor containing the current description of **TOXIC-SUBSTANCE**. The concept may be deleted altogether by pressing *Delete*, but only after the

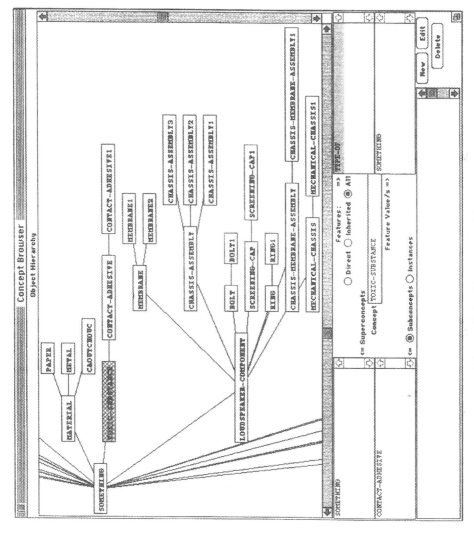

Figure 5.4   The concept browser.

description of **CONTACT-ADHESIVE** is modified to remove **TOXIC-SUBSTANCE** as one of its superconcepts. There may be other elements which are described in terms of **TOXIC-SUBSTANCE** and thus could also prevent the deletion of this concept until modified.

To create a new concept, the expert presses *New*. This also launches the concept editor but, in this case, initialized for the definition of a new concept.

5.3.3   Interactions with the concept editor

The expert interacts with the concept editor either to introduce descriptions of new concepts or to modify descriptions of existing concepts (see Figure 5.5). A concept description is created or modified via the three interaction buffers (*Concept/Instance, Relation/Property, Value*). The expert stores the appropriate elements in these buffers either directly or by selection from the elements which are listed in the panel below. The contents of the buffers are then moved up to the next line of the concept description. The radio buttons below the buffer specify which elements should be in the associated list. Notice that the layout and possible contents of the three interaction buffers correspond to the basic representation unit described in Section 3.1.2.4.

When the concept editor window is launched for a new concept, the concept description is clear and the interaction buffers are initialized for an **ISA** relation between two concepts.

The expert enters the name of the new concept in the *Concept/Instance* buffer and selects its superconcept from the *Values* list, then presses *Include* to move the contents of the interaction buffers into the concept description area located just above. Before the description is moved up, the agent ensures that the elements in the buffers are valid and appropriately used.

The remaining relations and properties of the concept and their values are similarly selected from the middle and right-hand lists and included in the main concept description. After the concept is completely described and the textual description of **SUPER-GLUE** is entered into the buffer at the top of the editor, the expert selects *Update* to add the new concept to the knowledge base.

The concept is added to the knowledge base only after the agent (knowledge base manager) ensures that the name of the new element is unique and the other elements in its description are currently defined and properly used. The knowledge base is also updated to ensure that the descriptions of the relations and concepts used by the new concept reflect their association with the new concept. Thus, the concept editor assists the expert to define concepts and instances in Disciple's knowledge representation and ensures the elicitation principles defined in Chapter 4 (dependency, integrity and consistency) are maintained.

Figure 5.5 shows the contents of the concept editor after **SUPER-GLUE** has been added to the knowledge base. Notice that middle list shows only those features which are included in **SUPER-GLUE**, and the right-hand list only contains the possible values for the selected relation **GLUES**. The interaction buffers contain the selected line of the complete description. This is also how the concept editor would look if it had been launched to edit the concept **SUPER-GLUE**.

5.3.4   Interactions with the Dictionary Browser/Editor

The dictionary browser/editor is a central, multi-purpose module of the agent (see Figure 5.6). Its main purpose is to give control over all of the knowledge elements currently existing

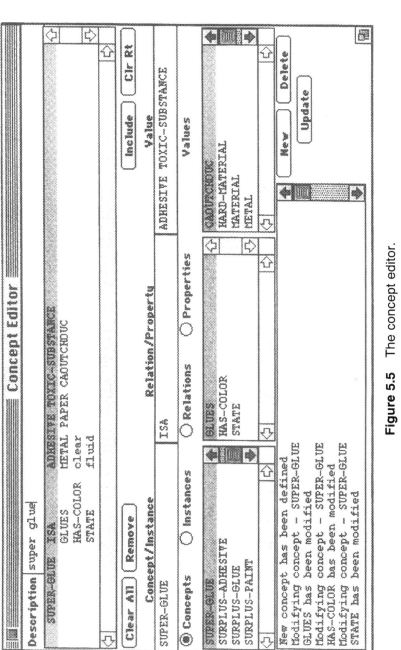

Figure 5.5 The concept editor.

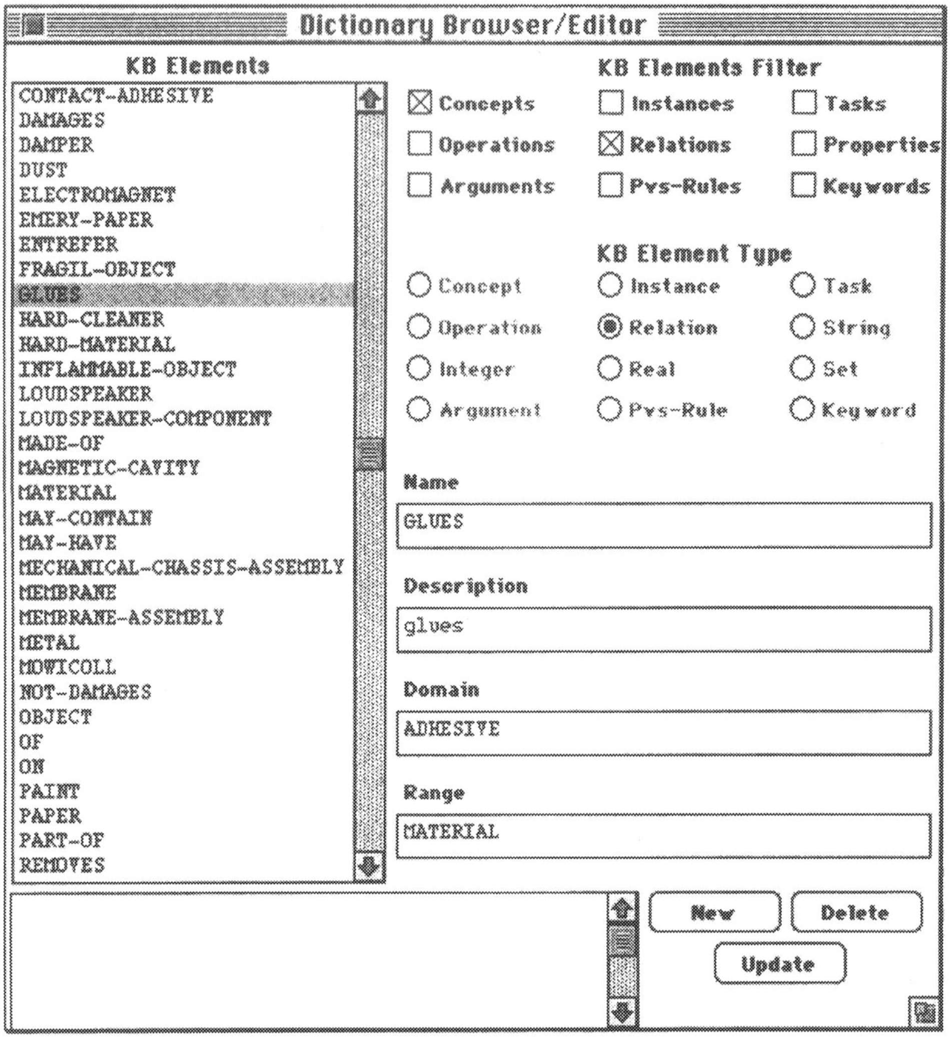

Figure 5.6   The dictionary browser/editor.

in the agent's knowledge base. With this interface, the expert can selectively display all the element names and their full or partial descriptions.

Another purpose of the dictionary browser/editor is that of an element editor for the elements whose descriptions are simple. The dictionary browser/editor can also be used to delete any knowledge element.

Figure 5.6 shows the description of the relation **GLUES**. All relations are similarly displayed. The contents of the list on the left, all the defined concepts and relations, is determined by the checked boxes in the *KB Elements Filter* pane. Notice that when more than one element type is selected, the list contains all selected elements sorted alphabetically in ascending order.

The selection of **GLUES** from the list causes its description to be stored in the four buffers to the right of the list. That **GLUES** is a relation is reflected in the selected radio button in the *KB Element Type* pane.

Every element in the knowledge base has a name and description, so the contents of the first two buffers are fixed. The contents of the remaining two buffers are specific to the type of element selected or being defined. In the case of a relation, such as **GLUES**, the third and fourth buffers contain the relation's domain (**ADHESIVE**) and range (**MATERIAL**), respectively. If we were examining a concept, such as **SUPER-GLUE**, the third buffer would list its superconcepts and the fourth would be unused. If we were examining a real property or an integer property, the third and fourth buffers would contain the bounds of its value, that is, its upper and lower values.

Once the element is displayed in the buffers, the description of the element can be modified by making the appropriate changes in the buffers then pressing *Update* in the lower right corner. The changes are made to the knowledge base only after the knowledge base manager ensures that they do not violate the knowledge elicitation principles cited in Section 4.1.

New elements can also be defined in the dictionary browser/editor. The expert presses *New*, selects the type of element being defined from the *KB Element Type* pane, then enters the appropriate information in the buffers. All elements except rules can be initially defined here. The descriptions of concepts and instances must be completed using the concept editor. The new element is added to the knowledge base only after the knowledge base manager verifies the contents of the buffers.

Selected elements can be deleted from the knowledge base by pressing *Delete*. If the element is not used as part of the description of another element, then it is deleted. If it is, the association browser indicates the associations between elements.

5.3.5   Interactions with the association browser

The association browser helps to track down the associations among knowledge elements (see Figure 5.7). As described in Chapter 2 and Section 5.1.3, associations exist between all knowledge elements and are monitored by the knowledge base manager. Figure 5.7 shows the associations of the concept **METAL** with all the other concepts listed under *Knowledge Element Associations*. For example, **METAL** and **MOWICOLL** are associated via the relation **GLUES** because **MOWICOLL GLUES METAL**, and **METAL** and **CHASSIS-ASSEMBLY** are associated via the relation **MADE-OF** because **CHASSIS-ASSEMBLY MADE-OF METAL**. The association browser is very useful for the maintenance of consistency in the knowledge base.

5.4   Expert–agent interactions during the rule-learning process

During the rule elicitation/learning process, the expert can either interact with the agent to explicitly define rules using Disciple's rule elicitation facilities, which include the rule browser and the rule editor, or can teach the agent how to solve domain-specific problems using the rule learning suite of interfaces, which include the rule learner, the example editor, and the explanation grapher. The learning tools were developed based on the plausible version space (PVS) representation described in Chapter 4.

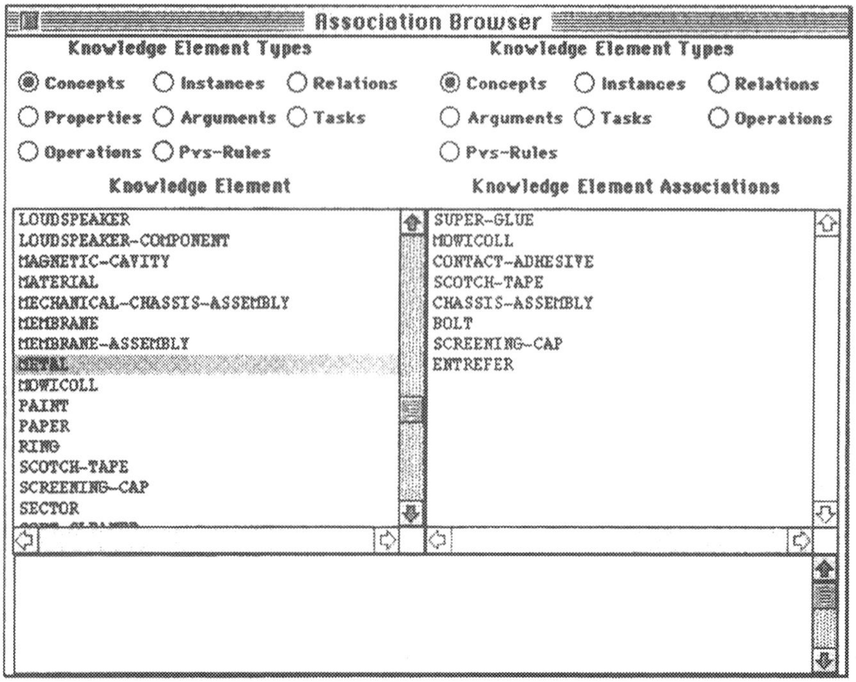

Figure 5.7   The association browser.

The process of directly introducing rule descriptions to the agent's knowledge seems to be easy from the agent's point of view – it is just told what to do in certain situations. However, it is not so easy from the expert's point of view. This is why the primary rule development facility in Disciple is its interactive rule-learning and refinement capabilities. And just in case the expert wishes to manually create some rules, an interface is also provided which assists the expert to specify rules which satisfy the agent's knowledge representation require-ments. The generic interfaces provided by the Disciple shell to support either method of developing rules are described in this section, starting with Disciple's rule-learning module.

5.4.1   Overview of the interaction process

Figure 5.8 illustrates the sequence of interfaces one can use in order for the agent to learn an initial rule. First, the rule learner is presented but no learning can take place until the first positive example is provided by the expert. The example can either come from a problem-solving operation or can be created interactively via the example editor. The example editor is launched directly from the rule learner.

After the example has been defined in the example editor, it is brought back for display in the rule learner where the expert can then select the types of explanations the agent should be looking for. These explanations are displayed in the rule learner. If one of the relevant explanations is either incomplete, due in part to the heuristics used in the agent's search, the expert can select that explanation for further editing. A relevant explanation may also be missing entirely.

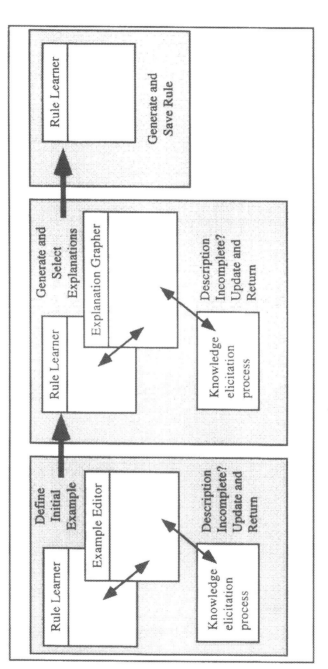

Figure 5.8   A process for learning a rule.

The creating and editing of explanations is done in the explanation grapher. Like the example editor, the explanation grapher is launched directly from the rule learner. When the explanation is complete, it is brought directly back into the rule learner and added to the list of accepted explanations. When the expert is satisfied with the example and the explanations, the initial rule is generated and displayed to the expert for inspection. If the expert is satisfied, the knowledge base is updated with the new rule.

In this section we describe the functionality of the rule learner and example editor while stepping through the creation of an initial rule for attaching one loudspeaker component to another, deferring the description of the explanation grapher until rule refinement. This is the same example used in Chapter 3 and Chapter 4 to describe both rule representation and the rule-learning and refinement process.

The rule-learning process essentially begins with the specification of a positive example of a problem-solving episode, so we shall ignore the rule learner for a moment and start with the example editor.

5.4.2   Interactions with the example editor

To introduce the description of a new learning example or to modify the description of an example generated by the agent, the expert uses the example editor shown in Figure 5.9. The example editor is generic, allowing the creation of example descriptions for any application domain in the format defined in Chapter 4. However, for some application domains there is a need to customize the example editor to the domain. This customization usually requires changes in the interface. An illustration of such a customization is the customized example editor of the assessment agent presented in Chapter 6 which replaces the generic example editor in that domain.

The example editor has similar functionality to the concept editor (Figure 5.5) and the rule editor (Figure 5.13).

The description of an example of a problem-solving episode is created by manually entering the appropriate elements into the three interaction buffers in the center of the example editor or by selecting the elements from the list panels which are located just below the buffers. Notice that the layout and possible contents of the three interaction buffers correspond to the clause representation described in Section 3.1.2.4 and, in particular, the representation of operations (Section 3.1.3.3). An example is created by selecting the elements of the example (in this case, the task **ATTACH** and its operation **DECOMPOSE**) and their related elements from the three lists below the interaction buffers. After each line of the description is formed in the interaction buffers, the expert presses *Include* to move it up to the example description.

During this process the agent assigns variables to abstract knowledge elements, such as concepts or tasks, and displays these variables as the example is being developed. The variables refer to specific instances of the elements from the particular problem-solving episode. For example, when the expert selects the task **ATTACH** from the list of tasks in the left pane, the agent displays the features of **ATTACH** in the middle pane for the expert's review. Once a feature is selected, its values are displayed in the right pane. After the expert has selected the appropriate value(s) of the feature and pressed *Include*, the agent generates a variable, **?A8** in our example, and assigns it to the task. Since there are additional features of the task to define, the agent places variable **?A8** in the left buffer and the expert continues to select from the listed features and values until the task is specified.

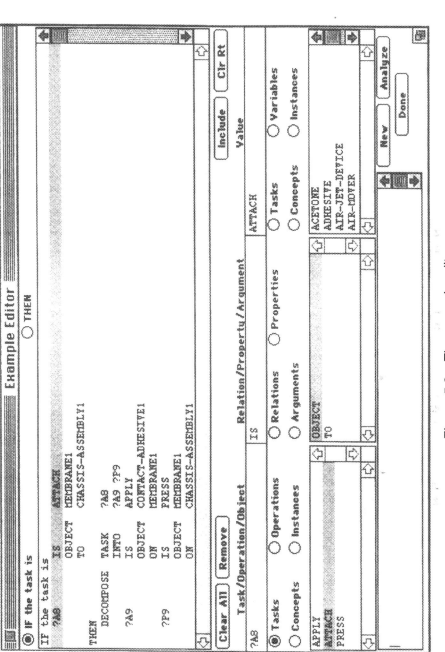

Figure 5.9 The example editor.

The output problem-solving operation is defined similarly, with the agent defining and providing the variables. Specific variables from the input task can be referenced in the output problem-solving operation by selecting *Variables* to list them under the right interaction buffer.

In this way, the expert chooses from the current elements of the agent's knowledge base to compose the learning example. When some components of the example description are missing, the expert uses the other editors (dictionary browser/editor, concept editor) to define them before continuing the development of the example description. When the description is complete, the expert presses *Done* to move the example back to the interface which launched the example editor, in this case the rule learner.

5.4.3   Interactions with the rule learner

The rule learner is used to create the initial PVS rule (see Figure 5.10), which is created from the description of the initial learning example and from the relevant explanations of the initial example. The initial example, a positive example of the rule to be learned, is created by the expert via the example editor and displayed in the top left pane of the window. The example is displayed either with or without variables depending upon whether the expert has selected the generic or customized view. The customized view is easier to read, but the generic view is necessary for correlating variables to abstract elements in the example. The customized view is shown in Figure 5.10.

The expert specifies the types of explanations to be generated by the agent via the checkboxes in the center of the interface. In this case, explanations of types *Association*, *Correlation* and *Relation* (described in Section 4.2.2.1) were specified. The explanations can be restricted to specific elements of the example by selecting one or two elements from the elements listed next to the checkboxes and checking the *Variable Filter* checkbox.

Generate was pressed to generate the explanations, which are shown in a list below the *Explanation Types* selection area. These explanations were reviewed and the two most relevant were selected (those marked with an asterisk). In this case, the expert decided that these explanations were sufficient for rule learning. Additional explanations can be generated by either directing the agent to generate explanations of type *Property* or by pressing *Edit*, which causes the explanation grapher (see Section 5.5.3) to be launched so that explanations can be created directly by the expert.

The initial rule, generated by the agent when *Create* is pressed, is shown in Figure 5.11. The rule is given a name by the expert, then is added to the knowledge base when the expert presses *Update*. The expert can review the rules contained in the knowledge base with the use of the rule browser.

5.4.4   Interactions with the rule browser

The expert interacts with the agent via the rule browser to investigate the agent's rule base (see Figure 5.12). The rule browser not only displays the descriptions of the rules but it also allows for easy scanning of the examples from which the rule was learned.

The displayed rule, **ATTACH–RULE–03**, was selected from the list of names at the top left side of the window. The rule is always displayed in generic format (with variables rather than element names) in order to distinguish between different elements of the same abstract

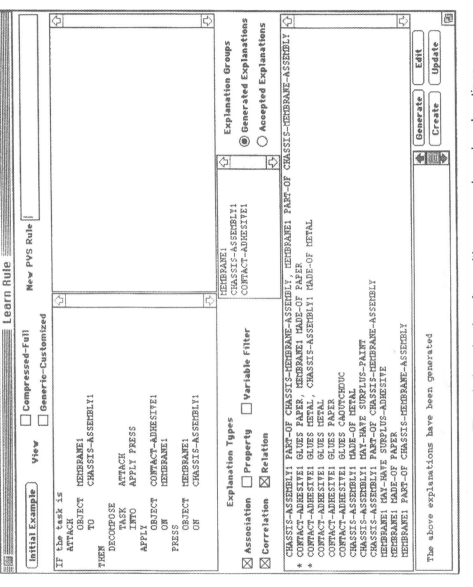

Figure 5.10 The rule-learning interface with an example and explanations.

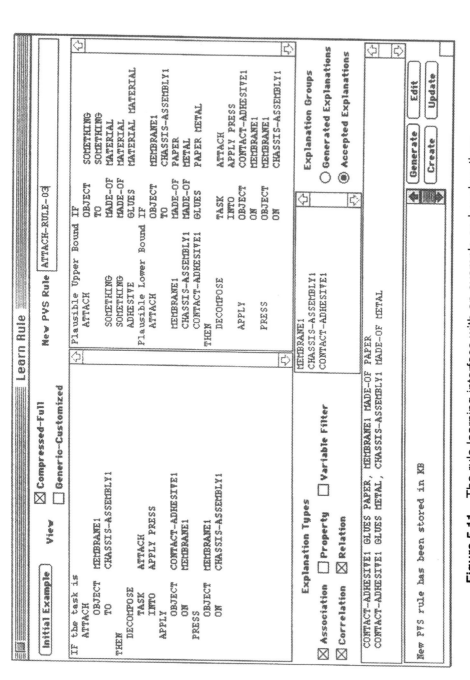

Figure 5.11   The rule-learning interface with an example and explanations.

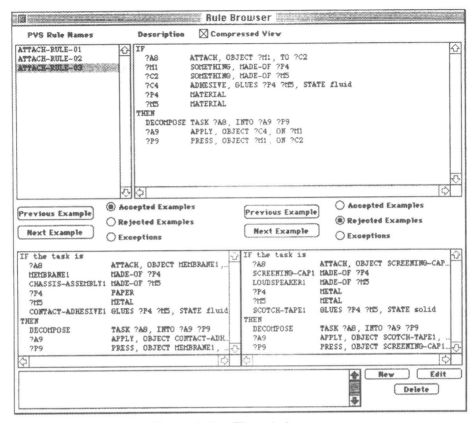

Figure 5.12   The rule browser.

type. In this case, it is also selected for display in the compressed form described in Chapter 3 and shown in Figure 3.10.

The examples used to create and refine the rule are displayed in the lower part of the rule browser. The expert can browse through two types of examples at the same time. In this case, the accepted (or positive) examples were selected for display on the left side of the window; the rejected (or negative) examples are being displayed on the lower right side of the window. If the displayed rule is invalid or redundant, it can be removed from the knowledge base via *Delete*.

The agent launches the rule editor if the expert decides either to create a new rule (presses *New*) or to modify the current rule (presses *Edit*).

5.4.5   Interactions with the rule editor

The expert interacts with the rule editor to manually define and refine rules (see Figure 5.13). Like the rule browser, all concepts, instances, and tasks included in the rule description are represented by corresponding variables in order to distinguish among different instantiations of the same abstract class. As described in Chapter 4, there are constraints to the representation of PVS rules which need to be enforced during knowledge elicitation.

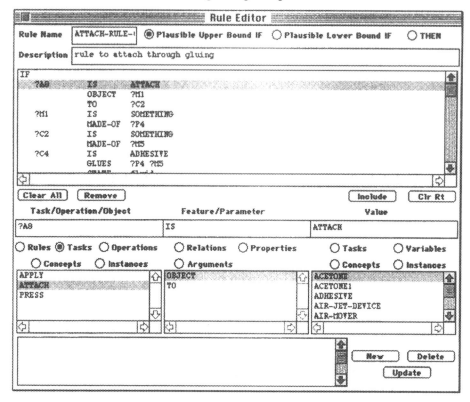

Figure 5.13   The rule editor.

The rule editor is structured in a similar manner to the other editors, with the components of the rule being built in the three interaction buffers in the center of the window and moved up to the rule description line by line. The possible contents of the buffers conform to the knowledge representation requirements.

5.5   Expert–agent interactions during the rule refinement process

The refinement of the knowledge base includes the refinement of rules and the elicitation of new knowledge which may be required as a result. During refinement of the knowledge base, the expert uses Disciple's principle rule refinement facility, the refine rule interface. Other interfaces which may be used during the rule refinement process are the rule editor (described in Section 5.4.5), the example editor (described in Section 5.4.2) and the explanation grapher.

During rule refinement, the agent generates examples and asks the expert whether they are correct. The expert can also give additional examples which satisfy conditions imposed by the PVS rule. These examples then need to be classified by the expert as positive (accepted) or negative (rejected). Accepted and rejected examples, together with the explanations introduced by the expert, create input to the rule refinement process conducted by the agent.

When the expert introduces additional examples or explanations, the need for definition of additional knowledge elements may require the use of the knowledge elicitation facilities. The domain-independent interfaces provided to support Disciple's rule refinement process, the rule refiner and the explanation grapher, are described in this section.

5.5.1   Overview of the interaction process

Once an initial PVS rule has been defined, the expert and the agent work together to refine the rule in a rule refinement process, described in Chapter 4. As illustrated in Figure 5.14, the rule refinement process begins with the rule refinement interface. The rule to be refined is designated and any constraints to be placed upon the variables of the rule are specified.

Either the agent can generate an example for rule refinement or the expert can choose to create an example. If the expert chooses to generate an example, the agent launches the example editor. When the expert is satisfied with the example, it is verified by the agent then returned to the rule refinement interface.

However the example is generated, the expert specifies whether it is a positive or negative example for the rule. If the example is accepted, the agent refines the rule and provides the modified rule for the expert's review. If the example is a negative example, the expert must then indicate the reason for the rejection. The failure explanation can either be an invalid component of the example or an explanation generated by the expert. An invalid component can be designated by the expert by pointing to it in the example description.

If the expert must generate the explanation, the agent launches the explanation grapher. The expert can designate the type of explanation to be generated and the explanation grapher interface assists in the construction of the explanation. When the expert is satisfied with the failure explanation, it is verified by the agent to ensure it satisfies the representation requirements before it is returned to the 'refine rule' interface.

Once the explanation is provided, the negative example is rejected by the expert and the agent uses the explanation to refine the rule. This process continues while there are still examples to process or until the bounds of the PVS rule coincide.

This section illustrates Disciple's rule refinement facilities in greater detail by refining the rule learned in the previous section.

5.5.2   Interactions with the rule refiner

The rule refiner, shown in Figure 5.15, employs learning by experimentation, inductive learning from examples and learning from explanations to refine rules. It is used to refine the PVS rule created either by using the rule editor or the rule learner (see Figure 5.13 and Figure 5.15).

The top pane of this window contains the current rule, **ATTACH–RULE–03**, selected by clicking on the name of the rule in the list to the left. The bottom left pane contains the current example generated by the agent, requested via *Generate*. The expert can apply certain constraints upon the variables before the agent generates the example. In this case, there were no constraints applied.

The pane next to the current example allows the expert to examine previous examples of the rule. The type of examples to be examined depends on the example types selected above.

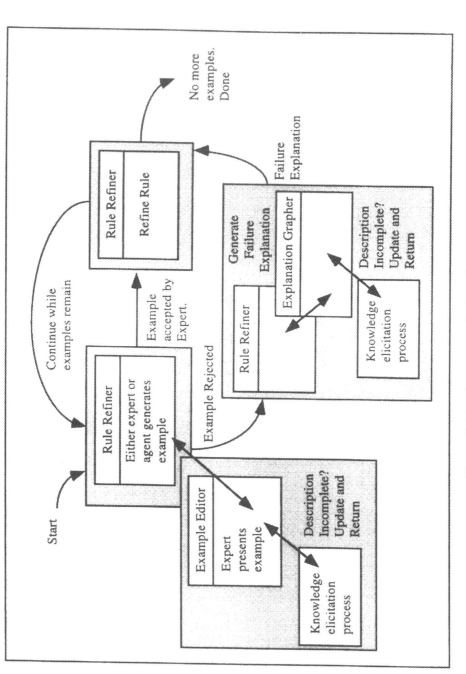

Figure 5.14 A process of rule refinement.

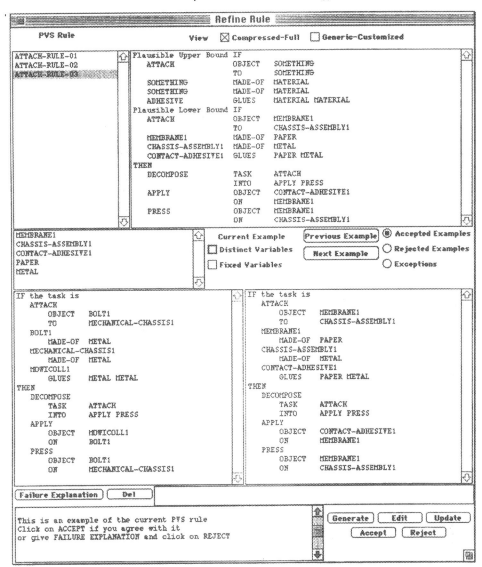

Figure 5.15 The refine rule interface showing the first example.

Currently, the only accepted example for this rule is the example from which the rule was learned. This is the example which is displayed.

The examples generated by the agent are accepted or rejected by the expert, being thus characterized as positive examples or as negative examples of the rule to be learned. The current example shown is a positive example, so the expert indicates acceptance via the *Accept* button, also located in the bottom right corner of the interface. Figure 5.16 shows the contents of the refine rule interface after the example is accepted. The generalizations made by the agent are itemized in the message pane at the bottom of the interface and the rule has been updated accordingly.

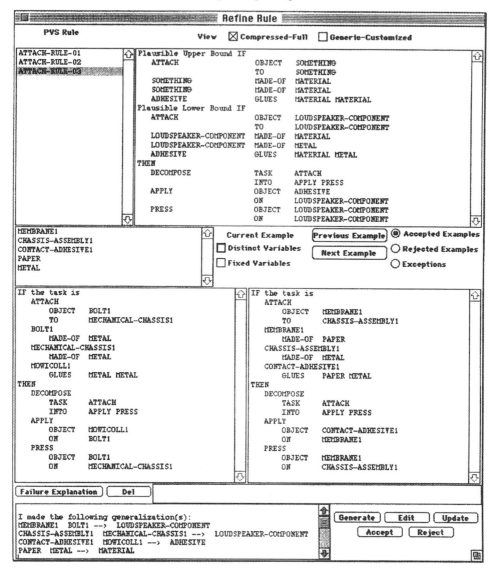

Figure 5.16   The refine rule interface with the first example accepted.

Some of the examples generated by the agent may be negative examples. If this is the case, an explanation must be provided by the expert. The expert may provide the explanation by selecting the line with the error in the current example or the expert can provide an explanation through the use of the explanation grapher, which is launched by the agent when *Failure Explanation* is pressed.

Figure 5.17 shows the refine rule interface with the negative example in the bottom left pane. On the right-hand side of this negative example is a positive example (in this case the example from which the rule was initially learned). The example on the left is negative because **SCOTCH-TAPE1** is solid while the corresponding object in the positive example,

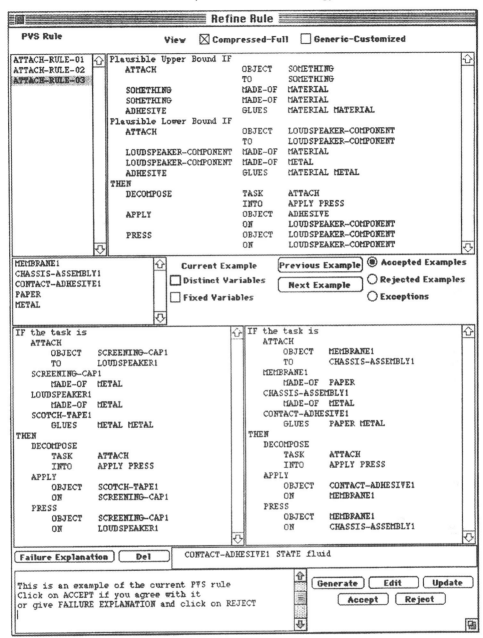

Figure 5.17 The refine rule interface with negative example and explanation.

CONTACT–ADHESIVE1, is fluid. The expert has used the explanation grapher to create the explanation **CONTACT–ADHESIVE1 STATE fluid**. After providing the explanation, the expert indicates rejection of the negative example and the agent refines the rule to not cover it.

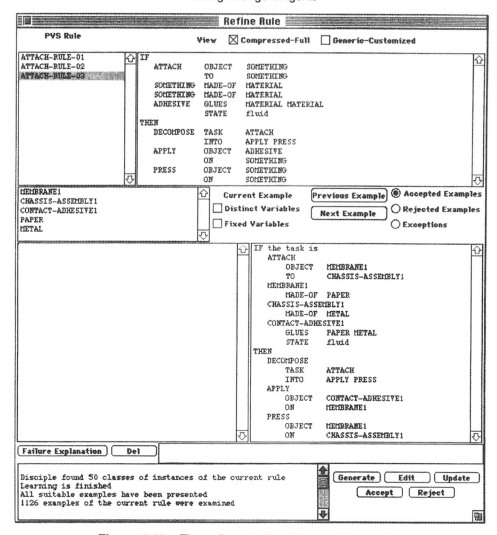

Figure 5.18   The refine rule interface with final rule.

Figure 5.18 shows the rule after all examples necessary for the completion of the rule refinement process have been presented to the expert and accepted or rejected. Of interest is that only eight examples needed to be presented to the expert for analysis from the 1126 examples examined internally by the agent.

5.5.3   Interactions with the explanation grapher

According to the rule-learning method and the rule refinement method described in Chapter 4, the expert should be able to introduce his or her own explanations. To implement this feature of the methods, Disciple is equipped with the explanation grapher shown in Figure 5.19. The explanation grapher is a specialized editor which allows the expert to build the

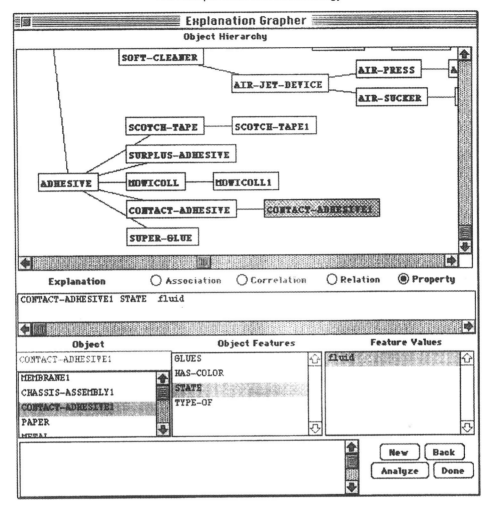

Figure 5.19 The explanation grapher.

types of explanations defined in Chapter 4. The explanation grapher can be launched from either the rule learner (via *Edit*) or from the refine rule interface (via *Failure Explanation*). Similar to the concept browser, the explanation grapher displays the semantic network hierarchy for the expert's review.

When entered from the refine rule interface, the objects listed in the left-most list near the bottom are the objects from the original example from which the rule was learned, not from the current example. The reasoning is as follows: since the current (negative) example was generated by analogy with the initial example, the explanations of initial example must be incomplete. Therefore, the expert may wish to provide additional ones.

The process of introducing the explanation begins by specifying the type of explanation to be created. In this case, the explanation is a property of the instance **CONTACT-ADHESIVE1**. When the expert selects **CONTACT-ADHESIVE1**, the agent positions the display of the semantic network so that **CONTACT-ADHESIVE1** and its superconcepts are visible. It also lists all the features of **CONTACT-ADHESIVE1** in the middle list pane.

Once the property **STATE** is selected, the agent lists its possible values in the third list. The explanation is built in the explanation buffer as the expert selects the relevant items in each list. When the explanation is complete (the expert selects *Done*), the agent returns the explanation to the refine rule interface.

If the object descriptions are not complete and correct enough to define the explanation needed, then the other editing tools should be used to refine them.

5.6   Final remarks

The current version of the Disciple learning agent shell has been developed by several members of the Learning Agents Laboratory. The core learning and problem-solving methods have been developed by Gheorghe Tecuci. Tomasz Dybala has developed all the browsers and the editors, as well as the knowledge base manager for the Sun version of Disciple. Kathryn Wright has ported Disciple onto the Macintosh, all the illustrations in this book being from her implementation. Michael Hieb, Harry Keeling, and Andrew Black have also contributed code to the Disciple shell.

6
Case Study: Assessment Agent for Higher-Order Thinking Skills in History

In this chapter we will present the case studies of building, training and using agents that generate history tests to assist in the assessment of students' understanding and use of higher-order thinking skills. Examples of specific higher-order thinking skills are: evaluation of historical sources for relevance, credibility, consistency, ambiguity, bias, and fact *vs* opinion; *analyzing* them for content, meaning and point of view; and synthesizing arguments in the form of conclusions, claims and assertions (Bloom, 1956; Beyer, 1987, 1988).

6.1   Characterization of two types of assessment agents

We have developed two agents that are representative of the class of agents built by an expert (an educational and history expert, in this case) for certain user groups (history teachers and students). One of the assessment agents is integrated with the MMTS (Multimedia and Thinking Skills) educational system (Fontana *et al.*, 1993), creating a system with expanded capabilities, called IMMTS (Intelligent Multimedia and Thinking Skills). We will refer to this agent as the integrated assessment agent. The other agent is a stand-alone agent that is used independently of the MMTS software. Figure 6.1 presents the general architecture of the IMMTS system.

The MMTS educational system is a computer software package that allows students to learn the higher-order thinking skills of evaluation, analysis and synthesis. Students learn as they explore and examine a set of multimedia historical sources (photos, drawings, text, audio segments, and videos) related to slavery and the experiences of African Americans around the period of the Civil War. To motivate the learner and to provide an element of game playing, the MMTS system employs a journalist metaphor. It enables students to play roles in a three-dimensional environment called the 'virtual office'. In this immersive environment, a student assumes the role of a novice journalist and learns on the job through both correspondence courses and by completing realistic assignments given to them by their boss, the Editor. An assignment could be, for instance, to write an article on the experiences of African American women during the Civil War. These assignments have editor's notes, a set of historical sources that might be considered and other support material. As exemplified above, the topics and other aspects of these assignments focus on historical issues. As the students play the role of reporters, they learn the trade by attempting to apply the skills of evaluation, analysis and synthesis to appropriate source material in much the way journalists do when they complete their assignments and prepare stories for publication.

The integrated assessment agent provides ongoing assessment of students' progress to both students and teachers. This agent expands the functionality of the MMTS system by

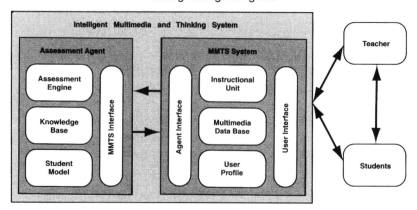

Figure 6.1   The general architecture of the IMMTS system.

generating a series of questions for students that test their higher-order thinking skills. As one can see in Figure 6.1, the assessment agent does not interact directly with the teachers and students. It interacts with them indirectly, through the MMTS system and its user interface. The integrated assessment agent receives a request for each test question from the MMTS system. In response, the agent dynamically generates test questions for a particular student and returns each question to the MMTS system. The MMTS system delivers the test to the student and records the student's answers.

This type of application of a Disciple-generated agent illustrates another possible role for these agents, that of enhancing the capabilities, generality, and usefulness of non-knowledge-based software.

As opposed to the integrated assessment agent, the stand-alone assessment agent interacts directly with students and teachers, and can be used independently of the MMTS system, as indicated in Figure 6.2. For instance, this agent can be used in a traditional classroom lecture on higher-order thinking skills to provide test questions.

Each of the two assessment agents has a knowledge base, an assessment engine and a student model. The stand-alone assessment agent also contains files with the historical sources

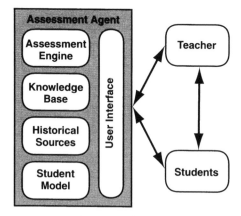

Figure 6.2   The general architecture of the stand-alone assessment agent.

while, in the case of the integrated agent, these sources are stored in the multimedia data base of the MMTS system. In addition, the stand-alone agent has its own user interface while the integrated agent only has an interface with the MMTS system. The stand-alone agent was developed first. This development effort proceeded in parallel with the development of the MMTS system. Then this agent was adapted and integrated with the MMTS system.

In the following sections we will present in detail the process of building, training and using the stand-alone assessment agent for higher-order thinking skills. Following, we will describe how the integrated agent was built starting from the stand-alone agent.

6.2   Developing a customized Disciple agent

The process of developing a customized Disciple agent begins with the analysis of the problem domain in order to define the requirements of the agent. During the course of this analysis, the significant objects and features of the application domain are identified. From this, the initial ontology of the domain can be developed. Likewise, the expected operation and behavior of the agent are defined and domain-dependent modules developed to support these capabilities.

6.2.1   Analyzing the problem domain and defining agent requirements

The general objective of the assessment agent is to generate tests that measure a learner's ability to use higher-order thinking skills after they have learned these skills by using the MMTS system, or another form of instruction. As a result of a testing interaction with a student, the agent produces a report containing the list of the test questions generated and the answers received, as well as various statistics such as the percentage of questions that were correctly answered. However, the actual grade is assigned by the teacher. Therefore, the assessment agent should be regarded as a tool to be used by teachers to provide assessment data about the students, or to be used by the students to provide assessment data about themselves.

The following are the main requirements that were established for the assessment agent.

6.2.1.1   *Application domain*
The agent should be compatible and consistent with the MMTS software. Subsequently, the application domain of the assessment agent should be the same as that of the MMTS system. The application domain of the MMTS system can be described as the domain of teaching the higher-order thinking skills of evaluation, analysis, and synthesis. The MMTS system has been used to teach these skills in the context of a history course about the period of slavery in America. Also, the MMTS system was concurrently developed to initially address only the evaluation skill of relevance. Consequently, the scope of the assessment agent was also initially limited to addressing this skill, and to be incrementally extended to address the other higher-order thinking skills.

6.2.1.2   *Use of the reporter paradigm*
It was decided that the tests generated by the agent should be consistent with the training approach taken in the MMTS system. As mentioned in Section 6.1, the students learn higher-

order thinking skills by playing the role of a novice journalist in the MMTS environment and they exercise these skills by completing writing assignments about the slavery period in American history. Therefore, the test questions generated by the assessment agent should also be formulated within the same reporter paradigm and should refer to the same type of topics.

6.2.1.3 Types of test questions

Each test question should ask the student to exercise higher-order thinking, and these questions should correspond to the tasks performed by the student while completing the MMTS writing assignments. To properly measure a student's ability, each test should only require the student to exercise one skill. The test questions should be of different levels of difficulty in order to provide a finer measurement. However, the level of difficulty of the tests should be appropriate for middle-school students, who are the target users of the agent.

These general requirements, together with the requirement to comply with the reporter paradigm, has led to the definition of the following four classes of test questions listed in ascending order of the level of difficulty:

- *If-relevant test question*: show the student a writing assignment and ask whether a particular historical source is relevant to that assignment.
- *Which-relevant test question*: show the student a writing assignment and three historical sources and ask the student to identify the relevant one.
- *Which-irrelevant test question*: show the student a writing assignment and three historical sources and ask the student to identify the irrelevant one.
- *Why-relevant test question*: show the student a writing assignment, a relevant historical source and three potential reasons why the source is relevant. Then ask the student to select the right reason.

Similar test questions could be generated for each evaluation skill, such as *If-credible test question* or *Why-credible test question*.

6.2.1.4 Modes of operation and feedback provided to the student

The assessment agent should have two modes of operation, final exam mode and self-assessment mode. In the final exam mode, the assessment agent generates an exam consisting of a set of test questions of different levels of difficulty. The student has to answer one test question at a time and, after each question, he/she receives the correct answer and an explanation of the answer. In the self-assessment mode, the assessment agent is used by a student to assess him/herself. In this mode, the student chooses the type of test question to solve, and will receive, on request, feedback in the form of hints to solve the test, the correct answer, and some or all the explanations of the answer. That is, in the self-assessment mode, the agent also tutors the student.

6.2.1.5 Dynamic and context-sensitive generation of tests

The test questions should be dynamically generated such that all students interacting with the assessment agent receive different tests even if they follow exactly the same interaction pattern. Moreover, the agent should build and maintain a student model and should use it in the process of test generation. For instance, to the extent possible, the assessment agent should try to generate test questions that involve historical sources that have not been investigated by the student while using the MMTS system, or historical sources that were not used in previous tests for that student.

6.2.2   Defining the top-level ontology of the knowledge base

An *ontology* is an explicit specification of a conceptualization of an application domain, where by a conceptualization we mean an abstract, simplified view of the world that we wish to represent for some purpose (Gruber, 1993a). An ontology contains the objects, concepts, and other entities that exist in an area of interest, as well as the relationships that hold them together (Genesereth and Nilsson, 1987).

The purpose of the assessment agent is to generate questions that require a student to determine whether a certain historical source is relevant (or credible, accurate, etc.) to a certain writing assignment on a topic related to slavery in America. It is therefore clear that the agent's ontology should include historical sources, historical concepts, and writing assignments. However, in order to further refine this ontology, we needed to first determine what kind of writing assignments should be included in the tests.

In the context of learning higher-order thinking skills with the MMTS system, the students receive various writing assignments from the Editor. The following are examples of such assignments:

1. 'You are a writer for *Harper's Weekly* in 1939 and you have been assigned to write and illustrate a feature article on the daily life of plantation slaves. Because *Harper's* is a national magazine and the issue of slavery is a sensitive one, try to avoid overt sectional bias and include such topics as work, food, clothing, housing, family, religion, and leisure activities.'
2. 'You are a writer in the year 2000 for a new magazine on the World Wide Web, and Black History Month and Women's History Month are approaching. You are assigned to develop a multimedia article about the experiences of African American women between 1820 and 1865.'
3. 'You are a reporter in the year 2000 for a new magazine on the World Wide Web. In commemoration of Martin Luther King's birthday, you are to develop a series of multimedia articles on the non-violent resistance to slavery and racism among African Americans in the ante-bellum south. Non-violent resistance to oppression was common and widespread, and took many forms, such as escape, "silent sabotage", and day-to-day resistance. It was also reflected in, and sustained by, African American religion, music, and folklore. Include all of these types of non-violent resistance in your series.'

Based on the MMTS assignments, we have defined templates for the writing assignments to be used in the test questions. The following are examples of such templates where each bold uppercase phrase is a concept representing the words to be used in its place:

- *Write-on-Topic*:
  You are a writer for a **PUBLICATION** and you have been assigned to write and illustrate a feature article on a **SLAVERY-TOPIC**.
- *Write-During-Period*:
  You are a writer for a **PUBLICATION** during a **HISTORICAL-PERIOD** and you have been assigned to write and illustrate a feature article on a **SLAVERY-TOPIC**.

- *Write-for-Audience*:
  You have been assigned to write an article for an **AUDIENCE** on a **SLAVERY-TOPIC** that supports the **ANGLE**.
- *Write-for-Occasion*:
  You are a reporter in the year 2000 for an **ELECTRONIC-MEDIA** and **OCCASION** is approaching. You are assigned to develop a multimedia presentation about **OCCASION-RELATED-SLAVERY-TOPIC**.

Each of these templates had to be formally represented as tasks in terms of a set of features and a description by using the dictionary editor. For example, the Write-During-Period template was defined in terms of three relations, **FOR**, **DURING**, and **ON**, and the following description:

'you are a writer **FOR** /for **DURING** /during period and you have been assigned to write and illustrate a feature article **ON**/on.'

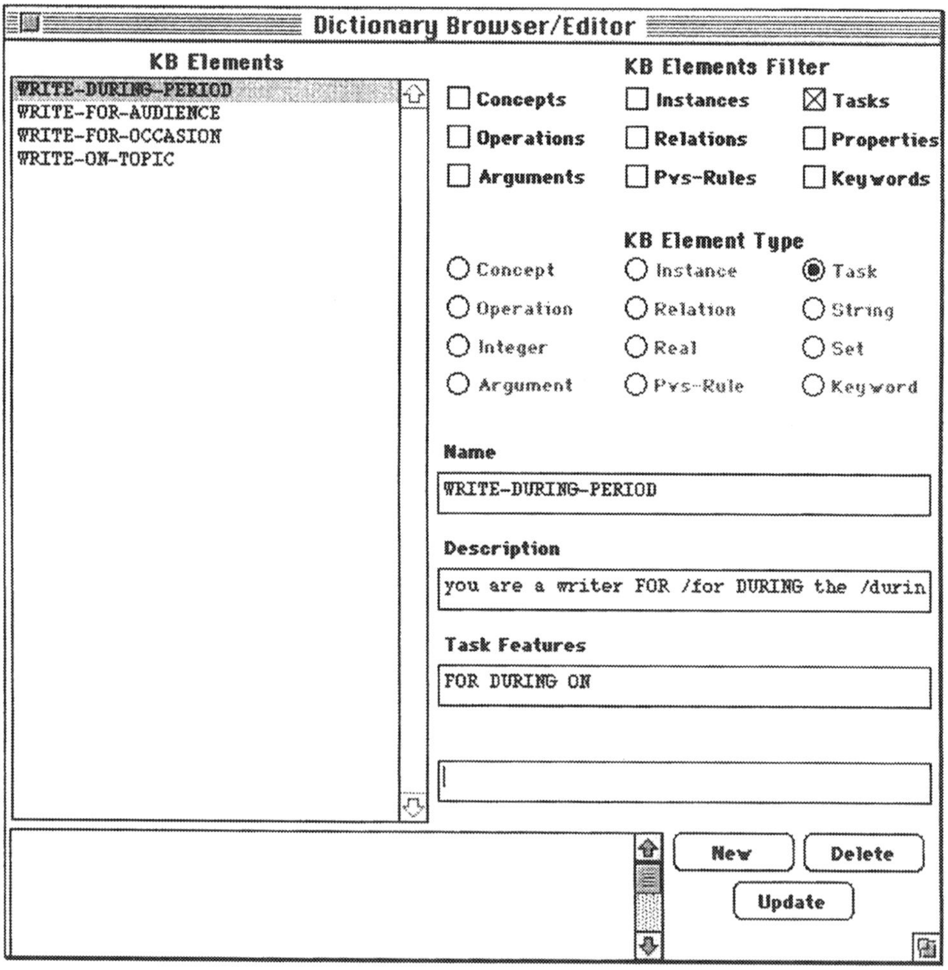

Figure 6.3   Definition of a task.

In this task template, the construct '/for' has been used to represent a value of the task feature **FOR**. The concepts and instances in the range of the relation **FOR** are the possible values that can be substituted for this construct in the template. For instance, the range of the task feature **FOR** is defined by the concept **PUBLICATION**. That is, any one of the publications in the agent's semantic network (e.g. *Harper's Weekly, Christian Recorder, Southern Illustrated News*) can be substituted for the construct '/for'. Also, in this task template, the constructs '/during' and '/on' represent values for the task features **DURING** and **ON**, respectively. This task template has been defined as illustrated in Figure 6.3.

Each of the features of the **WRITE–DURING–PERIOD** task also had to be defined as indicated in Table 6.1 and illustrated in Figure 6.4. This, in turn, led to the definition of the concepts representing the domains and ranges of these relations. As a result of the formal representation of each task, the agent's ontology was extended with additional concepts, as well as additional tasks and relations.

Having defined the task and its features, specific instantiations of the **WRITE–DURING–PERIOD** task could be represented as indicated in Section 6.4.1 and illustrated below:

?W1	IS	WRITE–DURING–PERIOD
	FOR	SOUTHERN–ILLUSTRATED–NEWS
	DURING	?P1
	ON	?S2
?P1	IS	POST–CIVIL–WAR
?S2	IS	SLAVE–LIFE

The 'Description' of this task instance is:

'You are a writer for *Southern Illustrated News* during the post Civil War period and you have been assigned to write and illustrate a feature article on slave life.'

This natural language description of the task instance is generated from the description of the **WRITE–DURING–PERIOD** task, the descriptions of the task features (which are 'for', 'during the' and 'on'), and the descriptions of the values of these features (which are '*Southern Illustrated News*', 'post Civil War', and 'slave life').

It was further decided that the agent should be able to judge the relevance of any historical source from its knowledge base with respect to any instantiation of the above tasks. Therefore, the agent's knowledge base had to contain relevancy rules of the form:

IF

the task is T

THEN

the source S is relevant.

Table 6.1 Definitions of **WRITE–DURING–PERIOD** task features

Relation	Description	Domain	Range
FOR	for	**WRITING–TASK**	**PUBLICATION**
DURING	during the	**WRITING–TASK**	**HISTORICAL–PERIOD**
ON	on	**WRITING–TASK**	**SLAVERY–TOPIC**

Since the relevance of a source is judged with respect to a given task, and there may be several types of relevancy, there will be families of relevancy rules, one family for each type of task. That is, there is a family of **WRITE–ON–TOPIC** relevancy rules, a family of **WRITE–DURING–PERIOD** relevancy rules, etc.

In addition to the relevancy rules, there will be credibility rules, accuracy rules, bias rules, etc. The right-hand side of each rule will contain an agent operation corresponding to that particular skill. Therefore, **RELEVANT**, **CREDIBLE**, **ACCURATE**, etc. were defined as agent

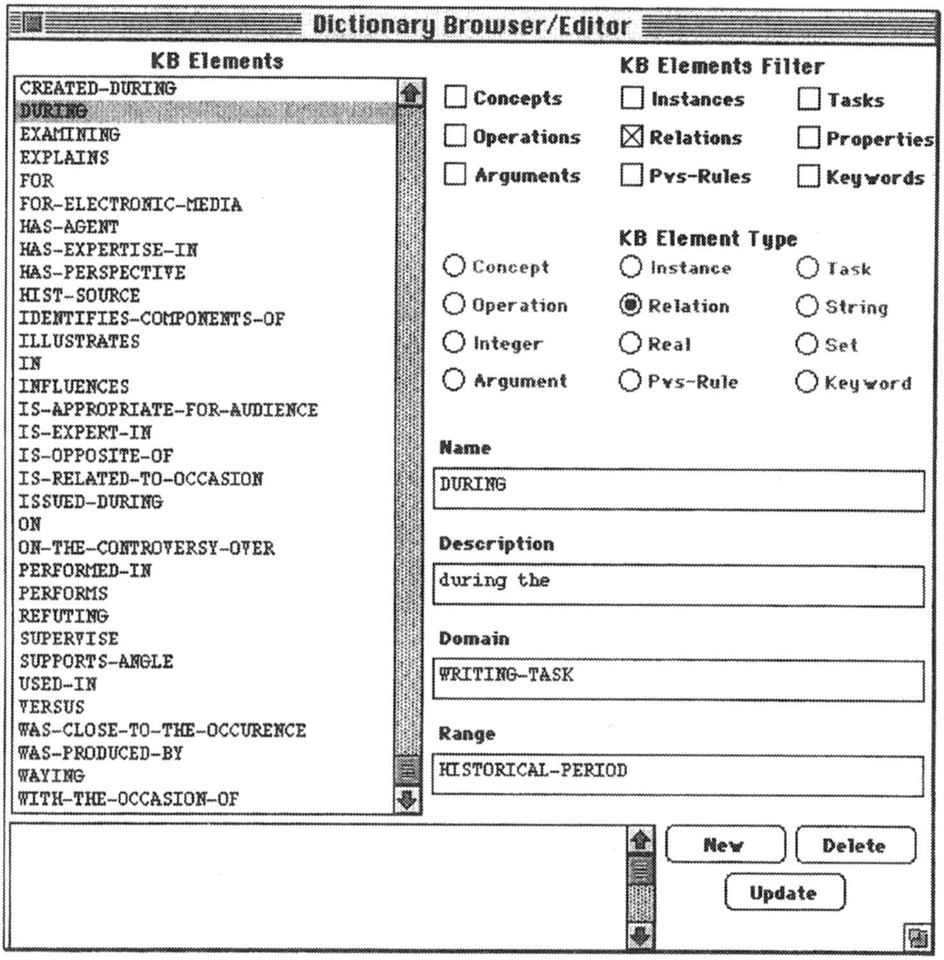

Figure 6.4 Definition of a relation.

operations used in these rules and in the generation of all the different types of test questions mentioned in the previous section.

Based on the above analysis and the resulting definitions, the top-level ontology in Figure 6.5 was created. The concepts contained in this ontology provided important guidance for defining the agent's initial semantic network of objects, instances and concepts, as will be described in Section 6.3.

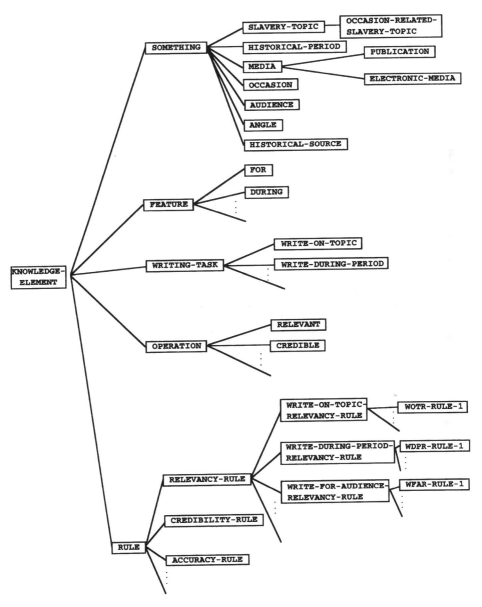

Figure 6.5   The top-level ontology of the assessment agent for higher-order thinking skills.

6.2.3   Developing domain dependent modules

Next, the Disciple learning agent shell had to be customized to facilitate the interaction between the learning agent and the domain expert. Functionality was added to make this interaction as natural as possible. This resulted in the creation of three domain specific interface modules, the *source viewer*, the *customized example editor* and the *example viewer*, to be presented in the following sections. The general-purpose problem-solving capabilities of the Disciple learning agent shell were used in the process of teaching the agent the evaluation skills and developing its knowledge base. These capabilities proved to be adequate and no specialized problem-solving engine needed to be developed at this point. However, the assessment engine was developed later, together with a graphical user interface. These will be described in Section 6.5.

6.2.3.1   The source viewer

The source viewer is a domain-dependent module that allows the agent to display a historical source. This module was needed both to facilitate the representation of the sources in the knowledge base by the domain expert (described in Section 6.3.2) and to display the sources during agent's teaching (described in Section 6.4). The source viewer automatically displays a historical source (see Figure 6.6) each time its name is used in any form of communication with the expert.

Figure 6.6   Source viewer window. (Picture reproduced from LC-USZ62-44265, Library of Congress, Prints and Photographs Division, Civil War Photographs.)

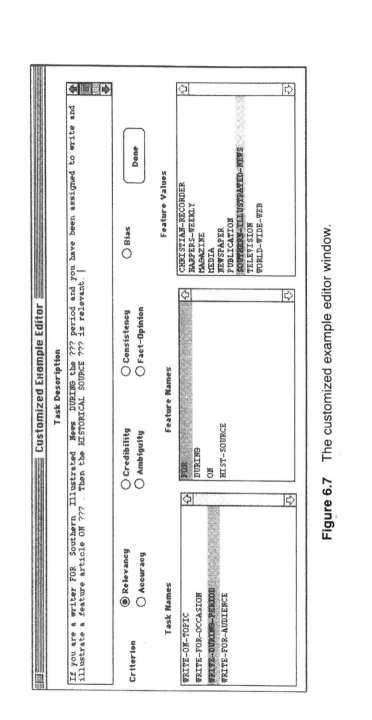

Figure 6.7 The customized example editor window.

Figure 6.8   Example viewer window.

6.2.3.2   The customized example editor

The customized example editor was developed to assist the history and educational expert in preparing the initial examples to teach the agent. It allows the expert to provide these examples by simply filling in a natural language pattern.

Figure 6.7 illustrates the use of the customized example editor. First, the expert selects the type of example to give to the agent (such as an example of determining the relevancy of a source, or an example of determining the credibility of a source), by selecting the appropriate button (i.e. **RELEVANCY** or **CREDIBILITY**). As a result, the names of the tasks associated with the selected evaluation skill will be displayed in the bottom left pane (see Figure 6.7). Next, the expert selects the type of task to be used in the example, by clicking on the task name in the bottom left pane of Figure 6.7. Once a task name is selected, this interface module displays the natural language pattern of the example to be defined by the expert. This pattern is instantiated by replacing each occurrence of '???' with a corresponding value indicated by the expert. The bottom middle pane displays the features of the example. Once the expert selects such a feature, the bottom right pane displays the possible values of this feature. The value selected by the expert will replace the corresponding '???' in the pattern.

6.2.3.3 The example viewer

The example viewer was developed to provide the expert with a natural English translation of the examples generated by the agent during rule refinement. This representation facilitates the expert's understanding of these examples. As other Disciple modules allow the expert to express his/her knowledge in the representation language of the agent, the example viewer (see Figure 6.8) allows the agent to express its knowledge in natural language.

6.3   Building the initial knowledge base

The organization of high-level concepts in Figure 6.5 provided a starting point for the creation of the agent's initial semantic network of object concepts and instances.

6.3.1   Defining the history curriculum

The abstract concept **SLAVERY-TOPIC**, from the ontology in Figure 6.5, is the top-level concept for all the topics that could be written about in the writing assignments. All these topics represent, at an abstract level, a part of the history curriculum to be taught to the students. Although it is not necessary to have all the topics that are part of the history curriculum on slavery, this set of topics should be complete enough to allow adequate testing of the students. The concepts corresponding to these topics were identified by the domain expert and were organized in an initial semantic network.

6.3.2   Selecting and representing historical sources

The next step in building the semantic network of the agent was to select and represent the historical sources to be used in test questions. These historical sources had to satisfy several constraints. Their content had to correspond to the history curriculum mentioned in the previous section. Their content also had to be appropriate for middle-school students for which the assessment agent was being developed. To ensure compatibility with MMTS, it was decided that these historical sources should be the same sources that were used in the MMTS software.

Before representing the historical sources, the domain expert had to define the source features that are necessary for applying the higher-order thinking skills of relevancy, credibility, etc. For instance, the expert concluded that the content of a source is relevant to some topic if it identifies components of the topic (or some historical concept related to this topic), or it illustrates the topic, or it explains the topic. As a consequence, the expert needed to represent the content of historical sources in terms of such features as **IDENTIFIES-COMPONENTS-OF**, **ILLUSTRATES**, or **EXPLAINS**. For instance, one could represent the source 'Am I Not a Man and a Brother' (see Figure 6.6) as **ILLUSTRATES** a **MALE-SLAVE**. Extending this analysis to other aspects of relevancy related to the writing tasks, and performing a similar analysis with respect to the other evaluation skills, led to the definitions of various other features for representing the historical sources, such as **IS-APPROPRI-ATE-FOR-AUDIENCE**, **CREATED-DURING**, **SUPPORTS-ANGLE**, etc.

Having identified and defined the features of the historical sources, the representation of a source became an easy process of describing it by using the concept editor and the source viewer, as illustrated in Figure 6.9. As shown, the source **STACKING-WHEAT** is described in terms of the features **INSTANCE-OF, ILLUSTRATES, IS-APPROPRIATE-FOR-AUDIENCE**, and **CREATED-DURING**. The content of this source is partially represented by the concepts: **SLAVE-FIELD-WORK**, **FIELD-SLAVE**, and **SLAVE-DAY**.

The representation of the sources must be guided by using the following 'projection' principle:

- *Any historical source must be completely described in terms of the concepts from the knowledge base.*

This means that if the knowledge base contains a certain historical concept, then any historical source describing that concept should contain the concept in the description of its content. Operationally, this simply means that if the expert decides to describe a source in terms of some topic, then any other source from the knowledge base that refers to that topic

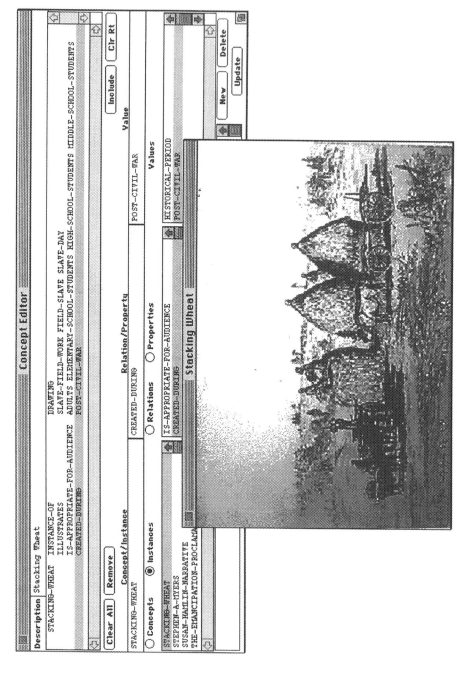

Figure 6.9 The concept editor and source viewer windows. (Picture reproduced from LC-USZ62-1073, Library of Congress, Prints and Photographs Division, Civil War Photographs.)

should also be described in terms of the same topic. This does not mean, however, that the contents of the historical sources are completely described (a task that would anyway be extremely hard because, for instance, 'a picture is worth a thousand words').

To describe a historical source, the expert had to sequentially select each appropriate feature of the source, from the bottom middle pane of the concept editor (see Figure 6.9). At that moment, the bottom right pane displayed all the historical concepts from the knowledge base in the range of the selected feature. Next, the expert chose the concepts that applied to the source. If the expert decided to introduce a new historical concept, then this new concept was first introduced in the semantic network by using the same concept editor. However, in this case, the expert needed to review the previously defined sources in order to determine whether they should also be described in terms of the newly defined concept. When this became necessary, the source viewer facilitated the inspection of each such historical source, and the concept editor was used to update the description of the source.

6.3.3   Populating the semantic network with other necessary concepts and instances

As indicated in the general ontology presented in Figure 6.5, there were other entities that needed to be represented in the semantic network of the agent such as the subconcepts and instances of media, historical period, audience, occasion and angle. The expert defined all of these entities by interacting with the agent through its editors and browsers. For instance, Figure 6.10 presents a part of the semantic network corresponding to the media concept.

The resulting initial semantic network of the knowledge base contained 80 historical sources, as well as other concepts and relationships. However, at this point there were no reasoning rules in the agent's knowledge base. The next step in building the assessment agent was to teach the agent how to apply the skills of judging relevancy or credibility, and to apply other thinking skills, as presented in the following sections.

6.4   Teaching the agent how to judge the relevancy of a source with respect to a task

After the initial semantic network was defined, the expert had to teach the assessment agent how to judge the relevancy (as well as the credibility, accuracy, etc.) of a source with respect to a certain task. This process followed the rule-learning and rule refinement processes described in Chapter 4. Therefore, in the following sections we will only illustrate and intuitively describe these processes and refer the reader to Chapter 4 for more details.

6.4.1   Giving the agent an example of a task and a relevant source

To teach the agent how to judge the relevance of a source to a given task, first the educational expert had to provide an example of a task and a source relevant to that task, by using the customized example editor as described in Section 6.2.3.2 and illustrated in Figure 6.7.

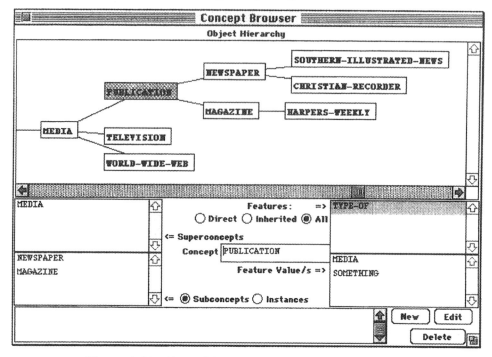

Figure 6.10   Part of the semantic network of the agent.

The bottom window of Figure 6.11 shows the natural language description of the example, as defined by the expert. The left-hand window of Figure 6.11 shows the simplified internal representation generated by the agent.

6.4.2   Helping the agent understand why the source is relevant

Next, the expert had to help the agent understand why this problem-solving episode in Figure 6.11 was correct. The expert guided the agent to propose explanations and the expert selected the relevant ones shown in the bottom pane of the window in Figure 6.12.

From these interactions, the agent concluded that the source **STACKING-WHEAT** is relevant to the task of writing an article for **SOUTHERN-ILLUSTRATED-NEWS**, during the **POST-CIVIL-WAR** period, on **SLAVE-LIFE**, because **STACKING-WHEAT** illustrates **SLAVE-FIELD-WORK**, which was a component of **SLAVE-LIFE** and was created during the **POST-CIVIL-WAR** period.

The expert was satisfied with the found explanation, and clicked on the *Create* button, asking the agent to create a rule from the example and the explanation. The generated rule is shown in a simplified form in the top-right pane of the window in Figure 6.12.

The internal representation of the rule is shown in Figure 6.13. It was obtained by applying the method described in Chapter 4. The comments on the right-hand side are introduced just to explain the rule to the reader and are not part of the internal representation.

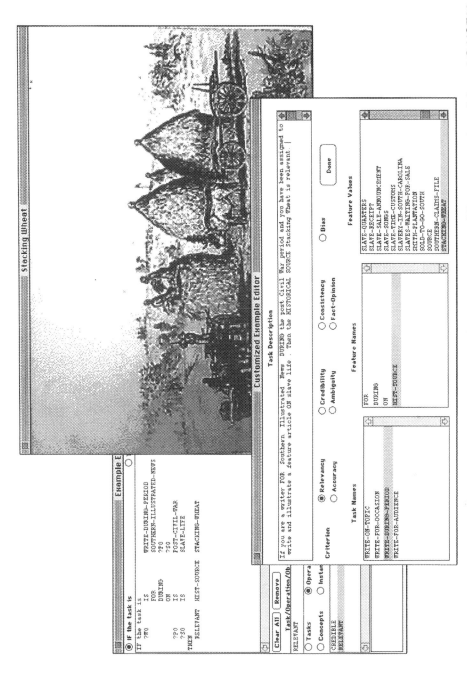

Figure 6.11 Creating the initial example using the customized example editor. (Picture reproduced from LC-USZ62-1073, Library of Congress, Prints and Photographs Division, Civil War Photographs.)

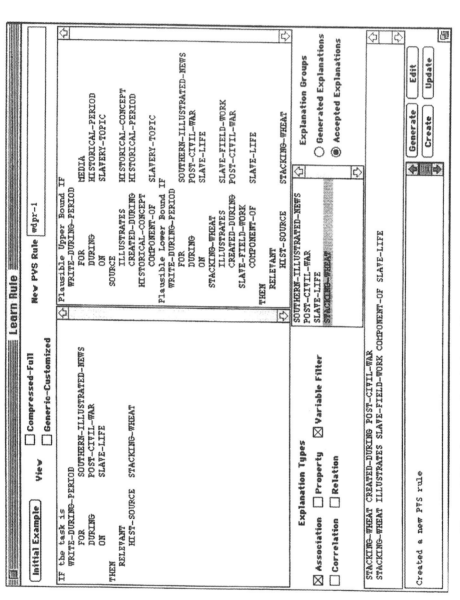

Figure 6.12 Learn rule window with accepted explanations and PVS rule.

```
Plausible Upper Bound    IF                                      ; IF (Plausible Upper Bound)
          ?W1            IS    WRITE-DURING-PERIOD                ; the task is to write an article
                         FOR   ?S1                               ; for a publication ?S1
                         DURING ?P1                              ; during the historical period ?P1
                         ON    ?S2                               ; on the slavery topic ?S2

          ?P1            IS    HISTORICAL-PERIOD

          ?S1            IS    PUBLICATION

          ?S2            IS    SLAVERY-TOPIC

          ?S3            IS    SOURCE                            ; and ?S3 is source that
                         ILLUSTRATES ?S4                         ; illustrates ?S4
                         CREATED-DURING ?P1                      ; and was created during ?P1

          ?S4            IS    HISTORICAL-CONCEPT                 ; and ?S4 is a historical concept
                         COMPONENT-OF ?S2                        ; that was a component of ?S2


Plausible Lower Bound    IF                                      ; IF (Plausible Lower Bound)
          ?W1            IS    WRITE-DURING-PERIOD                ; the task is to write an article
                         FOR   ?S1                               ; for Southern Illustrated News
                         DURING ?P1                              ; during the post Civil War period
                         ON    ?S2                               ; on slave life

          ?P1            IS    POST-CIVIL-WAR

          ?S1            IS    SOUTHERN-ILLUSTRATED-NEWS

          ?S2            IS    SLAVE-LIFE

          ?S3            IS    STACKING-WHEAT                    ; and ?S3 is 'Stacking Wheat' that
                         ILLUSTRATES ?S4                         ; illustrates slave field work and was
                         CREATED-DURING ?P1                      ; created during the post Civil War period

          ?S4            IS    SLAVE-FIELD-WORK                  ; and slave field work
                         COMPONENT-OF ?S2                        ; was a component of slave life

THEN                                                             ; THEN

          RELEVANT HIST-SOURCE ?S3                               ; the historical source ?S3 is relevant
```

Figure 6.13   The learned PVS rule.

6.4.3 Supervising the agent as it evaluates the relevance of other sources to similar tasks

Once the agent had learned a relevancy rule, the expert asked it to generate new tasks and sources that it considered to be relevant to these tasks. To do this, a rule refinement process was started with the agent. This allowed the expert to verify the agent's reasoning and to help it improve its understanding of relevance.

Figure 6.14 shows a task and source that were generated by the agent. They represent a new example of the relevancy rule from Figure 6.13. This example is covered by the plausible upper bound and is not covered by the plausible lower bound of the rule.

6.4.3.1 Confirming the agent's evaluation

Figure 6.15 shows the same example in the middle left pane of the refine rule window. The expert accepted the example generated by the agent as correct by clicking on the *Accept* button in the bottom right part of the window. Consequently, the plausible lower bound condition of the rule was generalized to cover it, as indicated in the bottom left pane of the window. The generalized plausible lower bound condition is shown in the top right pane of the window.

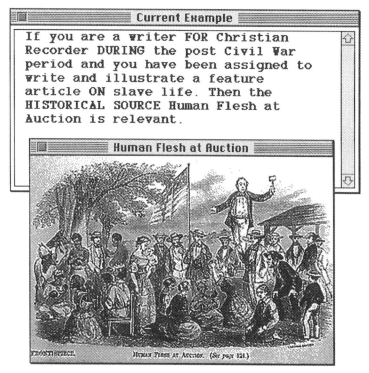

Figure 6.14 An example generated by the agent. (Picture reproduced from LC-USZ62-30797, Library of Congress, Prints and Photographs Division, Civil War Photographs.)

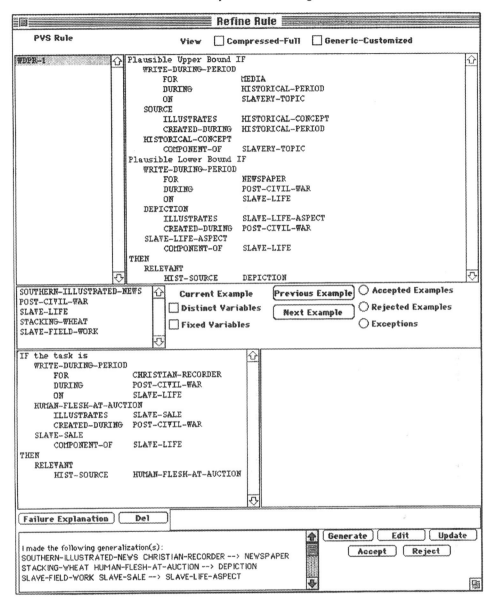

Figure 6.15 Refine rule window with the first example generated.

The next two examples generated by the agent are shown in Figure 6.16 and Figure 6.17. These examples were accepted by the expert and the lower bound condition of the rule was further generalized.

6.4.3.2 Rejecting agent's evaluation and helping it to understand its mistake

Figure 6.18 shows a new example generated by the agent. However, this example is obviously incorrect and was rejected by the expert. The window in Figure 6.19 shows both

Figure 6.16 The second example generated by the agent and accepted by the expert. (Picture reproduced from LC-USZ62-30796, Library of Congress, Prints and Photographs Division, Civil War Photographs.)

this negative example (in its bottom left pane) and the initial example provided by the expert (in its bottom right pane). The agent needed to understand why the generated example (which was generated by analogy with the initial example provided by the expert) is wrong, while the expert's example was correct. By comparing the two examples, the expert was able to explain that the generated example (in the lower left pane) is wrong because the **WORLD-WIDE-WEB** was not issued during the **CIVIL-WAR** period. On the contrary, the initial example (in the lower right pane) was correct because **SOUTHERN-ILLUSTRATED-NEWS** was issued during the **POST-CIVIL-WAR** period. To give this explanation, the expert invoked the explanation grapher (see Figure 6.20).

The expert chose the type of explanation to define which, in this case, is *Association*. The bottom left pane displays all the objects from the initial example. The expert chose **SOUTHERN-ILLUSTRATED-NEWS**. At this point, the agent displayed all the features of **SOUTHERN-ILLUSTRATED-NEWS** under *Object Features*. After the expert selected a feature, the agent displayed all the possible values of this feature in the right pane from which the expert had to choose the correct one. As the expert selected values in these panes, the explanation was extended. The final explanation, **SOUTHERN-ILLUSTRATED-NEWS ISSUED-DURING POST-CIVIL-WAR**, is displayed in the middle pane of Figure 6.20. This explanation given by the expert was used to specialize both bounds of the PVS rule by using the method presented in Section 4.3.3.4.

Figure 6.17 The third example generated by the agent and accepted by the expert. (Picture reproduced from LC-USZ62-30802, Library of Congress, Prints and Photographs Division, Civil War Photographs.)

The plausible upper bound shown in the top pane of the window in Figure 6.19 was specialized by adding to it the expression **MEDIA ISSUED-DURING HISTORICAL-PERIOD** and the plausible lower bound was specialized by adding to it the expression **PUBLICATION ISSUED-DURING HISTORICAL-PERIOD**. Notice that, in order to generate the new plausible lower bound, the agent had to check which of the previous positive examples of the rule are still covered by it, and to possibly elicit new knowledge about these examples, by using the method presented in Table 4.11. For instance, the agent had to check that **CHRISTIAN-RECORDER** was issued during the **POST-CIVIL-WAR** period. If some of the previous positive examples do not contain such information then the agent would ask the expert whether it is true or not and would update the semantic network accordingly.

The agent will continue the example generation process for as long as it can generate examples covered by the plausible upper bound condition of the rule without being covered by the plausible lower bound condition of the rule. In the case of the current rule, there are no such examples. Therefore, the final rule was the one in Figure 6.21 (represented in the condensed form).

Notice that several additional pieces of information are associated with the rule. The task description is the natural language description of the task. The operation description is the natural language description of the problem-solving operation from the right-hand side of the rule. The explanation pieces are the natural language patterns corresponding to the pieces of

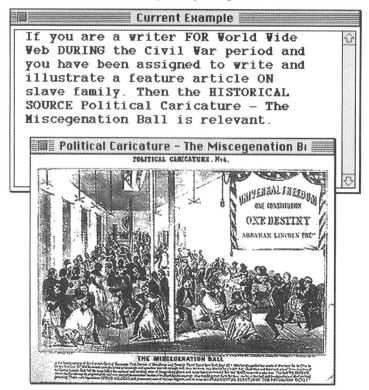

Figure 6.18 A new example generated by the agent. (Picture reproduced from LC-USZ62-14828, Library of Congress, Prints and Photographs Division, Civil War Photographs.)

explanations from which the rule was learned. All of these patterns are automatically generated from the description property of each component. In addition, there are the positive examples and the negative examples from which the rule was learned.

The training of the agent continued until a total of 54 rules were learned for the higher-order thinking skills of judging relevancy.

6.5   Developing the assessment engine and the graphical user interface

After the agent had been trained, an assessment engine that generates all the four types of test questions mentioned in Section 6.2.1 was developed. We will refer to these tests as follows: If relevant, Which relevant, Which irrelevant and Why relevant.

One of the agent's requirements was that it generate not only test questions, but also feedback for right and wrong answers, and hints to help the student in solving the tests, as well as explanations of the solutions. Moreover, agent's messages needed to be expressed in a natural language form. Although the rules learned by the agent contain almost all the

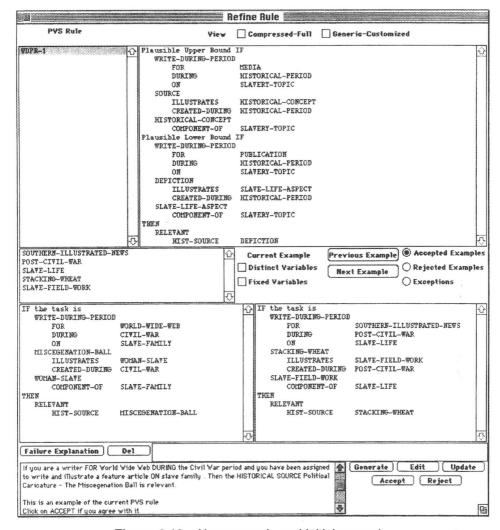

Figure 6.19   New example and initial example.

necessary information to achieve these goals, some small adjustments were necessary, as will be described in Section 6.5.1.

These rules were used by specialized methods to generate test questions from each of the above categories. The methods corresponding to each type of test question will be presented in the following sections.

Another requirement for the agent was that it generate test questions dynamically and randomly and that the tests should avoid, as much as possible, using the sources that have been investigated by the student or used in previous tests for the student. For this reason, each test generation method always uses a random function to choose an element from a set. For instance, the agent randomly chooses a rule from a set of applicable rules, and then randomly chooses an example of this rule from the set of the rule's examples. Also, when it becomes

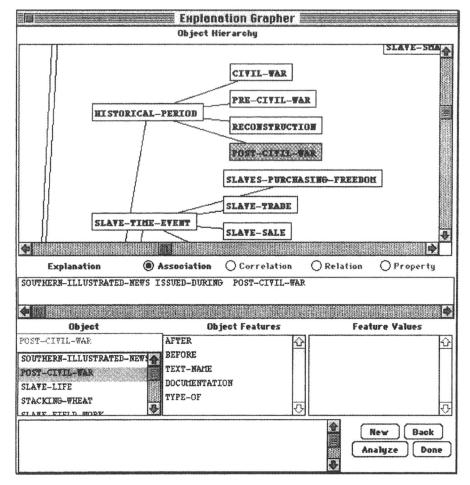

Figure 6.20 Explanation grapher window.

necessary to choose a historical source, the agent randomly chooses one from the set of sources to which the student has not been exposed. However, theoretically, this may not always be possible, in which case the agent may need to reuse a source that has been seen by the student before. In practice, this situation is very unlikely to appear because of the many available choices.

Therefore, in the presentations of the test generation methods, whenever a selection is made from a set, it should be understood to be a random one.

Let us also mention that the test generation methods rely heavily on a basic problem-solving operation of the agent: finding an example of a rule. The method corresponding to this operation was presented in Section 3.3.3.

Each of the test generation methods produces the elements of a test question that are passed to the agent interface to be displayed in a specific window. Also, this interface had to be developed at this point and its usage is illustrated in the following sections.

```
Plausible  Upper  Bound  IF
  ?W1    WRITE-DURING-PERIOD,   FOR   ?S1,   DURING   ?P1,   ON   ?S2
  ?P1    HISTORICAL-PERIOD
  ?S1    MEDIA,   ISSUED-DURING   ?P1
  ?S2    SLAVERY-TOPIC
  ?S3    SOURCE,   CREATED-DURING   ?P1,   ILLUSTRATES   ?S4
  ?S4    HISTORICAL-CONCEPT,   COMPONENT-OF   ?S2

Plausible  Lower  Bound  IF
  ?W1    WRITE-DURING-PERIOD,   FOR   ?S1,   DURING   ?P1,   ON   ?S2
  ?P1    HISTORICAL-PERIOD
  ?S1    PUBLICATION,   ISSUED-DURING   ?P1
  ?S2    SLAVERY-TOPIC
  ?S3    SOURCE,   CREATED-DURING   ?P1,   ILLUSTRATES   ?S4
  ?S4    SLAVE-LIFE-ASPECT,   COMPONENT-OF   ?S2
THEN
  RELEVANT   HIST-SOURCE   ?S3

Task  Description
  you  are  a  writer  for  ?S1  during  ?P1  and  you  have  been
  assigned  to  write  and  illustrate  a  feature  article  on   ?S2

Operation  Description
  ?S3  is  relevant

Explanation  Pieces
  ?S3  illustrates  ?S4  which  was  a  component  of   ?S2
  ?S3  was  created  during  ?P1
  ?S1  was  issued  during  ?P1

Positive  Examples
  ( ?S1  IS  SOUTHERN-ILLUSTRATED-NEWS,   ?P1  IS  POST-CIVIL-WAR,
    ?S2  IS  SLAVE-LIFE,   ?S3  IS  STACKING-WHEAT,
    ?S4  IS  SLAVE-FIELD-WORK)

  ( ?S1  IS  CHRISTIAN-RECORDER,   ?P1  IS  POST-CIVIL-WAR,
    ?S2  IS  SLAVE-LIFE,   ?S3  IS  HUMAN-FLESH-AT-AUCTION,
    ?S4  IS  SLAVE-SALE)

  ( ?S1  IS  HARPERS-WEEKLY,   ?P1  IS  CIVIL-WAR,
    ?S2  IS  SLAVE-CULTURE,   ?S3  IS  HALTING-AT-NOON,
    ?S4  IS  SLAVE-CLOTHES)

  ( ?S1  IS  CHRISTIAN-RECORDER,   ?P1  IS  POST-CIVIL-WAR,
    ?S2  IS  SLAVE-LIFE,   ?S3  IS  BLOODHOUND-BUSINESS,
    ?S4  IS  SLAVES-RUNNING-AWAY)

Negative  Example
  ( ?S1  IS  WORLD-WIDE-WEB,   ?P1  IS  POST-CIVIL-WAR,
    ?S2  IS  SLAVE-FAMILY,   ?S3  IS  MISCEGENATION-BALL,
    ?S4  IS  WOMAN-SLAVE)
```

Figure 6.21   The refined PVS rule.

6.5.1   Augmenting and adjusting the patterns associated with the learned rules

The patterns associated with the learned rules are used for the natural language generation of the test questions, answers, hints, and explanations. However, in order to generate all these natural language expressions, one may need to adjust some of the patterns, and to add others, depending on the desired form of the test questions.

In the case of the rule in Figure 6.21, the expert needed to define the patterns for the hint, right answer and wrong answer shown in Figure 6.22. The hint is the part of the explanation

```
IF
   ?W1     WRITE-DURING-PERIOD,   FOR   ?S1,   DURING   ?P1,   ON   ?S2
   ?P1     HISTORICAL-PERIOD
   ?S1     PUBLICATION,   ISSUED-DURING   ?P1
   ?S2     HISTORICAL-CONCEPT
   ?S3     SOURCE,   ILLUSTRATES   ?S4,   CREATED-DURING   ?P1
   ?S4     SLAVE-LIFE-ASPECT,   COMPONENT-OF   ?S2

THEN
   RELEVANT   HIST-SOURCE   ?S3

Task   Description
   You  are  a  writer  for  ?S1  during  the  ?P1  period  and  you
   have  been  assigned  to  write  and  illustrate  a  feature
   article  on  ?S2.

Operation   Description
   ?S3  is  relevant

Explanation
   ?S3  illustrates  ?S4  which  was  a  component  of  ?S2,  ?S3
   was  created  during  the  ?P1  period  and  ?S1  was  issued
   during  the  ?P1  period.

Hint
   To  determine  if  this  source  is  relevant  to  your  task
   investigate  if  it  illustrates  some  component  of  ?S2,
   check  when  was  it  created,  and  when  ?S1  was  issued.

Right   Answer
   The  source  ?S3  is  relevant  to  your  task  because  it
   illustrates  ?S4  which  was  a  component  of  ?S2,  ?S3  was
   created  during  the  ?P1  period  and  ?S1  was  issued  during
   the  ?P1  period.

Wrong   Answer
   Investigate  this  source  further  and  analyze  the  hints
   and  explanations  to  improve  your  understanding  of
   relevance.  You  may  consider  reviewing  the  material  on
   relevance.  Then  continue  testing  yourself.
```

Figure 6.22   The final adjusted relevancy rule.

that refers only to the variables used in the formulation of the test question. Some minor syntactic adjustments were required for correct punctuation and good sentence structure.

Finally, because the assessment agent was to be used to generate test questions without further learning itself, and because we wanted these test questions to be as correct as possible, the plausible upper bound conditions and the examples of the rules were no longer needed. Therefore, they were dropped from the rules that were transformed into single-condition IF–THEN rules.

In the following sections we will illustrate the interaction between the agent and the student during different types of test questions and then we will present the method used to generate each type of test question.

6.5.2 Sample agent–student interaction during an If-relevant test question

Figure 6.23 shows an If-relevant test question. The student is given a task and a historical source. He/she has to investigate the source and decide whether it is relevant to the task. In this case, the student did not know how to answer the question and requested a hint that is displayed in the lower right pane of the window.

Figure 6.24 illustrates the case where the student has responded correctly and the agent confirmed his/her answer and also indicated the reason why the source is relevant.

6.5.3 Generation of If-relevant test questions with relevant sources

To generate an If-relevant test question with a relevant source, the agent simply needs to generate an example of a relevancy rule. This rule example will contain a task T and a source S relevant to it, together with one hint and one explanation that will indicate one reason why S is relevant to T. However, if the student requires all the possible reasons for why the source S is relevant to the task T, then the agent will need to find all the examples containing the source S of all the relevancy rules corresponding to the task T. The method of generating an If-relevant test question and all the hints and explanations is presented in Table 6.2 and illustrated in the discussion that follows.

Let us briefly illustrate this method. First the agent randomly picks a relevancy rule R from the knowledge base. Let us assume that the picked rule is the one from Figure 6.22. Then the agent generates a random example of R. While generating this example, it attempts to randomly pick a source that satisfies the rule's condition and has not been used before. Let us suppose that the generated example is:

E = (?S1 IS SOUTHERN–ILLUSTRATED–NEWS, ?P1 IS POST–CIVIL–WAR,

?S2 IS SLAVE–CULTURE, ?S3 IS FIVE–GENERATIONS,

?S4 IS SLAVE–CLOTHING)

The partial example containing only the variables from the task is:

ET = (?S1 IS SOUTHERN–ILLUSTRATED–NEWS, ?P1 IS POST–CIVIL–WAR,

?S2 IS SLAVE–CULTURE, ?S3 IS FIVE–GENERATIONS)

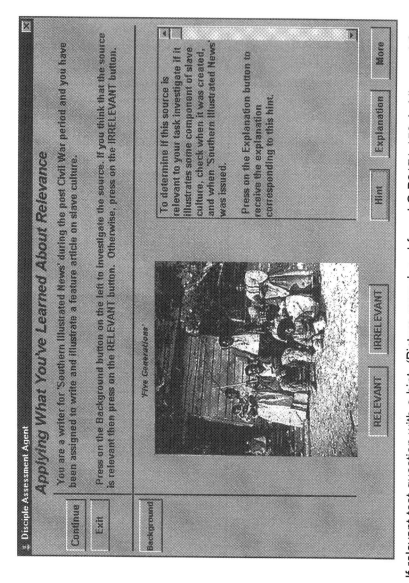

Figure 6.23 If-relevant test question with a hint. (Picture reproduced from LC-B8171-152-A, Library of Congress, Prints and Photographs Division, Civil War Photographs.)

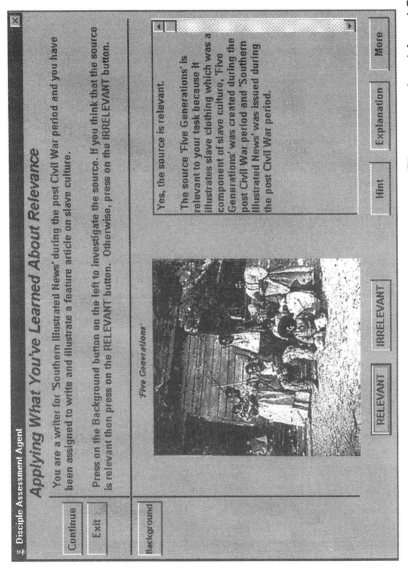

Figure 6.24 An if-relevant test question with the right answer and response. (Picture reproduced from LC-B8171-152-A, Library of Congress, Prints and Photographs Division, Civil War Photographs.)

Table 6.2   The method for generating If-relevant test questions when the source is relevant

---

- Choose a relevancy rule R from the KB.
- Generate an example E of R. Let S be the historical source from E, and let the partial example ET be the subset of E corresponding to the variables from the task description of R.
- Build the if question IFQ from the instances of the patterns of the task and operation of R, corresponding to ET and S.
- Build the feedback FR for the right answer.
- Build the feedback FW for the wrong answer.
- **Repeat** for each rule R_i of the family F of R:
  - Find all the examples E_{ij} of R_i starting from the partial example ET.
  - **Repeat** for each example E_{ij}:
    - Build the instance H_{ij} of the hint pattern corresponding to R_i and E_{ij}.
    - Build the instance EX_{ij} of the explanation pattern corresponding to R_i and E_{ij}.
    end
  end
- Return $(IFQ\ FR\ FW\ (H_{11}\ EX_{11})\ \dots\ (H_{nm}\ EX_{nm}))$, the list of the natural language descriptions of the test, feedback for the answers, hints and corresponding explanations.

The following is the same example where the values of the variables have been replaced by their descriptions:

$$E = (\textbf{?S1 IS} \text{ 'Southern Illustrated News', } \textbf{?P1 IS} \text{ post Civil War,}$$

$$\textbf{?S2 IS} \text{ slave culture, } \textbf{?s3 IS} \text{ 'Five Generations',}$$

$$\textbf{?S4 IS} \text{ slave clothing)}$$

The If question IFQ is:

> You are a writer for 'Southern Illustrated News' during the post Civil War period and you have been assigned to write and illustrate a feature article on slave culture.
>
> Press on the Background button on the left to investigate the source. If you think that the source is relevant then press on the RELEVANT button. Otherwise, press on the IRRELEVANT button.

Notice that the agent has generated this task by instantiating the task description pattern with the descriptions of the values of the variables from the pattern.

The feedback FR for the right answer is:

> Yes, the source is relevant.
>
> The source 'Five Generations' is relevant to your task because it illustrates slave clothing which was a component of slave culture, 'Five Generations' was created during the post Civil War period and 'Southern Illustrated News' was issued during the post Civil War period.

The feedback FW for the wrong answer is:

> No, the source is relevant.
>
> Investigate this source further and analyze the hints and explanations to improve your understanding of relevance. You may consider reviewing the material on relevance. Then continue testing yourself.

As illustrated in the previous section, a historical source S may be relevant to a certain task for several reasons. To generate all the explanations (and corresponding hints) of why the historical source 'Five Generations' is relevant to the task of writing a feature article for *Southern Illustrated News* during the post Civil War period on slave culture, the agent needs to find all the examples of all the relevancy rules from the Write-During-Period-Relevancy-Rule family, examples that cover ET. From each such rule and example a different hint and explanation is built.

We will only show here three of these examples. One is the above example. The corresponding hint H_{11} is generated by instantiating the Hint pattern of the rule:

> To determine if this source is relevant to your task investigate if it illustrates some component of slave life, check when it was created, and when 'Southern Illustrated News' was issued.

Similarly, the explanation EX_{11} is generated by instantiating the Explanation pattern:

> 'Five Generations' illustrates slave clothing which was a component of slave culture, 'Five Generations' was created during the post Civil War period, and 'Southern Illustrated News' was issued during the post Civil War period.

The second example E_{12} can be generated by the agent from the same rule that was used to generate the first example:

E_{12} = (?S1 IS SOUTHERN-ILLUSTRATED-NEWS, ?P1 IS POST-CIVIL-WAR,

     ?S2 IS SLAVE-CULTURE, ?S3 IS FIVE-GENERATIONS,

     ?S4 IS SLAVE-HOUSING)

The hint H_{12} for this example is the same as the hint in the first example. The explanation EX_{12} is:

> 'Five Generations' illustrates slave housing which was a component of slave culture, 'Five Generations' was created during the post Civil War period, and 'Southern Illustrated News' was issued during the post Civil War period.

To generate a third explanation and corresponding hint the agent selects another rule from the Write-During-Period-Relevancy-Rule family. One such rule is presented in Figure 2.17. Then the agent generates the following example of this rule, starting from ET:

E_{21} = (?S1 IS SOUTHERN-ILLUSTRATED-NEWS, ?P1 IS POST-CIVIL-WAR,

     ?S2 IS SLAVE-CULTURE, ?S3 IS FIVE-GENERATIONS,

     ?S4 IS SLAVE-FAMILY)

The following hint H_{21} is generated by instantiating the Hint pattern of the rule:

> To determine if this source is relevant to your task investigate if it identifies components of something that was a component of slave culture, check when it was created, and when 'Southern Illustrated News' was issued.

Similarly, the following explanation EX_{21} is generated by instantiating the Explanation pattern:

'Five Generations' identifies components of slave family which was a component of slave culture, 'Five Generations' was created during the post Civil War period, and 'Southern Illustrated News' was issued during the post Civil War period.

Let us mention that the method in Table 6.2 is presented at a conceptual level. However, it is implemented in such a way that it assures as fast a response time as possible. For instance, once an example of the initially selected rule is found, the test is generated together with a hint and an explanation. As the student is reading the test, hint or explanation, the agent is generating the next (hint–explanation) pair, in case the student asks for them, and so on.

In any case, the generation of an If-relevant test question where the source is relevant is a very efficient process, even if all the hints and explanations need to be generated. Although all the examples of several rules may need to be generated, the rule variables that appear in the test question already have values, and the only values that need to be determined are those corresponding to the other variables of the rules. For instance, in the case of the rule in Figure 6.22, the agent needed to find all the examples that contain:

$$ET = (?S1 \text{ IS SOUTHERN-ILLUSTRATED-NEWS}, ?P1 \text{ IS POST-CIVIL-WAR},$$

$$?S2 \text{ IS SLAVE-LIFE}, ?S3 \text{ IS FIVE-GENERATIONS})$$

That is, it only needed to find values of the variable $?S4$.

6.5.4   Generation of If-relevant test questions with irrelevant sources

As mentioned in the previous section, the generation of an If-relevant test question where the source is relevant is a very efficient process. The case where the source is not relevant is more computationally expensive. In such a case, the agent has to generate a valid task T by finding an example of a relevancy rule R, and then it has to find a historical source S such that the task and the source are not part of an example of any rule from the family of rules to which R belongs (including R). To do this the agent attempts to find examples of each of the rules from a certain family such as the Write-for-Audience-Relevancy-Rule family. Notice, however, that each time the agent starts with a partial example, only values for a few additional variables are needed. If such values are found for at least one rule of the family, then the current source is not good and the agent would need to try another one. This method is presented in Table 6.3.

Let us note that one can assure that a source is not relevant to a task only to the extent that the relevancy rules cover all the possible reasons why the source might be relevant to that task. This requirement was achieved in the case of the developed assessment agent.

6.5.5   Sample agent–student interaction during a Which-relevant test question

Figure 6.25 shows a Which-relevant test question. The student is given a task and three historical sources. He/she has to investigate the sources and decide which source(s) is (are) relevant to the task. The student is instructed to check the box next to the relevant source(s). No check means that the source is not relevant. In the presented example, the student has correctly indicated that only the source 'Rembert Slaves' is relevant and the other sources are not relevant. The agent has displayed feedback that evaluates the student's answers.

Table 6.3   The method for generating If-relevant test questions when the source is not relevant

---

Repeat
- Choose a family F of relevancy rules from the KB and choose a rule R from F.
- Generate an example E of R. Let S be the historical source from E, and let the partial example ET be the subset of E corresponding to the variables from the task description of R.
- **Repeat**
      Choose a new source NS different from S and from any other source in a previous step.
    Until a source NS is found such that there exists no example of any rule R_i from F that would contain ET and NS.
- Build the If question IFQ from the instances of the patterns of the task and operation of R, corresponding to ET and NS.
- Build the feedback FR for the right answer.
- Build the feedback FW for the wrong answer.
- Build the instance H of the hint description pattern of R corresponding to ET and NS.
- Build the explanation EX of the answer.
Until a test is generated.
Return (IFQ FR FW H EX), the list of the natural language descriptions of the question, feedback for the answers, hint and explanation.

6.5.6   Generation of Which-relevant test questions

The method for generating Which-relevant test questions is based on the methods for generating If-relevant tests, as indicated in Table 6.4. The method shown generates a Which-relevant test question where one source is relevant to the task and the other two are irrelevant. The actual method implemented in the assessment agent randomly selects how many relevant and irrelevant sources are in the test question.

6.5.7   Sample agent–student interaction during a Why-relevant test question

Figure 6.26 presents a Why-relevant test question. The student is given a task, a historical source and three possible reasons why the source is relevant to the task. He/she has to investigate the source and decide which reason(s) account for the fact that the source is relevant to the task. The student is instructed to check the box next to the correct reason(s). No check means that the reason is not correct. In the presented example, the student has correctly indicated that only reason 'C' explains why the source 'Christmas in Virginia' is relevant to the task. The agent has displayed feedback that evaluates the student's answer.

6.5.8   Generation of Why-relevant test questions

The method for generating Why-relevant test questions is presented in Table 6.5. First an example E_1 of a relevancy rule R_1 is generated. This example provides a correct task description T, a source S relevant to T, and a correct explanation EX_1 of why the source S is relevant to T. Then the agent chooses another rule that is not from the family of R_1. Let us suppose

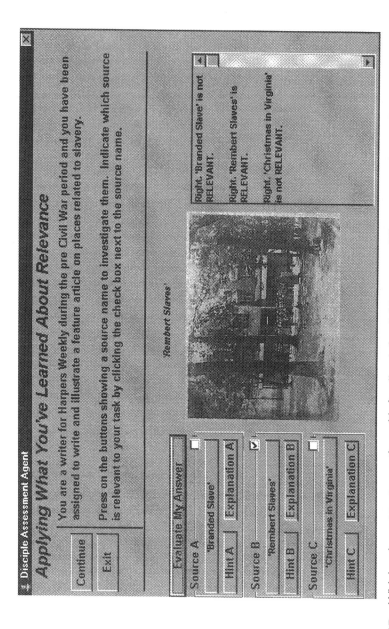

Figure 6.25 Which-relevant test question with feedback for right answer. (Picture reproduced from LC-USZ62-46606, Library of Congress, Prints and Photographs Division, Civil War Photographs.)

The screen content reads:

Disciple Assessment Agent

Applying What You've Learned About Relevance

You are a writer for Harpers Weekly during the pre Civil War period and you have been assigned to write and illustrate a feature article on places related to slavery.

Press on the buttons showing a source name to investigate them. Indicate which source is relevant to your task by clicking the check box next to the source name.

Continue

Exit

Evaluate My Answer

Source A
'Branded Slave'
Hint A   Explanation A

Source B
'Rembert Slaves'
Hint B   Explanation B

Source C
'Christmas in Virginia'
Hint C   Explanation C

'Rembert Slaves'

Right. 'Branded Slave' is not RELEVANT.

Right. 'Rembert Slaves' is RELEVANT.

Right. 'Christmas in Virginia' is not RELEVANT.

Table 6.4 The method for generating Which-relevant test questions

---

- Generate an If-relevant test IF_1 with a relevant source S_1:

$$IF_1 = (IFQ_1 \ FR_1 \ FW_1 \ (H_{11} \ EX_{11}) \ldots (H_{nm} \ EX_{nm})).$$

  Let T_1 be the task description from IFQ_1.

- Generate an If-relevant test IF_2 with an irrelevant source S_2, starting from T_1:

$$IF_2 = (IFQ_2 \ FR_2 \ FW_2 \ H_2 \ EX_2)$$

- Generate an If-relevant test IF_3 with an irrelevant source S_3, starting from T_1:

$$IF_3 = (IFQ_3 \ FR_3 \ FW_3 \ H_3 \ EX_3)$$

- Build the Which-relevant test question WRQ corresponding to T_1, S_1, S_2, and S_3, where the order of the sources is randomly generated.

- Return $(WRQ \ (IF_i, IF_j, IF_k)$ where $i, j, k \in \{1, 2, 3\}$ and $i \neq j \neq k$.

---

that the agent chooses a credibility rule R_2. It then generates an example E_2 of R_2, based on E_1 (that is, E_2 shares as many variables and values with E_1 as possible; certainly one of them is the source S). The agent also generates an explanation EX_2 of why S is credible. While this explanation is correct, it has nothing to do with why S is relevant to T. Then, the agent repeats this process to find another explanation that is true but explains something else, not why S is relevant to T.

One should notice that the false explanations presented in Figure 6.26 are based on relevancy rules that belong to other families of relevancy rules. Reasons 'A' and 'B' are correct explanations of why S is relevant to a Write-for-Audience task. These explanations mention the audience and the view point of the source. This has nothing to do with the task described in the test question. Of course, this method is correct only to the extent that the tasks to be used in the same test questions do not subsume one another.

6.6 Verifying, validating and maintaining the agent

The agent's knowledge base has been verified, validated and maintained as an integral part of the Disciple methodology. Disciple's learning approach verified the agent's knowledge (both the semantic network and the rules) by having the expert guide the learning process and evaluate the agent's problem-solving ability as the agent attempts to solve problems during rule refinement.

Further verification, validation and maintenance of the assessment agent was conducted in much the same manner as in traditional software engineering. Two testing tasks were performed. First, unit testing was conducted on each software module. Next, system-level testing was performed on the assessment agent using the test question interface.

The semantic network of the agent includes the description of 252 historical concepts, 80 historical sources and six publications. The knowledge base also contains 54 relevancy rules grouped in four families, each corresponding to one type of reporter task. Two of the families contain 18 relevancy rules and the other two contain nine relevancy rules.

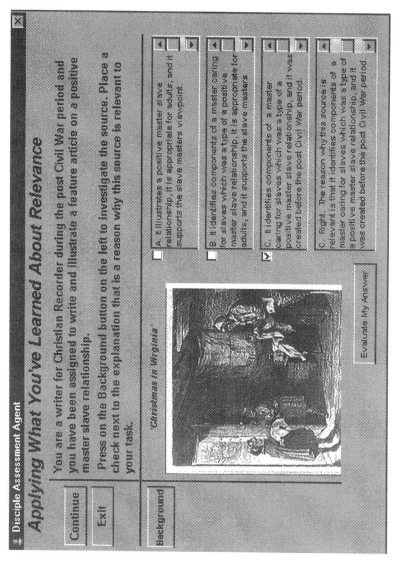

Figure 6.26 Why-relevant test question with feedback for right answer. (Picture reproduced from LC-USZ62-30813, Library of Congress, Prints and Photographs Division, Civil War Photographs.)

Table 6.5   The method for generating Why-relevant test questions

---

- Choose a family F of relevancy rules from the KB and choose a rule R_1 from F.
- Generate an example E_1 of R_1. Let S be the historical source from E_1, and let the partial example ET_1 be the subset of E_1 corresponding to the variables from the task description of R_1.
- Build the Why question WQ from the instances of the patterns of the task and operation of R_1, corresponding to ET_1 and S.
- Build the instance EX_1 of the explanation pattern of R_1, corresponding to E_1 and S.
- Choose a new rule R_2 that does not belong to the family F. R_2 may, but need not, be a relevancy rule. It may be a rule corresponding to another evaluation skill, such as credibility or accuracy.
- Generate an example E_2 of R_2 starting from the source S and the partial example ET_1.
- Build the instance EX_2 of the explanation pattern of R_2, corresponding to E_2.
- Choose a new rule R_3 that does not belong to the family F and is different from R_2. R_3 may, but need not, be a relevancy rule.
- Generate an example E_3 of R_3 starting from the source S and the partial example ET_1.
- Build the instance EX_3 of the explanation pattern of R_3, corresponding to E_3.
- Build the feedback FR for the right answer.
- Build the feedback FW for the wrong answer.
- Return (WQ S E_1 E_2 E_3 FR FW).

There are 40 930 instances of the 54 relevancy rules in the knowledge base. Each such instance corresponds to an If-relevant test question where the source is relevant. In principle, for each such test question the agent can generate several If-relevant test questions where the source is not relevant, as well as several Why-relevant, Which-relevant and Which-irrelevant test questions. Therefore, the agent can generate more than 10^5 different test questions.

The 54 relevancy rules have been learned from an average of 2.17 explanations (standard deviation 0.91) and 5.4 examples (standard deviation 1.37), which indicates a very efficient training process.

We have performed four types of experiments with the agent. The first experiment tested the correctness of the knowledge base, as judged by the domain expert who developed the agent. This was intended to clarify how well the developed agent represents the expertise of the teaching expert. The second experiment tested the correctness of the knowledge base, as judged by a domain expert who was not involved in its development. This was intended to test the generality of the agent, given that assessing relevance is, to a certain extent, a subjective judgment. The third and fourth experiments tested the quality of the agent, as judged by students and by teachers.

The results of the first two experiments are summarized in Table 6.6. To test the predictive accuracy of the knowledge base, 406 If-relevant test questions were randomly generated by the agent and answered by the developing expert, each time recording the agreement or the disagreement between the expert and the agent. They agreed in 89.16% of the cases.

We have performed a similar experiment with a domain expert who was not involved in the development of the agent. This independent expert has answered another 401 randomly generated If-relevant test questions. This time, the expert and the agent agreed in 85.76% of the cases and disagreed in 14.24% of the cases. These disagreements were analyzed by the developing expert and by the independent expert. There were cases where the two experts disagreed themselves, mainly because the independent expert had a broader interpretation of some general terms (such as slave culture, activities related to slavery, cruelty of slavery and

Table 6.6 Evaluation results

Reviewer	Number of reviewed questions	IF questions with relevant sources	IF questions with irrelevant sources	Time spent reviewing questions	Accuracy on relevant sources	Accuracy on irrelevant sources	Total accuracy
Developing expert	406	202	204	5 hours	96.53%	81.86%	89.16%
Independent expert	401	198	203	10 hours over 2 days	95.45%	76.35%	85.76%
Independent expert	1524	198 + 1326	—	22 hours for 1326 questions	96.19%	—	—

master slave relationships) than the developer of the knowledge base. However, the independent expert agreed that someone else could have a more restricted interpretation of those terms, and, in such a case, the answers of the agent could be considered correct. There were also 5 cases where the independent expert disagreed with the agent and then, upon further analysis of the test questions, agreed that the agent was right.

These two experiments have also revealed a much higher predictive accuracy in the case of If-relevant test questions where the source was relevant. This was 96.53% in the case of the developing expert and 95.53% in the case of the independent expert. The predictive accuracy in the case of irrelevant sources was only 81.86% in the case of the developing expert and 76.35% in the case of the independent expert.

The justification of the high predictive accuracy in the case of If-relevant test questions with relevant sources is as follows. Disciple learns one conjunctive rule at a time, starting from one example and its explanations, and is biased toward minimizing the errors of commission. Therefore, if an example is covered by the rule, one would expect it to be a positive example. On the other hand, if an example is not covered by this rule, it may or may not be covered by another rule. To more thoroughly test this hypothesis we have conducted a third experiment with the independent expert, who was shown an additional 1324 If-relevant test questions where all the sources were relevant (for a total of 1524 such questions). In this case, the predictive accuracy of the agent was 96.19%.

We have also analyzed in detail each case where both the developing expert and the independent expert agreed that the agent failed to recognize that a source was relevant to a certain task. In most cases it was concluded that the representation of the source was incomplete. This analysis suggested that the representation of the sources should be guided by the 'projection' principle mentioned in Section 6.3.2. If this principle had been followed more rigorously, then many of the agent's errors would have been avoided.

Table 6.6 also indicates the evaluation time because, unlike the automatic learning systems, the interactive learning systems require significant time from domain experts, and this factor should be taken into consideration when developing such systems. First of all, one could notice that it took twice as long for the independent expert to analyze 401 test questions than it took the developing expert. This is because the independent expert was not familiar with any of the 80 historical sources used in the questions, and he had to analyze each of them in detail in order to answer the questions. However, once the independent expert became familiar with the historical sources, he answered the new 1326 test questions much faster.

We have also conducted an experiment with a class of 21 students from the eighth grade at The Bridges Academy in Washington D.C. The students were first given a lecture on relevance and then were asked to answer 25 test questions that were dynamically generated by the agent. Students were also asked to investigate the hints and the explanations. To record their impressions, they were asked to respond to a set of 18 survey questions with one of the following phrases: very strongly agree, strongly agree, agree, indifferent, disagree, strongly disagree, and very strongly disagree. Figure 6.27 presents the results from seven of the most informative survey questions.

Finally, a user group experiment was conducted with eight teachers at The Public School 330 in the Bronx, New York City. This group of teachers had the opportunity to review the performance of the agent and was then asked to complete a questionnaire. Several of the most informative questions and a summary of the teacher's responses are presented in Figure 6.28.

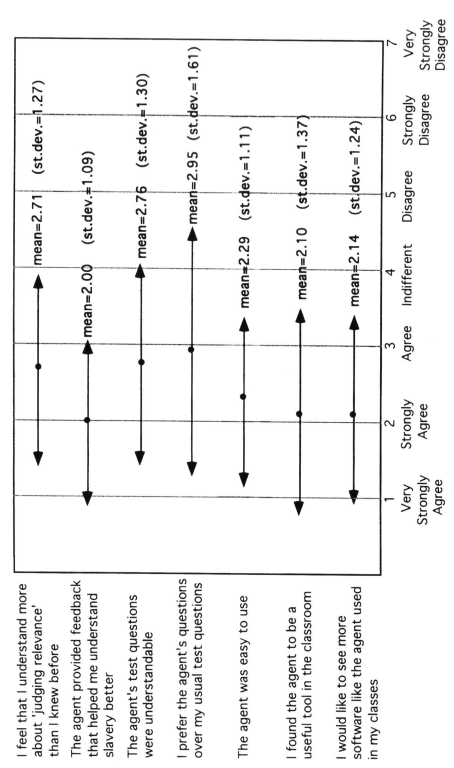

I feel that I understand more about 'judging relevance' than I knew before   mean=2.71   (st.dev.=1.27)

The agent provided feedback that helped me understand slavery better   mean=2.00   (st.dev.=1.09)

The agent's test questions were understandable   mean=2.76   (st.dev.=1.30)

I prefer the agent's questions over my usual test questions   mean=2.95   (st.dev.=1.61)

The agent was easy to use   mean=2.29   (st.dev.=1.11)

I found the agent to be a useful tool in the classroom   mean=2.10   (st.dev.=1.37)

I would like to see more software like the agent used in my classes   mean=2.14   (st.dev.=1.24)

1   Very Strongly Agree
2   Strongly Agree
3   Agree
4   Indifferent
5   Disagree
6   Strongly Disagree
7   Very Strongly Disagree

Figure 6.27   Student survey results.

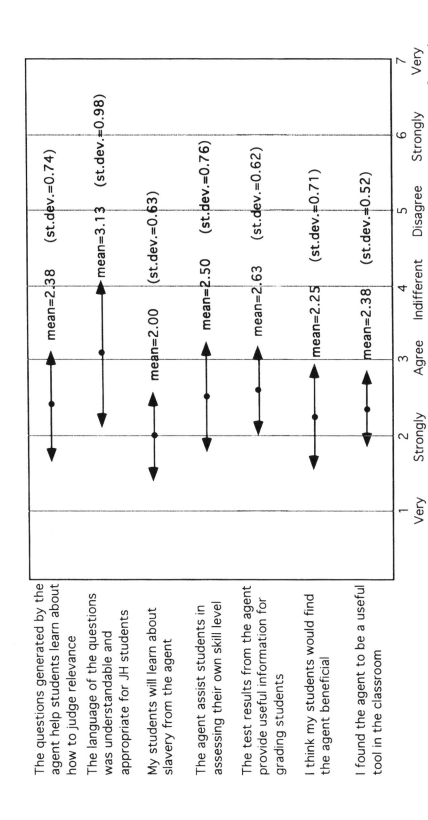

Figure 6.28 Teacher survey results.

6.7   Developing the integrated assessment agent

In the previous sections we presented in detail the methodology used to develop the stand-alone assessment agent. In this section we will describe how the integrated assessment agent (see Figure 6.1) was developed starting from the stand-alone assessment agent.

In principle, the knowledge bases, the assessment engines and even the user–agent interfaces of the two agents could have been identical, and only an interface with the MMTS system should have been developed. In this case, however, several changes needed to be made to each of these components in order to accommodate specific requirements from the developers of the MMTS system.

Instead of communicating directly with the student through its graphical user interface, the agent had to communicate only with the MMTS system. The MMTS system requests the agent to generate an exam, receives the test questions from the agent and then delivers the exam to the student. This required the development of a communication protocol and an interface between the MMTS system and the assessment agent. Both were based on a client/server approach where the assessment agent was the server and the MMTS software was the client sending requests for services. Secondly, the role of the integrated assessment agent was limited to that of dynamically generating final exams after a learning session with the MMTS system. These final exams consist of test questions, their answers, and explanations of the answers. Moreover, the agent was required to generate only one explanation of the answer even if there were more than one such explanation. However, if the student's answer to a question is not correct, then he/she receives a hint and is given the chance for a second try. Also, the Which-relevant test question described in Section 6.5.6 that generated any configuration of relevant and irrelevant sources was replaced by two simpler test questions: a Which-relevant test question (that contains one relevant source and two irrelevant ones) and a Which-irrelevant test question (that contains an irrelevant source and two relevant ones). These requirements concerning the form of the exam questions and the interactions with the students required an adaptation of the inference engine to no longer generate all the possible hints and explanations for a question, but only a single pair. Moreover, the patterns associated with each rule in the knowledge base had to be updated.

Figure 6.29 presents an updated Write-for-Audience relevancy rule. The patterns of all the rules in the agent's knowledge base had to be modified in the same manner to address the requirements of the integrated agent.

A final exam consists of eight test questions, each exam containing a variety of test question types. The student has to answer each question before receiving the next one. Figures 6.30–6.32 present three test questions from an agent-generated exam. The upper right pane of each window contains the *Assignment Sheet*, which corresponds to the types of assignments received by the students while using the MMTS system and playing the role of a reporter. The top part of each window contains instructions for the student. The middle pane displays the source and the pane under it contains the agent's feedback. None of the assignments and the feedback from the agent are canned texts, but are dynamically generated by the agent, and do not refer to sources that have been investigated by that particular student before.

Figure 6.30 presents an If-relevant test question that was correctly answered by the student. The student has selected the *RELEVANT* stamp from the left pane and dropped it onto the lower right pane. The agent has responded with the feedback shown in the bottom middle pane of the window.

```
IF
   ?W1   IS   WRITE-FOR-AUDIENCE
         IS-APPROPRIATE-FOR-AUDIENCE   ?A1
         ON   ?S1
         SUPPORTS-ANGLE   ?S2
   ?A1   IS   AUDIENCE
   ?S1   IS   SLAVERY-TOPIC
   ?S2   IS   ANGLE
   ?S3   IS   SOURCE
         IDENTIFIES-COMPONENTS-OF   ?S1
         IS-APPROPRIATE-FOR-AUDIENCE   ?A1
         SUPPORTS-ANGLE   ?S2

THEN

   RELEVANT   HIST-SOURCE   ?S3

Task   Description
   You have been assigned to write an article for ?A1 on
   ?S1 that supports ?S2.

Operator   Description
   ?S3 is relevant

Right   Answer
   The source is relevant because it identifies components
   of ?S1, it is appropriate for ?A1 and it supports the
   ?S2. This assignment calls for a source that identifies
   components of, explains, or illustrates ?S1, is
   appropriate for ?A1, and supports the ?S2.

Wrong   Answer
   The source ?S3 is not relevant because it does not meet
   the needs of this assignment. This assignment calls for a
   source that identifies components of, explains, or
   illustrates ?S1, is appropriate for ?A1, and supports the
   ?S2.

Try   Again   Answer
   ?S3 is relevant because it identifies components of
   ?S1, it is appropriate for ?A1 and it supports the ?S2.
```

Figure 6.29   An updated relevancy rule for the integrated assessment agent.

Figure 6.31 shows a Which-relevant test question. Initially, the left side of the window contained three sources from which the student was asked to choose the relevant one. When the student highlights one of these sources it is displayed in the middle pane. The selection of the relevant source is made by dragging its image from the left side of the window and dropping it into the lower right pane of the window. Figure 6.31 shows the interface after the student has incorrectly selected the source 'Smith's Plantation'. The bottom middle pane shows the feedback from the agent.

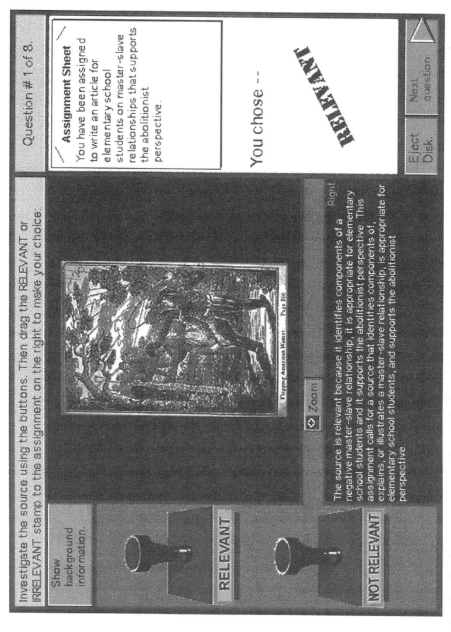

Figure 6.30 An If-relevant test question with the right answer and response. (Picture reproduced from LC-USZ62-30824, Library of Congress, Prints and Photographs Division, Civil War Photographs.)

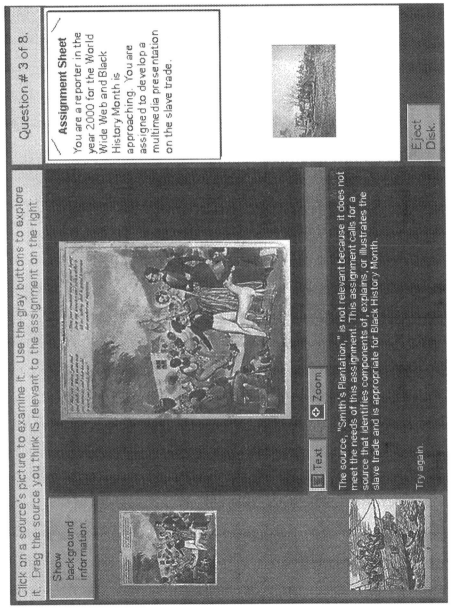

Figure 6.31 A Which-relevant test question with a wrong answer and response. (Picture reproduced from LC-USZ62-89745, Library of Congress, Prints and Photographs Division, Civil War Photographs.)

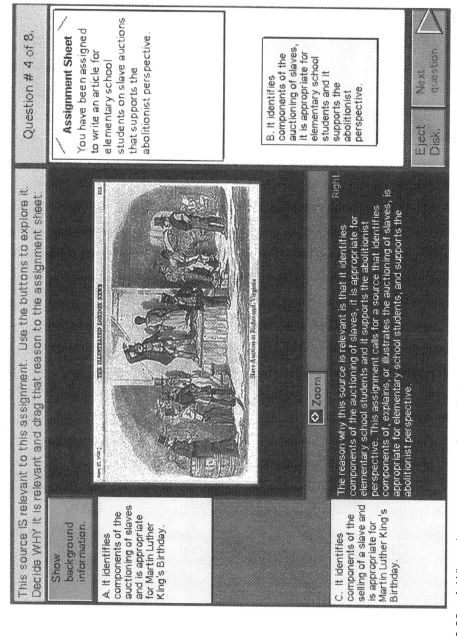

This source IS relevant to this assignment. Use the buttons to explore it. Decide WHY it is relevant and drag that reason to the assignment sheet.

Question # 4 of 8.

Show background information.

Assignment Sheet

You have been assigned to write an article for elementary school students on slave auctions that supports the abolitionist perspective.

A. It identifies components of the auctioning of slaves and is appropriate for Martin Luther King's Birthday.

Zoom

Right

B. It identifies components of the auctioning of slaves, it is appropriate for elementary school students and it supports the abolitionist perspective.

C. It identifies components of the selling of a slave and is appropriate for Martin Luther King's Birthday.

The reason why this source is relevant is that it identifies components of the auctioning of slaves, it is appropriate for elementary school students and it supports the abolitionist perspective. This assignment calls for a source that identifies components of, explains, or illustrates the auctioning of slaves, is appropriate for elementary school students, and supports the abolitionist perspective.

Eject Disk.

Next question.

Figure 6.32 A Why-relevant test question with the right answer and response. (Picture reproduced from LC-USZ62-15398, Library of Congress, Prints and Photographs Division, Civil War Photographs.)

Figure 6.32 shows a Why-relevant question. Initially, the left part of the window showed three potential reasons for why the displayed source is relevant to the assignment. The student has correctly moved the second reason to the pane under the assignment sheet and has received the agent's feedback in the bottom middle pane of the window.

6.8   Final remarks

In this chapter we have presented in detail the methodology for building a Disciple-based agent. We have also presented in detail the process of building the problem solver of the agent showing how a domain-specific problem solver could be built on top of the basic problem-solving operations of the Disciple Learning Agent shell. This case study shows an automated computer-based approach to the assessment of higher-order thinking skills (Beyer, 1987, 1988), as well as an assessment that involves multimedia documents. Both of these represent very important goals in the current education research. The IMMTS system was released for use in several US middle schools in Germany and Italy, in Spring 1997, and has been used by history teachers from these schools. The stand-alone assessment agent has been experimentally used, as described in Section 6.6.

The two developed assessment agents illustrate a new type of educational agent that serves as an asynchronous communication channel between one educator and an unlimited number of students, and yet provides one-to-one communication. Indeed, the educator can teach the agent in much the same way one might teach a human apprentice or student. Then the agent can test and tutor the students the way it was taught by the educator (see Figure 6.33). Because the educator instructs the agent as to what kind of tests to generate, and then the agent generates a wide variety of tests of that kind, the assessment agent provides the educator with a very flexible tool that lifts the burden of generating personalized tests for large classes, tests that do not repeat themselves and take into account the instruction received by each student. For instance, the developed assessment agents can generate over 100 000 relevancy test questions. Moreover, the ability of such agents to not only generate personalized tests, but also hints for solving them as well as detailed explanations, allows the students to be guided by the agent during their practice of the higher-order thinking skills as they would be directly guided by the educator.

The integration of the assessment agent and the MMTS software into the Intelligent Multimedia and Thinking Skill (IMMTS) system also illustrates another significant application of the Disciple approach and shell, that of developing intelligent agents that enhance the

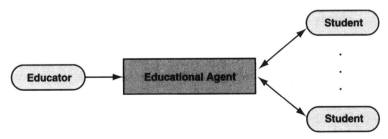

Figure 6.33   An educational agent tutored by an educator to tutor students.

capabilities, generality, and usefulness of non-knowledge-based software. In fact, this was the motivation for developing these assessment agents in the first place. Therefore, many of the features of the presented assessment agents are in response to requirements from the MMTS system, as discussed in Sections 6.2.1 and 6.7.

The MMTS system was developed by Lynn Fontana, Ward Cates, M.J. Bishop and Christopher M. vanNest. They and J. Mark Pullen have also contributed to the integration of MMTS and the assessment agent. Lynn Fontana, Lawrence Young, Rich Rosser, and Dave Webster served as our domain experts. The assessment agents were developed by Gheorghe Tecuci, Harry Keeling and Tomasz Dybala. This research was supported by the DARPA contract N66001-95-D-8653, as part of the Computer-Aided Education and Training Initiative, directed by Kirstie Bellman. Partial support was also provided by the NSF grant No. CDA-9616478, as part of the program Collaborative Research on Learning Technologies, directed by Caroline Wardle.

7
Case Study: Statistical Analysis Assessment and Support Agent

In this chapter we present the statistical analysis assessment and support agent, a Disciple agent that is integrated in an introductory, university level, science course and is accessed on the Internet through a web browser.

The course, The Natural World, introduces students to the world of science using collaborative assignments and problem-centered group projects that look at scientific issues that underlie public policy making and stimulate the development of students' analytic skills. The course focuses on three issues: diseases, evolutionary history, and nutrition. Most of the biology and statistics content material is available as web-based documents in the form of hyper-text tutorials, multimedia documents, and data sets. The focus of the course is on *analysis* of documents and data, and *integration* of biological knowledge, statistical methods, and historical and public health perspectives. This course is taught at George Mason University (GMU) and is part of an innovative curriculum designed for GMU's newest college, New Century College (NCC). The college's curriculum is organized around learning communities that focus on interdisciplinary themes and problem solving, student learning experiences which naturally incorporate collaborative projects, independent learning and the development of communication skills. In short, the goal of this course, and of the NCC curriculum, is to transform the culture of the classroom into a culture of thinking. The statistical analysis assessment and support agent is an integral part of this goal.

The statistical analysis assessment and support agent supports two aspects of students' learning: it assesses and supports students' knowledge and understanding of statistics, and students' analyses of issues related to statistics. The agent has three functions. First it can be used as a traditional test generator. This function is similar to the one for the history agent discussed in Chapter 6. Second the agent is integrated in the documents accessed by students. As students work through these documents they are asked periodically to 'interact' with them and to take an active role in the learning process. This takes the form of multimedia tasks to be performed and/or Disciple agent generated tasks to be completed. Finally, the agent can be used as an assistant to students as they work through their assigned projects. For most of these projects, students are required to include some data analysis to support their argument, and so students can use the statistical analysis assessment and support agent to help them extract all possible relevant information from a particular data set, and support their analysis of the data. This particular function, that of an assistant, is similar to the one of the manufacturing and the design assistants described in Chapter 2 and Chapter 8.

The course and the agent seek to support the development of students' higher order thinking skills, in particular, problem solving skills. While we recognize the importance of factual information, the course and the agent focus not only on the factual information, but also on the analysis of that information. In this sense, this agent, in its assessment mode, is similar to the history agent discussed in Chapter 6. But these two agents differ in many ways. First, the history agent is designed only as an assessment agent and it is not integrated into a course, and so it does not contribute to the learning process as intimately as the statistical analysis assessment and support agent does. Second, the history agent assesses whether students have acquired certain specific higher-order thinking skills, e.g. relevance or credibility of a historical document, and these skills are identified during the test, and are tested separately. So the history agent takes many documents and tests a single skill (see Chapter 6). In contrast, the statistical analysis assessment and support agent, in its assessment mode, handles the entire array of higher-order thinking skills, not identified to students, that are required for solving a particular problem. So the agent takes one problem and tests all the skills required for that problem, in turns, following closely the intuitive and natural problem-solving process. So while the history agent asks whether this or that document is relevant, the statistical analysis assessment and support agent, in its assessment mode, goes over the complete analysis of a data set. It tests:

1. whether the student has captured the type of the data set so as to determine the kind of questions that should be asked about that data set;
2. whether the student can formalize the questions in the form of hypothesis testing or statistical measures;
3. whether the student can identify the techniques and tools necessary to successfully complete the analysis.

We envision that ultimately the agent will also be able to test whether the student actually performs the analysis correctly, thereby completing the problem-solving methodology.

The development of the statistical analysis assessment and support agent follows the methodology described in Chapter 5 and illustrated in detail in Chapter 6. Therefore, in this chapter, rather than presenting in detail each step in the development process, we present an overview of this process, emphasizing those aspects that are characteristic of this agent. Moreover, we will not describe all three functions of the agent, rather we will focus on the assessment mode of the agent. The reader can imagine readily how the other modes are implemented. In the next section we give a description of the course and the innovative pedagogical approach behind the course and the NCC curriculum. Then, in Section 7.2, we discuss in more detail the functions of the statistical analysis assessment and support agent. In Section 7.3 we describe and illustrate the interactions between the agent and the student. Section 7.4 discusses the development of the higher-level ontology of the agent and of the semantic network. Section 7.5 presents a sample session of teaching the agent to generate tests. The last section summarizes this work.

7.1   The Natural World – the course

The recently established national standards for science education define a scientifically literate person as one who can identify the scientific issues underlying national and local decisions and express positions that are scientifically and technologically informed (National Academy of Sciences, 1996). Since new knowledge about science is being created each day, scientific literacy also requires lifelong learning (American Association for the Advancement of Science, 1993). The learning environment created by NCC seeks to implement these recent standards and to promote the development of lifelong learning skills by presenting students with models of learning science that they find both personally relevant and useful throughout their lives.

The NCC curriculum is organized around learning communities that focus on interdisciplinary themes and problem-solving, student learning experiences which incorporate collaborative projects, independent learning, and the development of communication skills. The NCC pedagogical approach is based on three broad themes: interdisciplinarity and team teaching, community building, and learners as lifelong constructors of understanding. This approach responds to a growing belief that the goal of education should no longer be dominated by the need to transfer information, but by the need to help students locate, retrieve, understand, analyze, construct, and use information (Naisbitt, 1990), and by the need to train lifelong learners. A sense of community in the NCC curriculum is built in the first year with a series of four units, required by all students in NCC. In this first year, students 'learn to learn', learn how to make distinctions, how to appreciate different perspectives, and how to find connections in what they learn. The Natural World (Figure 7.1) is the second unit and builds on the first unit, Community of Learners. After the first year, students take learning communities 'classes' to finish their degree program. The learning communities are interdisciplinary, team-taught, and address fundamental questions (e.g. Energy and Environment, Wealth-Power-Values, Religious myths and Scientific models).

Figure 7.1   The introductory screen of The Natural World course.

The Natural World has recently been transformed into a web-based course. At the same time, the statistical analysis assessment and support agent was introduced to further enhance the pedagogical goals of the course. Figure 7.1 is the introductory screen for the course. Students click on the icons to access the resources available for the course. As an inter-disciplinary science course, The Natural World focuses on the question of who we are as humans and how we analyze our human condition. The goal of The Natural World is to provide students with information about and experiences in exploring the world around them and in using these insights to formulate statements and solve problems.

This course emphasizes three issues: diseases (including their spread, their effect on who we are, the biology of diseases, etc.); evolutionary history (including population movement, links to the animal world, languages, diversity and commonality, genetics, race, etc.); and nutrition (including where our food comes from, how it affects who we are, and how we utilize it for energy, the molecular biology of the co-evolution of diet and digestive capa-bilities, etc.). The Natural World introduces students to the world of science using col-laborative assignments and problem-centered group projects that look at the scientific issues that underlie public policy making and stimulate the development of students' analytic and inquiry skills, among other higher-order thinking skills.

Most of the biology and statistics content material is available as web-based documents in the form of hyper-text tutorials, multimedia documents (such as digitized articles from scientific journals and popular press, digitized film clips and pictures, etc.), and data sets. This allows the class to be run without the traditional lectures, and contacts with students are mostly in the form of small group workshops where the focus is on *analysis* of the material accessed by the students, and on discussion of the group projects. Each group project involves the *integration* of biological knowledge, statistical methods, a historical perspec-tive, and a public health perspective. The course emphasizes the integration of statistical techniques throughout the biological content, in the learning process (in the tutorials and documents available to the students), but also in the assignments. So students are required to understand and communicate scientific knowledge (equivalent to an introductory biology course and an introductory statistics course) and how scientific ideas are involved in the decision-making process for important aspects of our lives. In this process they come to understand and appreciate scientific methods of inquiry and learn to employ these methods in seeking solutions to problems.

In such a class the role of the instructor is transformed drastically. The instructor does not act as the traditional 'well of knowledge' who passes knowledge to the students, but rather becomes an 'expert facilitator' of creation of knowledge – expert in the subject matter, but also expert in the dynamics involved in group learning and learning communities. In this new role, the instructor uses technology as a support for the learning process and for his/her expertise, and certainly not as a replacement of his/her expertise. This new role for the instructor and for technology calls for a new approach to assessment of students, and to students' interaction with the information technology available to them.

7.2   Assessment in The Natural World

Traditionally, assessment is done as a separate activity from the learning process, organized at discrete times during the course, sparsely distributed throughout the course (e.g. a

Table 7.1   Three levels of assessment or support

1. *Capture* the type of the data set and *determine* the kind of questions that should be asked about that data set.
2. *Formalize* these questions in the form of hypothesis testing (inferential statistics) or in the form of statistical measures (descriptive statistics).
3. *Identify* the techniques and tools necessary to successfully complete the analysis.

midterm, final exams, a few tests), with only one output, namely grades. The Natural World envisions assessment as an *integrated* part of the learning process, as an activity that is not discrete and sparse, but continuous and dense throughout the course, as an activity that outputs not only grades but also insights and guidance to the students. The assessment is also envisioned to be automated and 'intelligent', and built in the course information technology environment.

To demonstrate the feasibility of this approach and to make the description above more concrete, The Natural World has integrated the statistical analysis assessment and support agent to assess and support students' knowledge and understanding of statistics, and to assess and support students' analyses of issues related to statistics. More specifically, the agent is designed to test and support students' abilities defined in Table 7.1.

We will refer to these three parts as levels of assessment or support (level 1, level 2, and level 3 assessment or support). We note that these levels are sequential and describe precisely the problem-solving strategy involved in analyzing a data set. We envision that ultimately the agent will also be able to assess whether students actually perform the analysis correctly, thereby completing the problem-solving methodology.

We also envision that the role of the agent will be increased substantially to deal with other issues related to the course. For example we envision that the agent might assess and support students in their analysis of the *quality* of the data provided (e.g. this deals with the issue of sampling), or in the *design* of proper experimentation or data collecting for a particular study. Another direction we envision is to *integrate* the statistical analysis that the agent can presently assess and support with analysis of biological phenomena, e.g. of evolutionary forces, or of DNA material, etc. These are examples of our future work and they illustrate how the Disciple agent can be integrated very closely in the learning environment, and how it fits in the general pedagogical approach described earlier.

To achieve the three levels of assessment (or support) described in Table 7.1, the agent generates a sequence of three sets of questions, each set targeting one of the three levels of assessment (or support). Students answer these questions successively. For each set of questions, students are given the option of getting a hint. Moreover, if the agent is used for assessment, the agent generates explanations for correct and wrong answers to any question. So, for correct answers, the agent reinforces and verbalizes an explanation the student might have already generated him/herself, but maybe in an intuitive way, or without a clearly verbalized or complete reasoning. For incorrect answers, the agent gives an explanation of why the student chose a wrong answer and a description of what the student should have been looking for to obtain the correct answer (the correct answer is not given explicitly to the student), thereby giving feedback to the student and enhancing the learning process. The student is then asked to go back and retry answering the question. We will illustrate this process in three examples below, and we will make the description of this agent and how it works more precise in Sections 7.4 and 7.5.

Access to the agent is done through a web browser, just as the rest of the course's resources. Indeed we view the agent as yet another resource for the course. The agent resides locally on the student's computer's hard drive. The agent generates '.html' files that are displayed in the student's web browser whenever the student launches a session with the agent. Tracking of the student's interaction with the agent (and other activities, such as what documents the student is accessing, how much interaction is between the student and the documents, etc.) is recorded in a separate file that can then be given to the instructor, or sent electronically, for the purpose of assessment.

The Disciple agent is used in three different ways. First, the agent can be used as a traditional test generator. This can be launched directly from the computer icon in the introductory screen for the course (see Figure 7.1). A student might be asked by the instructor to take such a test, or might want to test his/her knowledge and understanding. Second the agent is integrated in the tutorials (both statistics tutorials and the biology tutorials) and the other documents. As students work through these resources, they are asked periodically to 'interact' with the documents. This takes the form of multimedia tasks to be performed, or agent-generated problems to be answered. These activities have two purposes:

- They are designed to help students become familiar with, understand and apply the concepts discussed in the particular section they are working on.
- They are also designed to make students take an active role (in a physical and intellectual way) in the learning process, by reflecting on the information they have accessed and by generating new knowledge and insight.

The web makes this non-linear learning process feasible. Finally, the Disciple agent is used as an assistant to students for the projects they have to work on. For most of these projects, students are required to include some data analysis to support their arguments, and so students can use the agent to help them extract all possible relevant information from a particular data set. Thus the Disciple agent also supports students' learning in a non-assessment mode.

In the following sections we illustrate how the Disciple agent works only in the assessment mode, but it is easy to imagine how this mode can be adapted to using the agent in a support mode.

7.3   Sample interactions between the agent and the student

In this section we illustrate the type of the test questions generated by the agent and the interactions with the student during answering these questions.

7.3.1   Example 1 – analysis of cigarette data

In this example students are asked to analyze a data set* which contains measurements of weight, tar, nicotine, and carbon monoxide contents for 25 brands of domestic cigarettes. The data are available to students (directly from the web) as an Excel worksheet if the student desires to manipulate the data and/or perform any sort of computations with the data.

* Obtained from the StatLib web site of the American Statistical Association (http://lib.stat.cmu.edu).

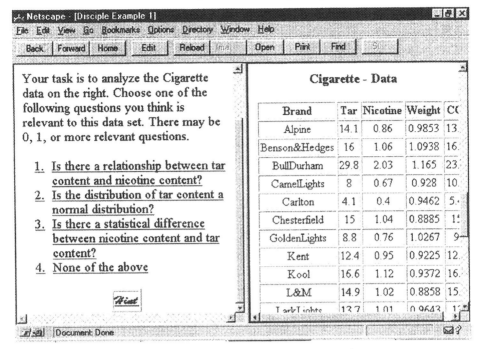

Figure 7.2 Analysis of cigarette data – level 1 of assessment.

This first set of questions (see Figure 7.2) attempts to assess whether the student has captured the type of the data set and whether he/she can determine the kind of questions that should be asked about the cigarette data (this is the first level of the assessment, see Table 7.1). The underlined text indicates that students can click there and be linked to a new page. Each set of questions of this type may have one or more correct answers, and possibly no correct answer at all. This makes the task more realistic and interesting, in the sense that, in a real situation, one may go in many wrong directions before finding the right way, and there may be more than one way to analyze a particular data set.

This is an example of what we have called *case* data. Such data are a collection of measurements where each row in the data table is an individual case. For such data there are two issues that are relevant (Questions 1 and 2 in Figure 7.2): the relationship between measurement variables, and the nature of the distribution of any one of the measurement variables (we will discuss these variables in Section 7.4).

In general we have a correspondence between the relevant issues (or questions) and the type of data. There are a number of types of data sets (see Section 7.4), case data being one of them. This data set is fairly simple because it is only case data. Other data sets may not be only case data, but also other types (such as categorical or time data). Therefore a more complex set of relevant questions might be asked about them. Example 2 will illustrate this fact.

So, Questions 1 and 2 in Figure 7.2 are relevant questions, and when the student clicks on one of Questions 1 or 2 he/she is led to a new set of questions (see Figures 7.4 and 7.5), and if the student clicks on Question 3 he/she is led to Figure 7.3. The screen that the student links to when he/she clicks on Question 4 gives the same information as the one in Figure 7.3.

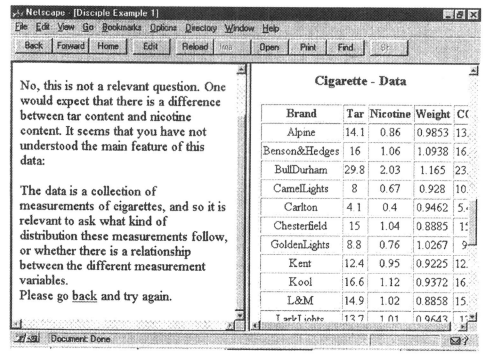

Figure 7.3   Analysis of cigarette data – feedback for wrong answer.

As indicated earlier, this page gives an explanation of why the student chose a wrong answer (one would expect a difference between tar and nicotine contents) and points to the kind of information the student should have looked at (it basically tells the student that this is case data and it tells what the relevant issues are for such data). Note that the correct answer is not given. The information given at the bottom of Figure 7.3 is the same as the one that the student would have obtained by clicking on the *Hint* button on the first page (Figure 7.2).

Question 1 (in Figure 7.2) is a relevant question and if the student clicks on it, it leads to a new screen (Figure 7.4) which contains an explanation of why this is a good question. As mentioned before it reinforces the student's conceptualization of the type of data (case data) and the relevant issues concerning the data. That page also contains three other questions. These new questions attempt to assess whether the student can formalize the question he/she has identified as a relevant question (this is the second-level assessment, see Table 7.1). Again, students have access to a hint. Since the relevant question the student has chosen is the one about a relationship between tar content and nicotine content, the correct formalization of this questions is the investigation of the correlation between these two variables (this is Point 1 in Figure 7.4).

Question 2 from Figure 7.2 links to a similar screen (see Figure 7.5). Again the questions try to determine whether the student can formalize the question he/she has identified as a relevant question.

We now illustrate the third-level assessment (see Table 7.1). We go back to Figure 7.4. If the student clicks on the second or third point, he/she is linked to a page that explains why

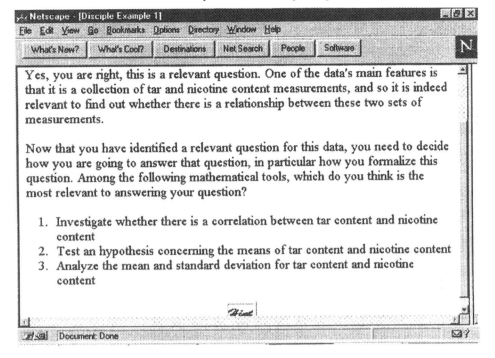

Figure 7.4   Analysis of cigarette data – level 2 of assessment.

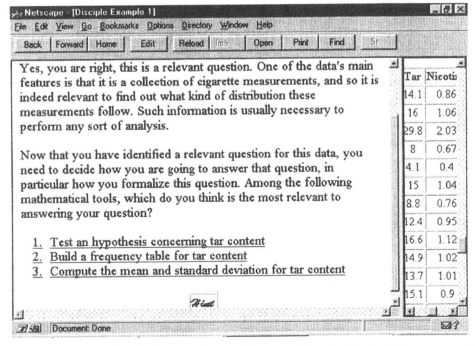

Figure 7.5   Analysis of cigarette data – other questions for level 2 of assessment.

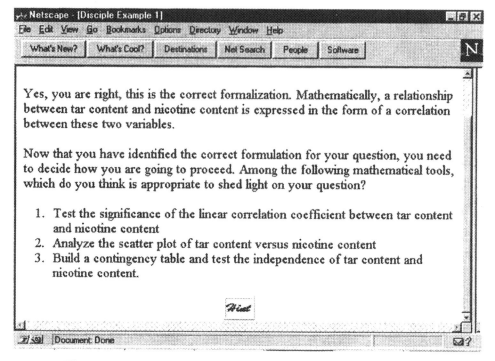

Figure 7.6   Analysis of cigarette data – level 3 of assessment.

he/she is wrong and gives some more information, the same as the information given with the *Hint* button, about how to find the right formalization of Question 1 in Figure 7.2. If the student clicks on the first point in Figure 7.4, then he/she is linked to Figure 7.6. Again, the student is given a reason why this is the correct answer. To analyze the correlation between the tar and nicotine contents, the student has two choices, first he/she may analyze the scatter plot of tar content *vs* nicotine content (this is Point 2 in Figure 7.6), but this will only give subjective information (but at least a place to start), or he/she can compute the correlation coefficient and test its significance (this is Point 1 in Figure 7.6). From Figure 7.6 the student will be linked to the appropriate page that will indicate this information.

We have now gone over the three levels of assessment for this particular data set. The student started by asking general questions about the kind of information he/she could extract from this data set (see Figure 7.2) and finished by identifying the appropriate mathematical tool and technique to answer the question he/she asked. As mentioned before, it would now be appropriate to ask the student to actually perform the analysis, compute what is needed and interpret the results as needed.

7.3.2   Example 2 – analysis of iris data

The iris data example is different from the cigarette data example because of the type of the data. In this example students are asked to analyze a data set* which contains measurements

---

* Obtained from the StatLib web site of the American Statistical Association (http://lib.stat.cmu.edu).

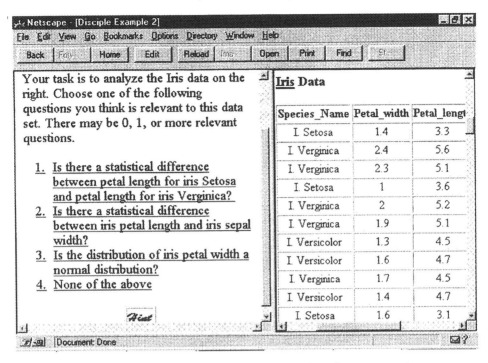

Figure 7.7   Analysis of iris data – level 1 of assessment.

of the petals and sepals for 150 irises. In addition, the species each iris belongs to is recorded. The first set of questions attempts to assess whether the student has captured the type of the data set and whether he/she can determine the kind of questions that should be asked about the iris data (this is the first level of the assessment, see Table 7.1).

This is another example of what we called case data, since it is a collection of measurements; each row in the data table is an individual case. So we can ask the case data questions (Question 3 in the picture), as we did for the cigarette data in Figure 7.2. However, the data are also of a different type; the irises are separated in three species. This is an example of what we have called *categorical* data. Such data have a 'categorical' variable (in this case it is the species name), usually a symbolic variable, which has a small number of values (categories), and the values for each measurement variable in the data can be separated according to the different categories. For example, in this case, the data can be separated according to the three species of iris (*Iris Setosa*, *Iris Verginica*, and *Iris Versicolor*). There is one relevant question for such data (Question 1 in Figure 7.7): whether there is a difference between two categories, i.e. whether there is a difference between the values of a measurement variable for one category and the values of the same measurement variable for another category (in Figure 7.7 this measurement variable is petal length, and the categories are *Setosa* and *Verginica*).

Another typical example of categorical data might be data that describe certain features of human beings where gender is reported for each individual (here gender becomes the categorical variable, and the data can be separated, and hence compared, according to gender). Another example might be clinical data that describe certain features of patients

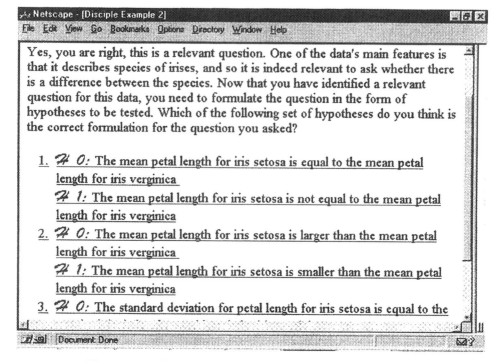

Figure 7.8   Analysis of iris data – level 2 of assessment.

with and without a certain medical procedure (here whether a patient has received the procedure is the categorical variable, the data can then be separated, and hence compared, according to whether a patient received the treatment or not).

The iris data therefore are of two types, case data and categorical data, and hence one could ask relevant questions corresponding to either of these types. When the student clicks on Question 1 in Figure 7.7 (the question is relevant because the data are categorical), he/she is linked to Figure 7.8.

The correct formulation of the question is in the form of a pair of hypotheses (the first pair in Figure 7.8). The assessment proceeds as in the cigarette data example leading the student to one more level of assessment where he/she is asked what tools he/she would use to test the hypotheses. It is now easy to imagine how the agent proceeds in this example.

We should note that the second-level assessment in this example is quite different from that of Example 1. In the present case, the second-level assessment deals with inferential statistics (hypothesis testing) while the first example dealt with descriptive statistics. This difference is due to the type of the data captured by first-level assessment: the first-level assessment for the cigarette data captured the case type of the data, while the first-level assessment of the iris data captured the categorical type of the data.

7.3.3   Example 3 – analysis of flies data

The last example we give is that of a data set which is neither case nor categorical data. It is an example of the third type of data, namely *time data*. The flies data (Carey *et al.*, 1992) test Gompertz's 1825 theory of mortality rates with Mediterranean fruit flies: the population for the experiment comprises 1 203 646 fruit flies, and the number of flies found dead each day is recorded. The data consist of the number of surviving flies on each day, until day 171, when the last two flies died. So the number of surviving flies varies over time, and it is this feature that makes this data set different from the other two examples.

The first set of questions, in Figure 7.9, attempts to assess whether the student has captured the type of the data set and whether he/she can determine the kind of questions that should be asked about the flies data (this is the first level of the assessment, see Table 7.1). There are two relevant questions for this type of data: whether there is a pattern over time for one of the variables, or whether there is significant difference over time. Which of the two questions is relevant for a particular data set depends on what is 'expected'. For the flies data, we expect great variation, so we are looking for a pattern or model of the data over time, hence Question 3 in Figure 7.9 is the correct question. For other data sets where it is expected that the data will not vary much over time, the relevant question would be whether there is significance difference over time.

We note that this situation is quite different than the two previous examples. In those examples only the type of the data set, case or categorical, determined the first-level assessment. In the present situation, i.e. for time data, the first-level assessment is determined not only by the type of the data, but also by a 'property' that we connect to the data in the form of an 'expectation', i.e. whether the expectation is that the data vary greatly over time, or whether the expectation is that the data do not vary much over time.

The next two sections will present how the Disciple agent has been developed to interact with the student as illustrated above.

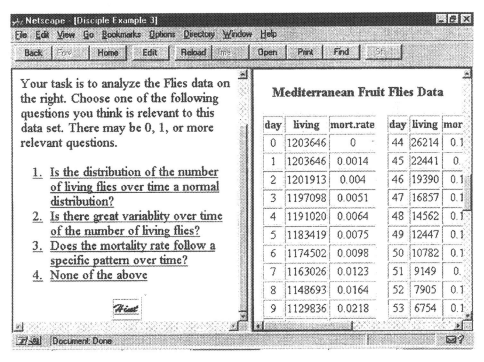

Figure 7.9 Analysis of flies data – level 1 of assessment.

7.4   Building of the initial knowledge base

In this section we describe how the semantic network of the agent has been developed.

The starting point in building the knowledge base is the three levels of assessment we want to develop, as described in Table 7.1. These levels describe precisely the steps which one would take to analyze a data set. We emphasize again that this assessment mode can easily be transformed to become a support or assistant mode when the agent is used in a non-assessment mode (to help students with their projects).

There are two concepts needed in order to have a 'qualitative' description of a data set: the type of the data and the variables contained in the data set. The type of the data can be described by the concept **TYPE–OF–DATA** (see Figure 7.10) There are three types of data:

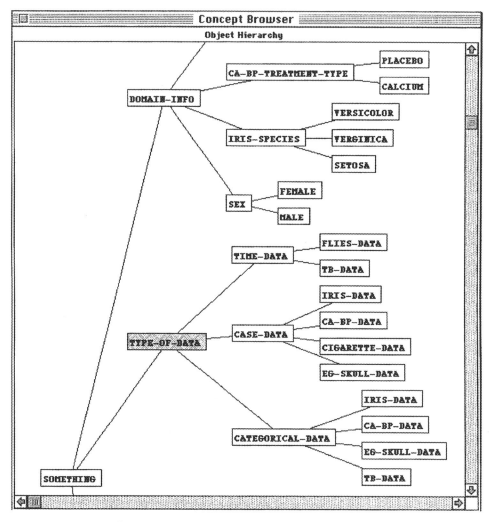

Figure 7.10   Sample of the agent's ontology.

CASE-DATA, **CATEGORICAL-DATA**, and **TIME-DATA**. These are concepts in our knowledge base, subconcepts of the concept **TYPE-OF-DATA**. We have already briefly discussed these types of data in Section 7.3.

Recall that we define **CASE-DATA** to be a collection of measurements where each row in the data table is an individual case. **IRIS-DATA**, **CA-BP-DATA**, **CIGARETTE-DATA** and **EG-SKULL-DATA** are instances of the concept **CASE-DATA** (see Figure 7.10). Recall that we define **CATEGORICAL-DATA** to be a data set that has a 'categorical' variable, usually a symbolic variable which has a small number of values (the categories), and the values for each measurement variable in the data can be separated according to the different categories. For example, **IRIS-DATA** is an instance of the concept **CATEGORICAL-DATA** where the values of the variable 'petal length' can be separated according to the species of iris. Recall that we define **TIME-DATA** to be a data set which contains one or several measurement variables that vary over time. **FLIES-DATA** is an instance of the concept **TIME-DATA**. Other instances of **TIME-DATA** might be the number of cases of a disease over time, birthrate over time, measurements of evolutionary traits over time, etc.

The type of the data set then indicates the kind of questions that are relevant to ask about that data set. This one-to-one relationship between the type(s) of the data set and the relevant questions will be the basis for teaching the agent rules for generating questions (which we will describe in Section 7.5). For **CASE-DATA** we might ask two types of questions: the relationship between two measurement variables, and the nature of the distribution of any one of the measurement variables. For **CATEGORICAL-DATA** there is one relevant question: whether there is a difference between the categories, i.e. whether there is a difference between the values of one measurement variable for one category and the values of the same measurement variable for a second category. For **TIME-DATA** there are two relevant questions: whether there is a pattern over time for a measurement variable, or whether there is significant difference over time (which question is relevant depends on what is expected; e.g. for the **FLIES-DATA**, we expect great variation, so we are looking for a pattern or model of the data over time, but for the birthrate data, the relevant question will be whether there is difference over time).

The concept **VAR** will encode the information about the kinds of variables a data set contains, the relations among them, and the properties they may have. There are three types of variables: **SYMBOLIC-VARIABLE**, **MEASUREMENT-VARIABLE**, and **SEQUENCE-VARIABLE** (indicated in Figure 7.11). Symbolic variables are further divided into **CATEGORICAL-VARIABLE** and **NON-CATEGORICAL-VARIABLE**. Each data set contains measurement and/or symbolic variables, and thus to each data set we associate a concept of the data set's measurement and symbolic variables. For example, **IRIS-MEASUREMENT-VAR** is the concept whose instances are the actual measurement variables for **IRIS-DATA**, such as **IRIS-PETAL-LENGTH**. **CIG-MEASUREMENT-VAR** is the concept whose instances are the actual measurement variables for **CIGARETTE-DATA**, such as **CIG-WEIGHT**. The variables **IRIS-SPECIES-NAME**, a variable contained in **IRIS-DATA**, and **CIG-BRAND-NAME**, a variable contained in **CIGARETTE-DATA**, are symbolic variables; the first one is categorical, while the second is non-categorical.

Now that we have defined the data's type and variables, we define relations that encode information about the data set and its variables. For each data set we have the relation **CONTAINS-VARIABLE**, which relates the data set to its variables, and conversely, for each variable we have the relation **IS-A-VARIABLE-OF**, which relates the variable to the data set it belongs to.

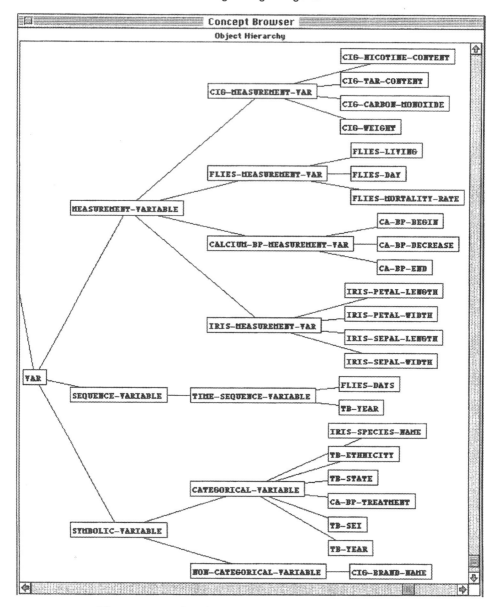

Figure 7.11 Another sample of the agent's ontology.

Since the types of the data sets are linked to particular variables in the data, we define relations that encode that information. So for a data set that is **CASE-DATA**, we define the relation **CASE-WITH-RESPECT-TO** which links the data set to the variable(s) that make it case data. For a data set that is **CATEGORICAL-DATA**, we define the relation **CATEGORICAL-WITH-RESPECT-TO** which links the data set to the variable(s) that make it categorical data. Finally, for a data set that is **TIME-DATA**, we define the relation **TIME-WITH-RESPECT-TO** which links the data set to the variable(s) that make it time data. For

Table 7.2   Definitions of relations between objects

Name	Description	Domain	Range
CATEGORICAL-WITH-RESPECT-TO	is categorical data with respect to	CATEGORICAL-DATA	CATEGORICAL-VARIABLE
CASE-WITH-RESPECT-TO	is case data with respect to	CASE-DATA	CASE-VARIABLE
TIME-WITH-RESPECT-TO	is time data with respect to	TIME-DATA	TIME-VARIABLE
CONTAINS-VARIABLE	contains the variable	DATA	VAR
HAS-VALUE-FOR	has value for	VAR	DOMAIN-INFO

example, **CIGARETTE-DATA** is case data and it is **CASE-WITH-RESPECT-TO** the variable **CIG-WEIGHT**, but it is not **CASE-WITH-RESPECT-TO** the variable **CIG-BRAND-NAME** (in fact there is no special relationship between the data and this variable, except for the two relations that indicate that it is a variable of the data). The data set **IRIS-DATA** is case data and it is **CASE-WITH-RESPECT-TO** the variable **IRIS-PETAL-LENGTH**. On the other hand it is also categorical data and it is **CATEGORICAL-WITH-RESPECT-TO** the variable **IRIS-SPECIES-NAME**. The data set **FLIES-DATA** is **TIME-DATA** and it is **TIME-WITH-RESPECT-TO** the variable **FLIES-DAYS**. Table 7.2 contains the definitions of some of these relations.

Finally, we turn our attention to a part of the semantic network describing the concept **DOMAIN-INFO** and its subconcepts and their instances (see Figure 7.10). This will encode the values of certain categorical variables for certain data sets. The reason we need this information for categorical variables is that certain measurement variables do not have values for certain categories (i.e. values of a categorical variable), and this is information we need to encode in our semantic network. So, for example, we have the concept **IRIS-SPECIES**, a subconcept of **DOMAIN-INFO**, and the instances of this concept are the actual species of irises, namely **SETOSA**, **VERGINICA**, and **VERSICOLOR**. So we define a relation **POSSIBLE-VALUES** that indicates that the variable **IRIS-SPECIES-NAME** has **POSSIBLE-VALUES** the values **SETOSA**, **VERGINICA**, and **VERSICOLOR**. Moreover we define a relation **HAS-VALUES-FOR** which indicates that a measurement variable has values for this or that value of the categorical variable. For example, **IRIS-SEPAL-LENGTH** is a measurement variable in **IRIS-DATA**, and it **HAS-VALUES-FOR** the categories **VERSICOLOR** and **VERGINICA**, but not for the category **SETOSA**.

7.5   Teaching the agent to generate test questions

In this section we discuss how the agent has been taught to generate questions. The teaching of the agent follows the rule learning and rule refinement processes described in Chapter 4 and illustrated by the history agent in Chapter 6. Therefore here we are only going to briefly illustrate this process from the point of view of the expert and focus on how the expert interacts with the agent to perform the teaching.

The starting point, following the general Disciple approach, is an example which, in this case, should be some data set and a relevant question to ask about that data. We first consider the **IRIS-DATA** and focus our attention on the first level assessment (see Table 7.1). The task at hand is to analyze the **IRIS-DATA** by asking a relevant question for the data. Such a relevant question is whether there is a statistical difference between the **IRIS-PETAL-LENGTH** for iris **SETOSA** and the **IRIS-PETAL-LENGTH** for iris **VERGINICA** (see Figure 7.7). The interaction with the agent is done through the customized example editor (see Figure 7.12). The bottom window in Figure 7.12 is the **IRIS-DATA** to be analyzed (displayed as a web document, the same way the students or the expert would see it) and the window above it is the customized example editor that displays the natural language representation of the initial example. This customized example editor is the same as the one used in the case of the History Assessment Agent described in Chapter 6. The top window of Figure 7.12 shows the internal formal representation of the initial example, in a simplified form, that was automatically generated by the agent.

The operation in the right-hand side of the example is called **RELEVANT-QUESTION** to indicate that its argument is a relevant question about the **IRIS-DATA**. Another operator is

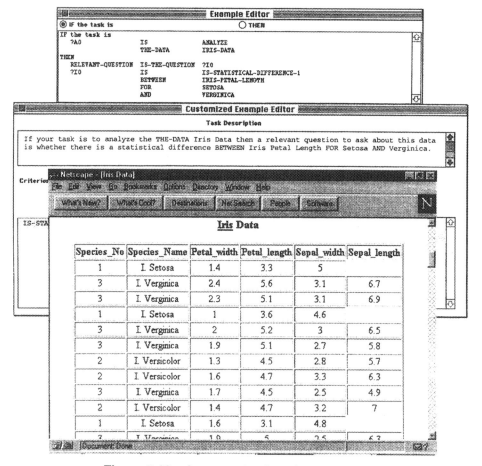

Figure 7.12   An example of a relevant question.

IRRELEVANT-QUESTION that indicates that its argument is a question that is not relevant to the data set being analyzed. This operator is necessary because the agent needs also to generate irrelevant questions, as will be discussed latter.

In our example the relevant question is whether there is a statistical difference between **IRIS-PETAL-LENGTH** for **SETOSA** and **IRIS-PETAL-LENGTH** for **VERGINICA**. Now we have to help the agent understand why this question is a relevant question for this data set. We first ask it to generate all the plausible explanations of type Association or Relation. The generated explanations are displayed in the bottom pane of the window of Figure 7.13. From them we select those that really explain why this is a relevant question. As we click on a good explanation piece, it becomes marked with an asterisk. These explanations are very intuitive and easy to understand. They indicate that **IRIS-DATA** contains the variable **IRIS-PETAL-LENGTH** that has values both for **SETOSA** and for **VERGINICA**. Moreover, **IRIS-DATA** is categorical with respect to **IRIS-SPECIES-NAME**. Now we click on the *Create* button asking the agent to create a rule from this example and its explanations, as explained in Chapter 4. The learned rule is shown in the upper right pane of the window in Figure 7.13. The rule is shown in the customized internal form that does not include the rule's variable, but only their classes. The generic form of the rule is shown in Figure 7.14. On the right-hand side of Figure 7.14 are our comments that explain the rule to the reader.

The upper and lower bounds of the rule are also very intuitive, as can be seen from the right-hand side of Figure 7.14. The upper bound corresponds to the highest level concepts that can possibly fit the example given, and the lower bound is exactly the example given. The exact rule is somewhere in between these two bounds and we are going to interact with the agent to refine the current rule. We are therefore asking the agent to generate new relevant questions similar to the one we have formulated. Figure 7.15 shows a question generated by the agent for the data set **CA-BP-DATA*** that describes the effect of calcium on blood pressure. A treatment group of 10 men have received a calcium supplement for 12 weeks, and a control group of 11 men have received a placebo during the same period. All subjects have had their blood pressure tested before and after the 12-week period (these measurements are denoted by **CA-BP-BEGIN** and **CA-BP-END**). The question generated by the agent is whether the two groups (treatment and control) for the study are similar before the experiment begins. This is a very relevant issue if one wants to draw conclusions about the effect of the treatment after the experiment is completed.

We accept the example generated by the agent by clicking on the *Accept* button in the lower right corner of the window in Figure 7.16. As a consequence, the agent generalizes the plausible lower bound condition of the rule, as indicated in the bottom left pane. The new plausible lower bound condition is shown in the upper right pane.

The next question generated by the agent is shown in Figure 7.17. The tuberculosis data set* used in this question consists of 11 338 individual cases of tuberculosis sampled randomly from 113 417 cases reported to the Center for Disease Control (CDC) during 1985–1989. Each record includes seven fields, the state code, the year the case was counted by CDC, the month the case was counted by CDC, as well as the age, sex, race and ethnicity of the persons.

The question generated by the agent is not relevant because the year a case was reported to the CDC is not related to gender. Therefore, we point to the wrong variable used by the agent (which is **TB-YEAR**), and reject this example. As a consequence, the agent specializes

---

* Obtained from the StatLib web site of the American Statistical Association (http://lib.stat.cmu.edu).

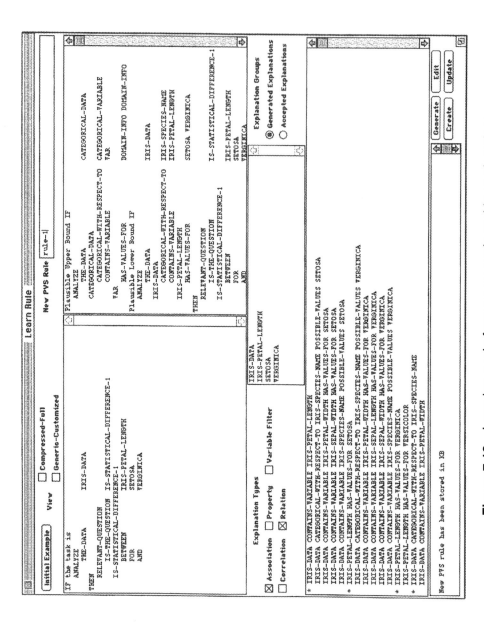

Figure 7.13 Learning a rule from an example and its explanations.

```
Plausible Upper Bound IF                              ; IF (Plausible Upper Bound)
  ?W1 IS      ANALYZE                                 ; the task is to analyze
    THE-DATA ?S1                                      ; the data ?S1 and

  ?S1 IS      CATEGORICAL-DATA                        ; ?S1 is a CATEGORICAL-DATA that
    CONTAINS-VARIABLE ?V1                             ; contains the variable ?V1 and
    CATEGORICAL-WITH-RESPECT-TO ?V2                   ; is categorical with respect to ?V2 and

  ?V1 IS      VAR                                     ; ?V1 is a variable
    HAS-VALUES-FOR ?H1  ?H2                           ; that has values for ?H1 and ?H2 and

  ?V2 IS      CATEGORICAL-VARIABLE                    ; ?V2 is a categorical variable and

  ?H1 IS      DOMAIN-INFO                             ; ?H1 is DOMAIN-INFO and

  ?H2 IS      DOMAIN-INFO                             ; ?H2 is DOMAIN-INFO


Plausible Lower Bound IF                              ; IF (Plausible Upper Bound)
  ?W1 IS      ANALYZE                                 ; the task is to analyze
    THE-DATA ?S1                                      ; the data ?S1 and

  ?S1 IS      IRIS-DATA                               ; ?S1 is the IRIS-DATA that
    CONTAINS-VARIABLE  ?V1                            ; contains the variable ?V1 and
    CATEGORICAL-WITH-RESPECT-TO ?V2                   ; is categorical with respect to ?V2 and

  ?V1 IS      IRIS-PETAL-LENGTH                       ; ?V1 is IRIS-PETAL-LENGTH
    HAS-VALUES-FOR ?H1  ?H2                           ; that has values for ?H1 and ?H2 and

  ?V2 IS      IRIS-SPECIES-NAME                       ; ?V2 is IRIS-SPECIES-NAME and

  ?H1 IS      SETOSA                                  ; ?H1 is SETOSA and

  ?H2 IS      VERGINICA                               ; ?H2 is VERGINICA

THEN                                                  ; THEN
  RELEVANT-QUESTION  IS-THE-QUESTION  ?Q1             ; a relevant question is:
  ?Q1 IS   IS-STATISTICAL-DIFFERENCE-1                ; "Is there a statistical difference
    BETWEEN  ?V1                                      ; between ?V1
    FOR      ?H1                                      ; for ?H1
    AND      ?H2                                      ; and ?H2 ?"
```

Figure 7.14 The initial PVS rule learned.

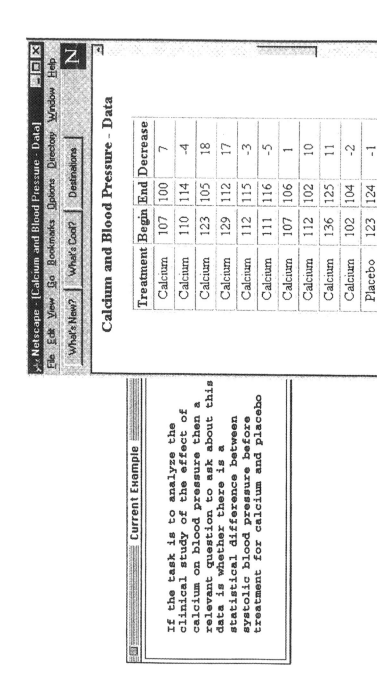

Figure 7.15 An example generated by the agent.

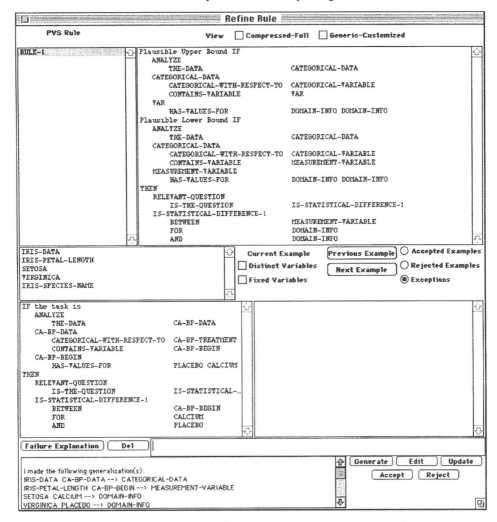

Figure 7.16   Generalization of the rule to cover the generated example.

the concept **VAR** from the plausible upper bound to **MEASUREMENT-VARIABLE** which no longer covers **TB-YEAR**, but is still more general than or as general as the concept from the plausible lower bound (**MEASUREMENT-VARIABLE**). At this point the plausible upper bound has become identical with the plausible lower bound and the agent has learned an exact rule, shown in the upper right pane of Figure 7.18.

Figure 7.19 shows the final rule learned with the patterns that the agent has automatically associated with it, as explained in Chapter 6.

In a similar way the agent is taught to generate irrelevant questions, i.e. questions that are not relevant to the data set, for the purpose of assessment. The generation of irrelevant questions is a significant aspect of our approach, since these questions are generated using the same techniques as the correct questions, and are designed to 'make sense'. These irrelevant questions are based on pedagogical experience, reflecting the mistakes students

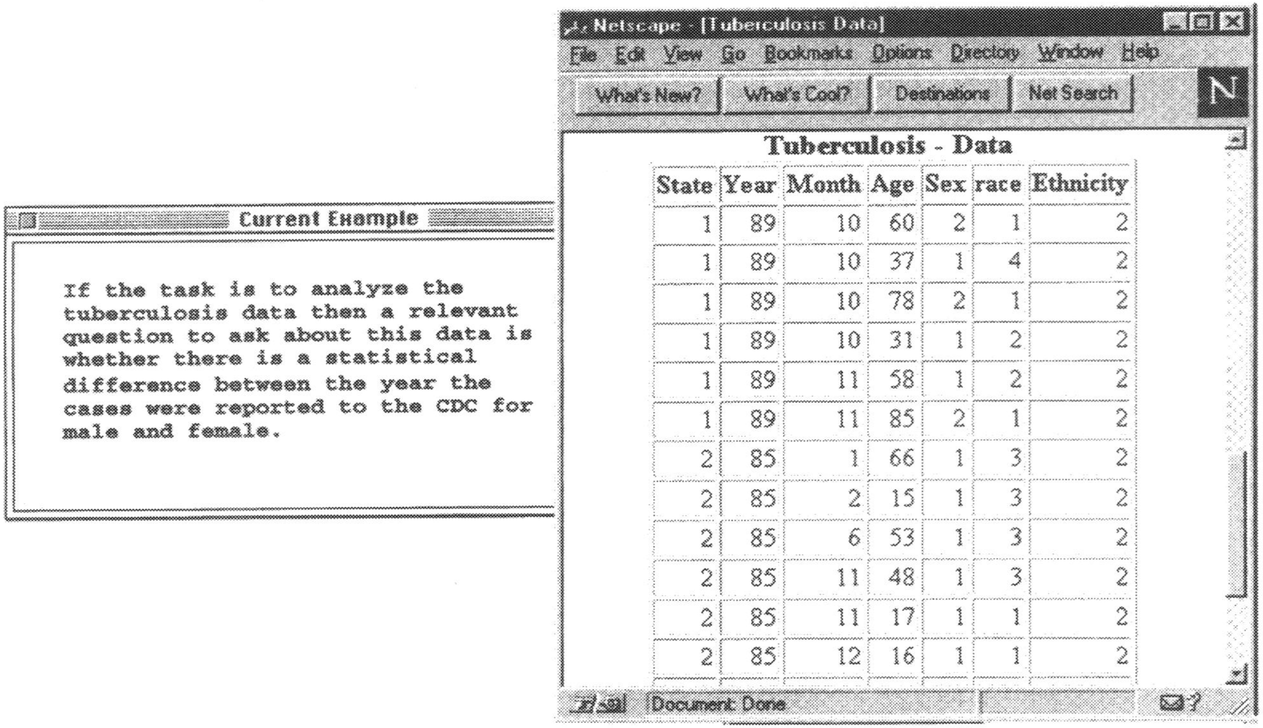
Current Example
If the task is to analyze the tuberculosis data then a relevant question to ask about this data is whether there is a statistical difference between the year the cases were reported to the CDC for male and female.
Netscape - [Tuberculosis Data]
File Edit View Go Bookmarks Options Directory Window Help
What's New? What's Cool? Destinations Net Search
Tuberculosis - Data
State Year Month Age Sex race Ethnicity
1 89 10 60 2 1 2
1 89 10 37 1 4 2
1 89 10 78 2 1 2
1 89 10 31 1 2 2
1 89 11 58 1 2 2
1 89 11 85 2 1 2
2 85 1 66 1 3 2
2 85 2 15 1 3 2
2 85 6 53 1 3 2
2 85 11 48 1 3 2
2 85 11 17 1 1 2
2 85 12 16 1 1 2
Document: Done

Figure 7.17 Another example generated by the agent.

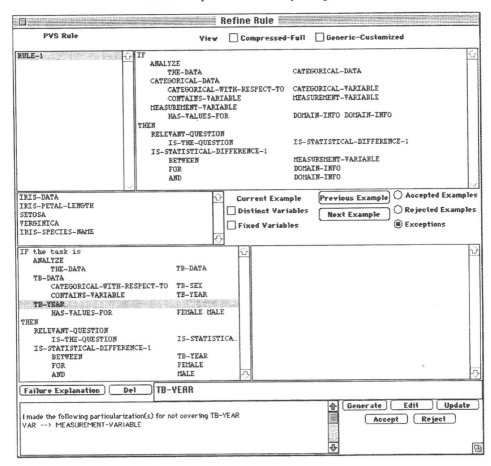

Figure 7.18   Specialization of the rule to uncover a negative example.

might make, and/or reflecting the subtle points of a statistical technique. Notice that this is different from the case study of the history assessment agent. In that case the agent had only rules for determining whether a source was relevant to a task and was concluding that the source was not relevant when no relevancy rule for that task applied. This was possible in that case because any source that was not relevant to that task was irrelevant. For this agent, however, it is not the case that any question that is not a 'relevant question' can be considered a good 'irrelevant question', as has been mentioned above.

7.6   Final remarks

Rapid technology-driven changes in subject matter, both in school and on the job, have created a need for educational systems that are rapidly produced and easily updated and yet retain the high pedagogical quality of the best intelligent tutoring systems and interactive

```
IF      ?W1   IS    ANALYZE
              THE-DATA ?S1

        ?S1   IS    CATEGORICAL-DATA
              CONTAINS-VARIABLE ?V1,
              CATEGORICAL-WITH-RESPECT-TO ?V2

        ?V1   MEASUREMENT-VARIABLE
              HAS-VALUES-FOR ?H1 ?H2

        ?V2   CATEGORICAL-VARIABLE

        ?H1   DOMAIN-INFO

        ?H2   DOMAIN-INFO

THEN
        RELEVANT-QUESTION IS-THE-QUESTION ?Q1

        ?Q1   IS-STATISTICAL-DIFFERENCE-1,
              BETWEEN ?V1
              FOR ?H1
              AND ?H2

Task Description
        the task is to analyze the ?S1

Operator Description
        a relevant question to ask about this data is whether there is a
        statistical difference between ?V1 for ?H1 and ?H2

Explanation Pieces
        ?S1 contains the variable ?V1
        ?S1 is categorical with respect to ?V2
        ?V1 has values for ?H1
        ?V1 has values for ?H2
```

Figure 7.19 The learned rule.

learning environments (Khuwaja *et al.*, 1996). This case study and the one presented in the previous chapter show that the Disciple learning agent shell is able to provide precisely such a capability by significantly facilitating the process of building intelligent educational agents to the point where such agents can be built by educators themselves, with little or no assistance from a knowledge engineer.

The statistical analysis assessment and support agent is used as a multipurpose assistant to the student, either to test the student or to provide expertise and guidance to the student. It acts as a one-on-one teacher/tutor, using the same processes as the ones by which it was taught by the expert. The agent therefore can replicate the usual teacher–student or mentor–student interactions. As in the case of the history assessment agent, the expert instructs the agent what kind of test questions to generate, and then the agent generates a wide variety of test questions of that kind. Therefore, the agent provides the educator with a very helpful tool for generating personalized tests for large classes, tests that do not repeat themselves and take into account the instruction received by each student. Moreover, the ability of such agents to generate hints and explanations allows the student to be guided by the agent during the learning process.

The statistical analysis assessment and support agent also fits very closely with a traditional statistics or biology course, and can be scaled to take more and more 'responsibilities', as mentioned throughout the chapter. For example, the agent can be expanded to assess final stages of the statistical analysis (computations and analysis of the results). It can also be expanded to incorporate biological issues with the statistical ones (e.g., in connection with environmental issues or evolutionary processes). The agent can also be expanded to support students' development of projects and presentations in a way similar to the manufacturing agent presented in the previous chapters and the design assistant presented in the next chapter. So the agent is scalable, but in the sense of the conceptual content it can handle. Moreover, this scalability of the agent can be handled by the expert him/herself with little help from the knowledge engineer.

The statistical analysis assessment and support agent was developed by Philippe Loustaunau (who was also the domain expert), Gheorghe Tecuci, Harry Keeling and Kathryn Wright, for the course 'The Natural World'. This course is taken each year by 250 students of the GMU's New Century College. This web-based course was developed by Luther Brown, Jonathan Hughes, and Philippe Loustaunau. This research was supported by the NSF grant No. CDA-9616478, as part of the program Collaborative Research on Learning Technologies, directed by Caroline Wardle. Partial support was also provided by the DARPA contract N66001-95-D-8653, as part of the Computer-Aided Education and Training Initiative, directed by Kirstie Bellman.

8
Case Study: Design Assistant for Configuring Computer Systems

This chapter presents the development of a design assistant for configuring computer systems. The main feature of this agent is that it is under constant supervision of the user who is both the developer of the agent and the beneficiary of its services. The agent is continuously supervised and customized by the user according to the changing practices in the user's domain, as well as the needs and the preferences of the user.

8.1   Disciple methodology applied to engineering domains

The growing complexity of contemporary engineering designs requires design tools that can play a more active role over the whole design process (Sriram and Tong, 1991). One way to make the design tools more active is to have them behave as assistants and collaborators in the design process. Figure 8.1 presents the cooperation between the designer and his/her computer-based assistant. The human designer and the assistant create designs together. The level of interaction between them depends on the quantity and quality of the assistant's knowledge. Initially, the assistant will behave as a novice that does not contribute significantly to the design process. In this case the designs are created by the designer who is also teaching the agent. Gradually, the assistant learns to design, becoming a more useful assistant. This process constantly increases the capabilities of the design team consisting of the human designer and the computer assistant (Mitchell *et al.*, 1985; Tecuci and Kodratoff, 1990; Dybala, 1996).

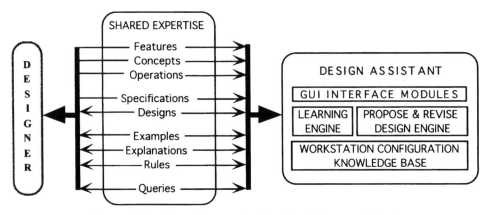

Figure 8.1   Teaching and using the design assistant.

Engineering design is a process by which products or systems are created to perform desired functions (Coyne *et al.*, 1990). The configuration problem, so often encountered in contemporary engineering design, is defined as the problem of designing a product or a system from ready-to-use components. The result of the design activity is a design description. The purpose of such a description is to represent sufficient information about the artifact so that it can be constructed. There seems to be a general agreement on the classification of designs into routine, innovative, and creative (Gero, 1990).

The process of designing configurations of computer systems (McDermott, 1982) is an example of a configuration problem. Contemporary computing devices are highly customized with respect to the needed memory, computing power, disk space, input/output devices, etc. The space of all possible configurations is very large. The knowledge needed to make such configurations is mainly heuristic, but it also includes procedures to calculate quantitative design parameters. Most of this knowledge is embedded only in the minds of the human designers. Therefore the design assistant has not only to support a human designer in the configuration process but also to build a model of his/her design knowledge.

An input to the configuration process is a functional specification of an artifact to be designed. This specification characterizes the functional units of the product and their features. For instance, Figure 8.2 shows the functional specification of a computer workstation that needs to be configured for a user who requires moderate power for text processing, moderate graphics and communication capabilities.

The main task of the design process is to convert the input functional specification into an operational description of the designed product. This describes the configuration of the final product in terms of partial configurations, connections among them, and lists of components used to create these configurations. For instance, the functional requirements of the product described in Figure 8.2 are satisfied by a Macintosh workstation, **QUADRA-840AV**, in the configuration described in Figure 8.3. The configuration is described in terms of the components used and their quantities.

There is an inherent order in which designers approach a configuration problem. Typically, they start with an analysis of the functional specifications of a product, acquire new knowledge if necessary, synthesize one or more potential designs, choose one of them, and then evaluate the design constraints based on the original problem specification. If the configuration is not satisfactory, the entire process is repeated.

The cooperation between the designer and the assistant is carried out during these phases of the design process. The analysis is made by the assistant but the designer serves as a source of the needed information. The synthesis is made by the assistant or by the designer, depending on the problem's complexity. The evaluation is made by the designer.

We will illustrate the learning capabilities of the Disciple-based design assistant by using an example of memory configuration which is a part of the computer workstation configuration process. Table 8.1 illustrates the heuristic knowledge used by a system integrator to configure the memory of a computer workstation. Section 8.4 shows how the design assistant is able to learn such heuristic knowledge.

The configuration of a workstation's memory is an illustrative example of the propose-and-revise design method (Marcus and McDermott, 1989). According to this method the design synthesis is carried out through a sequence of propositions and revisions of the values of the design parameters. A value for each design parameter is proposed, based on the input functional specification and the values of other design parameters. Once proposed, the

```
?C01  IS       CONFIGURE              ; the task is to configure
      PRODUCT  ?C02                   ; a computer workstation for a moderate user
      FOR      ?M01                   ; who will  perform text processing,
?C02  IS       COMPUTER-WORKSTATION   ; moderate graphics and network communication
?M01  IS       MODERATE-USER
      ACTIVITY TEXT-PROCESSING  MODERATE-GRAPHICS  COMMUNICATION
```

Figure 8.2 Functional specification of a workstation.

```
?C02  IS           QUADRA-840AV          ; the configured product is a  QUADRA 840AV
      CAT-NUMBER   M6710LLA              ; with catalogue  number M6710LLA,
      MEMORY       ?S01                  ; it is composed of memory,
      HARD-DISK    ?S02                  ; a hard disk,
      MONITOR      ?A01                  ; and a monitor.
      QUANTITY     1                     ; one workstation is needed.

?S01  IS           SIMM-8MB-60NS-32B-72P
      CAT-NUMBER   11027                 ; the memory is configured of 4 SIMM modules
      CAPACITY     8-MB                  ; of 8 MB each
      QUANTITY     4                     ; with catalogue number 11027.

S02   IS           SCSI-DISK             ; the SCSI disk is an internal
      CAT-NUMBER   INTERNAL-COMPONENT    ; preinstalled component
      CAPACITY     525-MB                ; with capacity  525 MB
      ACCESS-TIME  13MS                  ; and access time 13 ms.
      QUANTITY     1                     ; one SCSI  disk is needed.

?A01  IS           APPLE-17-MONITOR      ; the monitor is APPLE display
      CAT-NUMBER   M2612LLA              ; with catalogue number M2612LLA
      SIZE         17"                   ; and size 17".
      QUALITY      MODERATE              ; it is a moderate quality monitor.
      QUANTITY     1                     ; one monitor is needed.
```

Figure 8.3 Operational description of the workstation specified in Figure 8.2.

Table 8.1   Heuristic knowledge used to configure the memory of a workstation

- The random access memory (RAM) of contemporary computer workstations is composed of memory banks.
- A memory bank contains a certain number of RAM slots which can be populated with memory modules.
- Memory modules can be plugged into memory slots that have the same number of pins.
- The memory module can be either single interface memory module (SIMM) or dual in-line memory module (DIMM).
- The workstation can accept either SIMMs or DIMMs, but not both.
- The memory modules placed in one memory bank should be of the same type (access time, word length) and capacity.
- Memory modules in different banks can be of different types or capacity.
- The total number of the memory modules placed in the RAM slots of all the memory banks give the required memory capacity which should be within a required range.

value is checked against the design constraints and possibly revised to maintain design consistency.

The propose-and-revise design method creates a validated design extension in two steps:

- *Propose design extension* – a value for a design parameter is either determined by evaluating a procedure which calculates the value, or the value is a component name selected from a set of available components;
- *Revise design extension* – additional constraints on the proposed value are checked. If constraints are violated, the design extension proposed is retracted and revisions to it and/or to other parameters are made. The crucial information to be included in the revision description is the specification of the parameter to change, how to change it and some idea of the expert's preference for this revision over others that may be tried.

For example, the memory of a workstation is configured as follows:

- Propose memory modules that fit into the workstation's memory slots.
- Calculate the resulted capacities of the memory banks and the total memory capacity.
- Check constraints related to the minimum and maximum required memory, and propose memory modules with different capacities if the constraints are violated.

8.2   Defining the initial knowledge base of the design assistant

To transform a functional specification into an operational description, the assistant needs to have two types of knowledge: descriptions of components and heuristic design rules. The components are described as objects with features and are hierarchically organized in abstract classes. This hierarchical organization of objects provides the generalization language for learning. For example, the description of the Macintosh workstation **QUADRA–840AV** and its relative place in the hierarchy is shown in Figure 8.4.

The heuristic rules are PVS rules that describe the conditions under which the steps of the propose-and-revise design method are performed. The 'propose' rules usually indicate

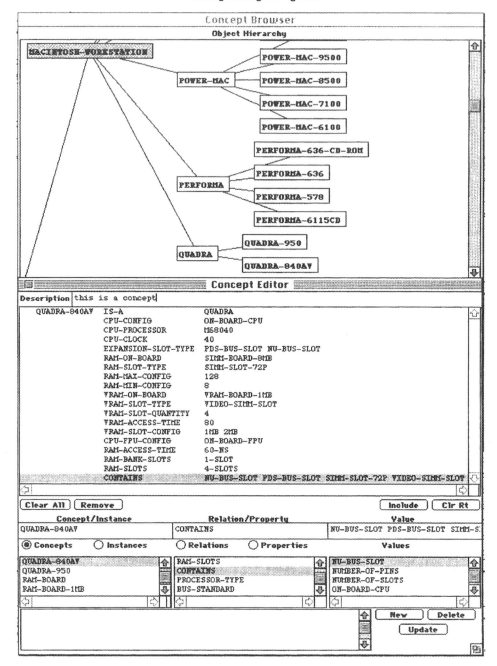

Figure 8.4   Semantic network of engineering concepts.

the applicability conditions for procedures which calculate the values of some design parameters. A typical procedure retrieves information about relevant decisions made and some initial conditions and uses mathematical expressions to calculate a parameter value. Another type of 'propose' rule is a rule that indicates the applicability conditions for

```
Plausible Upper Bound IF
    ?R34                    IS                      REVISE-DESIGN-EXTENSION
                            ARTIFACT-DESIGN         ?M36
                            MAX-MEM-LIMIT           ?X43
                            MM-IN-BANK1             ?S37
                            BANK-1-CAPACITY         ?X44

    ?X43                    IS                      MB-CAPACITY

    ?X44                    IS                      MB-CAPACITY
                            GREATER-THAN            ?X43

    ?M36                    IS                      COMPUTER
                            CONTAINS                ?S64

    ?S37                    IS                      MEMORY-MODULE
                            PLUGS-INTO              ?S64
                            MEMORY-CAPACITY         ?X62

    ?S41                    IS                      MEMORY-MODULE
                            PLUGS-INTO              ?S64
                            MEMORY-CAPACITY         ?X63

    ?X63                    IS                      MB-CAPACITY
                            LESS-THAN               ?X62

    ?X62                    IS                      MB-CAPACITY

    ?S64                    IS                      EXPANSION-SLOT

Plausible Lower Bound IF
    ?R34                    IS                      REVISE-DESIGN-EXTENSION
                            ARTIFACT-DESIGN         ?M36
                            MAX-MEM-LIMIT           ?X43
                            MM-IN-BANK1             ?S37
                            BANK-1-CAPACITY         ?X44

    ?X43                    IS                      32-MB

    ?X44                    IS                      64-MB
                            GREATER-THAN            ?X43

    ?M36                    IS                      WORKSTATION
                            CONTAINS                ?S64

    ?S37                    IS                      RAM-MODULE
                            PLUGS-INTO              ?S64
                            MEMORY-CAPACITY         ?X62

    ?S41                    IS                      RAM-MODULE
                            PLUGS-INTO              ?S64
                            MEMORY-CAPACITY         ?X63

    ?X63                    IS                      MB-CAPACITY
                            LESS-THAN               ?X62

    ?X62                    IS                      MB-CAPACITY

    ?S64                    IS                      EXPANSION-SLOT
THEN
    RETRACT                 PARAMETER               ?P38
                            ARTIFACT-DESIGN         ?M36

    ?P38                    IS                      PARAMETER-DESCRIPTION
                            MM-IN-BANK1             ?S37
                            BANK-1-CAPACITY         ?X44

    EXTEND-BY               PARAMETER               ?P39
                            ARTIFACT-DESIGN         ?M36

    ?P39                    IS                      PARAMETER-DESCRIPTION
                            MM-IN-BANK1             ?S41
```

Figure 8.5   A memory configuration rule.

selection of components that create partial or final configurations. The 'revise' rules are fired when values of the design parameters violate some design constraints or the initial specification. They retract values assigned to the design parameters and replace them with other values that improve design consistency.

An instance of a PVS rule with two conditions is shown in Figure 8.5. The rule's informal description is shown in Figure 8.6. The rule in Figure 8.5 is a 'revise design extension' rule. When fired, this rule makes revisions of the design parameters describing the memory module selected for the first memory bank, and the parameter describing the resulted capacity of the first memory bank.

Figure 8.7 shows an instance of the rule in Figure 8.5 that reads:

IF the task is

to revise the design extension of a Macintosh model **MAC-IIVX** that should have a maximum memory capacity of **32-MB**, and the design extension is described by the following parameters:

- the memory module selected for the first memory bank is **SIMM-16MB-80NS-8B-30P**,
- the capacity of the first memory bank is **64-MB**,

and

   MAC-IIVX contains a **SIMM-SLOT-30P**

and

   the **SIMM-16MB-80NS-8B-30P** memory module with capacity **16-MB** can be plugged into the **SIMM-SLOT-30P** memory slot.

THEN

retract the values of the following parameters of the **MAC-IIVX**:

- the memory module selected for the first memory bank (**MM-IN-BANK1**),
- the capacity of the first memory bank (**BANK-1-CAPACITY**),

and

   extend the design of **MAC-IIVX** Macintosh with the following parameter description:

- the memory module selected for the first memory bank (**MM-IN-BANK1**) is **SIMM-8MB-80NS-8B-30P**.

Because this solution is generated by using the plausible lower-bound condition of the rule, it is considered correct. However, if the problem were to revise the same parameters for a file server which cannot be classified as a workstation, but can be classified as a computer, then the plausible lower bound condition would no longer be satisfied. However, the plausible upper bound condition would still be satisfied and the solution indicated by the rule would be considered only plausible.

Such PVS rules are learned by the design assistant from the expert designer by applying the Disciple approach. During learning, the two bounds progressively converge toward the exact applicability condition of the rule. However, due to the incompleteness and the partial incorrectness of the assistant's knowledge, there is no guarantee that the two bounds will become identical. This feature allows Disciple agents to perform plausible inferences and to continuously improve their knowledge, as we will show in the remaining sections of this chapter. Moreover, this type of rule representation supports a natural cooperation between the expert and the assistant in design and learning. The use of PVS rules for

Plausible Upper Bound IF

the problem is to revise a design extension of a partially configured
computer (**?M36**) that should have a maximum memory capacity (**MAX-MEM-
LIMIT**) of the order of MB (**?X43**), and the design extension is described
by the following parameters:
- the memory module selected for the first memory bank (**MM-IN-BANK1**),
- the capacity of the first memory bank (**BANK-1-CAPACITY**),
and the capacity of the first memory bank (**?X44**) is greater than the
maximum memory capacity (**?X43**),
and the computer contains an expansion slot (**?S64**),
and there is a memory module (**?S37**) that can be plugged into the
expansion slot (**?S64**), and its capacity is of the order of MB (**?X62**),
and there is another memory module (**?S41**) that can be plugged into the
same expansion slot (**?S64**), and its capacity is of the order of MB
(**?X63**),
and the capacity of the second memory module (**?X63**) is less than the
capacity of the first memory module (**?X62**)

Plausible Lower Bound IF

the problem is to revise a design extension of a partially configured
workstation (**?M36**) that should have a maximum memory capacity (**MAX-MEM-
LIMIT**) of 32 MB (**?X43**), and the design extension is described by the
following parameters:
- the memory module selected for the first memory bank (**MM-IN-BANK1**),
- the capacity of the first memory bank (**BANK-1-CAPACITY**),
and the capacity of the first memory bank (**?X44**), which is 64 MB, is
greater than the maximum memory capacity (**?X43**), which is 32 MB,
and the workstation contains a RAM memory slot (**?S64**),
and there is a RAM memory module (**?S37**) that can be plugged into the
RAM memory slot (**?S64**), and its capacity is of the order of MB (**?X62**),
and there is another RAM memory module (**?S41**) that can be plugged into
the same RAM memory slot (**?S64**), and its capacity is of the order of
MB (**?X63**),
and the capacity of the second memory module (**?X63**) is less than the
capacity of the first memory module (**?X62**)

THEN
retract the values of the following parameters of the designed
artifact (**?M36**):
- the memory module selected for the first memory bank (**MM-IN-BANK1**),
- the capacity of the first memory bank (**BANK-1-CAPACITY**),
and
extend the designed artifact (**?M36**) with the following parameter
description:
- the memory module selected for the first memory bank (**MM-IN-BANK1**)
  has a value described by the variable **?S41**.

Figure 8.6   The informal description of the rule in Figure 8.5.

```
IF  the  task  is
    ?R34              IS                    REVISE-DESIGN-EXTENSION
                      ARTIFACT-DESIGN       ?M36
                      MAX-MEM-LIMIT         32-MB
                      MM-IN-BANK1           ?S37
                      BANK-1-CAPACITY       64-MB
    64-MB             GREATER-THAN          32-MB
    ?M36              IS                    MAC-IIVX
                      CONTAINS              ?S64
    ?S37              IS                    SIMM-16MB-80NS-8B-30P
                      PLUGS-INTO            ?S64
                      MEMORY-CAPACITY       16-MB
    ?S41              IS                    SIMM-8MB-80NS-8B-30P
                      PLUGS-INTO            ?S64
                      MEMORY-CAPACITY       8-MB
    8-MB              LESS-THAN             16-MB
    ?S64              IS                    SIMM-SLOT-30P
THEN
    RETRACT           PARAMETER             ?P38
                      ARTIFACT-DESIGN       ?M36
    ?P38              IS                    PARAMETER-DESCRIPTION
                      MM-IN-BANK1           ?S37
                      BANK-1-CAPACITY       64-MB
    EXTEND-BY         PARAMETER             ?P39
                      ARTIFACT-DESIGN       ?M36
    ?P39              IS                    PARAMETER-DESCRIPTION
                      MM-IN-BANK1           ?S41
```

Figure 8.7   An instance of the memory configuration rule in Figure 8.5.

interactive engineering design is controlled by the shared expertise method discussed in the next section.

8.3   The shared expertise method of problem solving and learning

The use of PVS rules allows the design assistant to use not only deductive reasoning but also plausible reasoning while comparing designs. This, in turn leads to the consideration of several design spaces as shown in Figure 8.8.

Let us consider, for instance, a design assistant with a complete and correct knowledge base. The set of designs which can be constructed using such a hypothetical knowledge base is called the *target design space* (TDS). An expert's design space is a good approximation of a TDS. The learning goal of our design assistant is to improve its knowledge base so as to approximate as accurately as possible a complete and correct knowledge base.

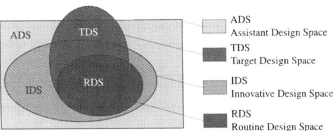

Figure 8.8   Types of design spaces.

The typical knowledge base of a design assistant is most likely incomplete and partially incorrect. All the design descriptions which could be constructed by using the assistant's representation language represent the *assistant design space* (ADS). As one could notice, some of the designs from the TDS cannot be represented in the agent's representation language. Therefore this language would need to be extended.

The *routine design space* (RDS) represents the set of designs which can be deductively constructed using the current knowledge base. Deductive derivations are made when exact rule conditions or plausible lower bound conditions are satisfied when the rules are applied. As indicated in Figure 8.9, some deductively constructed designs are wrong (those within the set $RDS - TDS$), because the current knowledge base is partially incorrect. Other correct designs are not deductively derivable (those within the set $IDS - RDS$), because the knowledge base is incomplete. Designs composed within the RDS are called *routine designs* ($D_R \in RDS$).

The *innovative design space* (IDS) represents the set of designs that can be composed from the knowledge base by using either deductive or plausible inferences. Plausible inferences are made by using the rules from knowledge base when only their plausible upper bound conditions are satisfied. Designs composed by making at least one plausible inference are called *innovative designs* ($D_I \in IDS - RDS$). The design assistant can create an innovative design based solely on the existing knowledge base and its plausible reasoning capabilities.

Designs composed outside the IDS but within the TDS are called *creative designs* ($D_C \in TDS - IDS$). A creative design is composed by the expert designer who introduces new object descriptions and design parameters and therefore extends the ADS. The heuristic and procedural knowledge needed for a creative design is not derivable from the existing knowledge base.

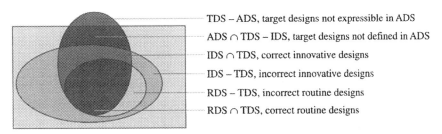

Figure 8.9   Design subspaces.

The set union $TDS \cup RDS \cup IDS$ represents the shared expertise space where both the human designer and the design assistant compose designs. The assistant proposes routine designs or innovative designs, and the designer composes creative designs. This gradation of designs helps them to exchange experience and expertise by mutual evaluations and explanations of the collaborative designs. We say that the designer and the assistant *share their expertise* in the interactive design process. Composing designs from imperfect knowledge creates opportunities to acquire new knowledge and to improve existing knowledge, as will be shown in Section 8.4.

The design and the learning processes as well as the communication between the designer and his Disciple-based assistant are controlled by the shared expertise method (Dybala *et al.*, 1996; Dybala, 1996) shown in Table 8.2.

After receiving design specifications the assistant analyzes them, and if some of the terms used are new, their definitions are elicited from the human designer. After the analysis phase, the design is synthesized by the assistant (in the case of a routine or an innovative design) or elicited from the expert (in the case of a creative design). During the synthesis phase the assistant may also apply various constraints and preference criteria that are part of its meta-knowledge to choose between competing designs. The evaluation of the proposed design extension is an opportunity for improving assistant's knowledge. There are two basic learning scenarios: learning from success, and learning from failure.

The behavior of the assistant is dependent upon the quality and the quantity of its knowledge base. Initially, the process of knowledge elicitation is dominant. Over time, knowledge

Table 8.2 An outline of the shared expertise method

---

repeat
    Analyze design specifications S_D
        if S_D contains new terms
            then Elicit new term definitions from the designer
        end
    Synthesize design: propose and revise design extension
        if S_D is covered by plausible lower bound condition
            then compose *routine design* D_R
            else if S_D is covered by plausible upper bound condition
                then compose *innovative design* D_I
                else S_D is not covered by any rule
                    Elicit *creative design* D_C from the designer
            end
        end
    Evaluate design
        if the design extension is accepted
            then Learn from success
                if D_I **then** Generalize plausible lower bound condition **end**
                if D_C **then** Learn new PVS rule **end**
            else Learn from Failure
                if D_R **then** Define exception **end**
                if D_I **then** Specialize plausible upper bound condition **end**
        end
    until satisfactory design

---

refinement occurs more and more often. Ultimately, the assistant's knowledge base becomes good enough so that knowledge acquisition is rarely needed. However, at any time the assistant is able to properly react to unknown input, and to learn from it.

8.4 Training and using the design assistant

The design assistant reacts to new information obtained from the human expert with the goal of extending, updating, and improving the knowledge base to integrate the new input. For instance, each design task and its correct solution (indicated by the expert or generated by the agent) is regarded by the agent as a positive example of a general heuristic design rule, and incorrect solutions proposed by the agent will be treated as negative examples. The agent learns and modifies PVS rules based on the examples of the design episodes that it encounters and the explanations received from the designer. These modifications cause an evolution of the RDS and IDS of the assistant toward the TDS. The following subsections illustrate the different learning processes invoked during design and learning.

8.4.1 Learning a heuristic design rule from a new creative design

The creative design tasks belong to the set difference $TDS - IDS$ (see Figures 8.8 and 8.9) which represents the set of the designs that cannot be derived by the assistant even by performing plausible inferences. Solutions to these tasks are provided by the expert and recorded by the assistant. Each new creative design task is an opportunity for the assistant to learn a new PVS rule. Adding this rule to the knowledge base enlarges both the RDS and IDS.

Let us assume that during the task analysis phase the assistant needed to solve the task shown in the top left pane of Figure 8.10. This is the task of proposing a design extension for a Macintosh **QUADRA-840AV** that should have a maximum memory of 32 MB. However, the assistant was not able to solve this task even by applying its plausible reasoning capabilities. In such a case, the assistant treats this as a creative design task and elicits the solution from the designer. The designer chooses to extend the description of **QUADRA-840AV** with **SIMM-8MB-60NS-32B-72P**, which is the memory module selected for the first memory bank. The problem and solution described by the designer become an initial example for a rule-learning session. When the expert specifies a new creative design like the one in Figure 8.10, he must also explain it to the assistant. The explanation is expressed in terms of the properties and relations between the objects from the creative design. There are various techniques to facilitate the process of defining these explanations (see Chapter 4). For instance, the assistant can search the semantic network representing the objects from the creative design, proposing their relationships or properties as plausible explanations. The expert has to choose from them the relevant ones, and can also define additional explanations using the explanation grapher.

For the given example of creative design, several explanations are proposed by the assistant (see the bottom of Figure 8.10). Two of them are accepted by the expert as relevant (those marked by '*'):

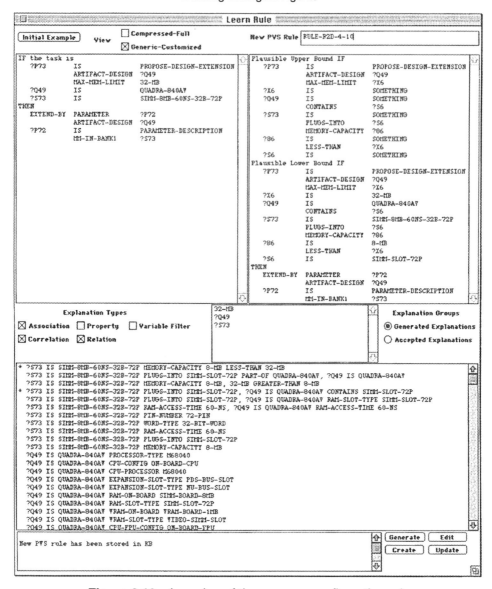

Figure 8.10   Learning of the memory configuration rule.

- The selected memory module **SIMM-8MB-60NS-32B-72P** has capacity **8-MB** which is less than **32-MB**, the required maximum memory limit.
- The selected memory module **SIMM-8MB-60NS-32B-72P** can be plugged into the **SIMM-SLOT-72P** slot with 72-pin layout, and the workstation **QUADRA-840AV** contains the same type of memory expansion slot.

The PVS rule learned from the creative design and these explanations is shown in the top right pane of Figure 8.10. The plausible lower bound of this rule is simply a reformulation of the object descriptions from the creative design, in terms of rule variables. For instance, the following component of the lower bound (see the top right pane of Figure 8.10)

?Q49   **IS**   **QUADRA-840AV**

            CONTAINS   **?S6**

?S6   **IS**   **SIMM-SLOT-72P**

indicates that the variable **?Q49** can only take the value **QUADRA-840AV**, and the variable **?S6** can only take the value **SIMM-SLOT-72P**. Therefore, the plausible lower bound can only match the current design specification.

The plausible upper bound is an analogy-based generalization of the plausible lower bound obtained by applying the method in Table 4.4. For instance, the above component of the plausible lower bound was generalized to

?Q49   **IS**   **SOMETHING**

            CONTAINS   **?S6**

?S6   **IS**   **SOMETHING**

The meaning of the above condition is that the variable **?Q49** can be any object that contains another object.

The purpose of the plausible upper bound is to allow the agent to propose innovative designs for specifications that are similar to the current one. Examples of such cases are presented in the following subsections.

8.4.2   Extending the RDS

The rule from the right-hand side of Figure 8.10 is later applied to generate designs that are analogous with the creative design from the top left pane of Figure 8.10. These designs are accepted or rejected by the designer. For instance, the assistant will apply this rule to the new task specified in the bottom left pane of Figure 8.11 because the plausible upper bound of this rule is satisfied. The corresponding design is an innovative design.

The problem is to propose a design extension for another Macintosh workstation, **MAC-IIVX**, which should have a maximum memory of **8-MB**, and contains the memory slot **SIMM-SLOT-30P**. The extension proposed by the assistant is a memory module for the first memory bank. This is the **SIMM-4MB-80NS-8B-30P** module with **4-MB** capacity. This solution proposed by the assistant is correct and is therefore accepted by the designer. The capacity of the selected module is less than the maximum memory limit, and the module can be plugged into the 30-pin **SIMM** slot that is part of the **MAC-IIVX** mother board.

In general, if an innovative design proposed by the agent is accepted by the designer, then the plausible lower bound of the rule is generalized as little as possible so as to cover the current design and to remain less general than the plausible upper bound. These generalizations (indicated in the bottom part of the window in Figure 8.11) are made by climbing generalization hierarchies similar to the one shown in Figure 8.4. The new PVS rule, with a generalized plausible lower bound condition, is shown in the top part of Figure 8.11. As a result of generalizing the rule's plausible lower bound the RDS was extended to cover more

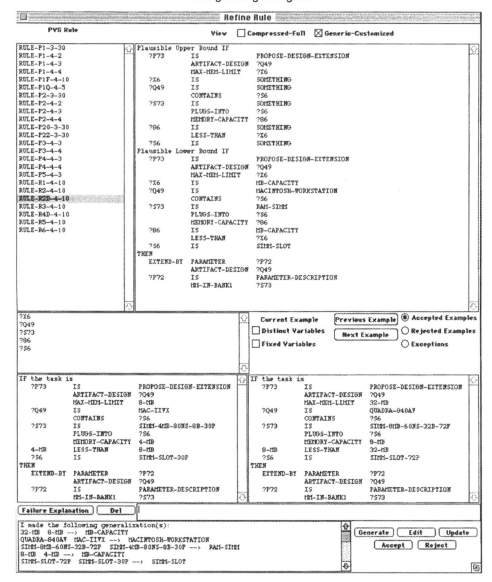

Figure 8.11   Memory configuration rule refinement — accepted example.

of the intersection *IDS* ∩ *TDS* (see Figure 8.9). As a result of this learning process, designs that used to be innovative have become routine designs.

8.4.3   Improving the IDS

Let us now consider another configuration problem faced by the assistant. This is the problem of proposing a design extension for a video controller board called **DIAMOND-STEALTH-64-ID**, as shown in Figure 8.12. The rule from Figure 8.11 has been refined,

```
IF  the  task  is
     ?P73              IS                    PROPOSE-DESIGN-EXTENSION
                       ARTIFACT-DESIGN       ?Q49
                       MAX-MEM-LIMIT         730-MB
     ?Q49              IS                    DIAMOND-STEALTH-64-ID
                       CONTAINS              ?S6
     ?S73              IS                    VDIMM-1000K-70NS
                       PLUGS-INTO            ?S6
                       MEMORY-CAPACITY       1-MB
     1-MB              LESS-THAN             730-MB
     ?S6               IS                    VIDEO-DIMM-SLOT
THEN
     EXTEND-BY         PARAMETER             ?P72
                       ARTIFACT-DESIGN       ?Q49
     ?P72              IS                    PARAMETER-DESCRIPTION
                       MM-IN-BANK1           ?S73
```

Figure 8.12   Rejected example of a memory configuration rule.

based on learning from other configuration problems and then has been applied to the problem shown in Figure 8.12. The assistant applied this PVS rule because its plausible upper bound was satisfied. Consequently, it proposed to plug a video memory module into the video memory slot of the controller board. However, this innovative solution is rejected by the designer who explains that the memory module selection rule can be applied to computers, but not to electronic control devices such as the video controller boards.

In this case, the assistant will use the explanation provided by the expert and will specialize the plausible upper bound of the rule to no longer cover this situation and to remain more general than the plausible lower bound. Because the expert pointed to the wrong object from the design (**?Q49 IS DIAMOND-STEALTH-64-ID**) the agent needs only to specialize the corresponding concept from the plausible upper bound (**?Q49 IS SOMETHING**) to no longer cover **DIAMOND-STEALTH-64-ID** while still covering the corresponding concept from the plausible lower bound (**?Q49 IS COMPUTER**). As a result of this specialization the value of the variable **?Q49** is constrained to be a computer. This updated rule is shown in Figure 8.13.

A rejected innovative design falls into the set $IDS - TDS$ (see Figure 8.9). The described process of specializing the plausible upper bound of a PVS rule will cause the IDS to be specialized, removing wrong designs from it.

8.4.4   Handling exceptions

When the assistant proposes a routine design which is rejected by the designer, the corresponding problem specification and solution description are defined in the knowledge base as an exception. For instance, if the assistant faced a problem of selecting the type of a memory module for an old XT model of a PC, it would apply the rule from Figure 8.13 because the rule's plausible lower bound condition is satisfied (PC-XT, one of the first versions of IBM-PC compatible computers, can be classified as a computer workstation).

```
Plausible  Upper  Bound  IF
    ?P73              IS               PROPOSE-DESIGN-EXTENSION
                      ARTIFACT-DESIGN  ?Q49
                      MAX-MEM-LIMIT    ?X6
    ?X6               IS               MB-CAPACITY
    ?Q49              IS               COMPUTER
                      CONTAINS         ?S6
    ?S73              IS               MEMORY-MODULE
                      PLUGS-INTO       ?S6
                      MEMORY-CAPACITY  ?86
    ?86               IS               SOMETHING
                      LESS-THEN        ?X6
    ?S6               IS               SOMETHING
Plausible  Lower  Bound  IF
    ?P73              IS               PROPOSE-DESIGN-EXTENSION
                      ARTIFACT-DESIGN  ?Q49
                      MAX-MEM-LIMIT    ?X6
    ?X6               IS               MB-CAPACITY
    ?Q49              IS               COMPUTER
                      CONTAINS         ?S6
    ?S73              IS               MEMORY-MODULE
                      PLUGS-INTO       ?S6
                      MEMORY-CAPACITY  ?86
    ?S86              IS               SOMETHING
                      LESS-THEN        ?X6
    ?S6               IS               SOMETHING
THEN
    EXTEND-BY         PARAMETER        ?P72
                      ARTIFACT-DESIGN  ?Q49
    ?P72              IS               PARAMETER-DESCRIPTION
                      MM-IN-BANK1      ?S73
```

Figure 8.13   The refined memory configuration rule.

However, the rule would not produce a correct solution because the memory of PC-XT was not configured from memory modules. Instead it used to be configured from memory chips plugged directly onto the mother board. In such a case, the plausible lower bound condition cannot be specialized to uncover this negative example because such a specialization will cause the uncovering of some positive examples of the rule. The described exception falls into the set difference $RDS - TDS$ (see Figure 8.9). This is the set of designs that are deductively derivable from the knowledge base but are incorrect. This shows that there are errors in the set of facts and deductive rules in the knowledge base which will generate incorrect solutions to routine problems. This is one of the most dangerous situations in the assistant's behavior in which deductive inferences produce incorrect results. Such covered negative examples point to the incompleteness of the assistant knowledge, and are used to guide the elicitation of new concepts and features, by using the knowledge elicitation methods described in Chapter 4.

8.5 Final remarks

In this chapter we have presented the process of developing a design assistant and have illustrated it with examples related to the memory configuration subproblem of the workstation configuration problem. A more detailed description of this Disciple application is given in Dybala (1996). This application showed that the knowledge elicitation and learning methods developed are capable to efficiently operate in formally defined engineering domains. Moreover, the distinction between the knowledge formulation process and the knowledge refinement process proves to be very useful for domains where the expert's knowledge has to follow a rapidly changing technology.

A distinctive feature of the design assistant is the close integration of its problem-solving and learning methods, as well as its ability to not only learn from the expert but to also actively participate to the design process. As a consequence, the agent learns not only during teaching sessions but also during its normal operation when it encounters novel design problems.

Hadi Rezazad contributed to the development of the design assistant with his expert knowledge. Partial support for this research was provided by the School of Information Technology and Engineering of George Mason University.

9
Case Study: Virtual Agent for Distributed Interactive Simulations

This chapter presents the application of the Disciple approach to build automated agents for distributed interactive simulations of military exercises. We will first present the application domain. Then we will define the virtual agent to be built and will illustrate the main stages of the agent development process.

9.1 Distributed interactive simulations

Distributed interactive simulation (DIS) is a set of standards and supporting methodology for creating a synthetic environment for multi-agent interactions (Miller, 1994; Pullen, 1994). It enables human participants at various locations to enter a virtual world containing all the essential elements of their application domain. A DIS system consists of a number of simulators which use a computer network to exchange messages about their behavior in the virtual world being simulated. The data must be exchanged in real time, with level of fidelity sufficient to permit their acceptable representation by all the participating simulators. There are many potential applications of the DIS concept, among the most important being that of training individuals who need to interact with each other. For instance, training of emergency and fire-fighting personnel, air traffic controllers, customs officers, anti-drug or counter-terrorist agents, and even high-school student drivers are possible in DIS.

Military DIS applications provide a system for collective training of military forces that uses no fuel, causes no training injuries, and avoids much of the cost associated with travel to far-away training grounds. While virtual simulation will not completely replace field training, it will significantly reduce the total cost of training. An important class of DIS simulators are computer-generated forces (CGFs). These are software agents that emit real-time information equivalent to that produced by the 'real' simulators. Most experience with CGFs has been in the area of semi-automated forces (SAFORs) where a human commander directs the performance of a group of software agents representing simulated tanks, aircraft, etc. CGFs allow DIS to represent larger forces in order to improve the richness of the training experience for small numbers of human participants.

Modular Semi-Automated Forces, or ModSAF for short (Ceranowicz, 1994), is one of the most successful virtual and real-time military simulation environments. Its development was sponsored by the US Army Simulation, Training and Instrumentation Command. Figure 9.1 presents the overall architecture of a ModSAF-based distributed interactive simulation that creates a virtual world where live forces, vehicle simulators and synthetic forces (CGFs) interact. The left-hand side of Figure 9.1 represents a real M1A1 tank operated by a tank crew on the ground. Second from the left is an M1A1 tank simulator, also operated by a

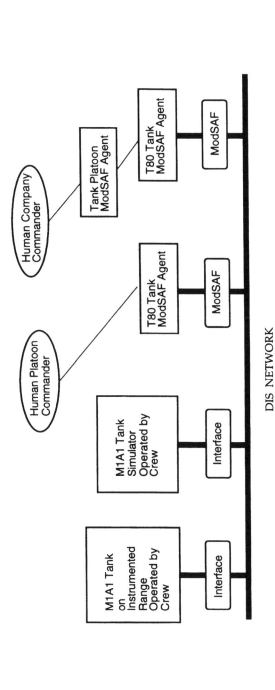

Figure 9.1 Interactions between instrumented live forces, vehicle simulators and semi-automated forces in DIS.

crew. Third from the left is a ModSAF software agent acting as a T80 tank. This agent receives orders from a human platoon commander. Finally, on the right is another semi-automated agent that also behaves as a T80 tank, but receives orders from another ModSAF agent behaving as a platoon commander. This agent, in turn, receives orders from a human company commander. Each human participant in this simulation has access to a computer display that shows the virtual environment consisting of a terrain view and the agents positioned in different places. Of course, each participant will observe as much of the situation as would be available to it in a real exercise. For instance, an agent that is well covered and concealed will not be observed by other agents. The representation of the environment is based on a common terrain database that is used by all the participants in the simulation. Each agent takes actions in this environment and these actions, as well as their effects, are represented in this environment. For instance, if the tank crew in the simulator fires at one of the ModSAF entities, DIS-compliant algorithms calculate if there is a hit, and the subsequent amount of damage, and modifies the behavior of the ModSAF entity accordingly. ModSAF is a very complex real-time application (consisting of over 300 source code libraries in C) which simulates military operations. It provides facilities for creating and controlling virtual agents (such as virtual tanks, planes, ships, or infantry) acting in a realistic representation of the physical environment that utilizes digital terrain databases. The ModSAF agents use a task-level architecture, similar to a robotic subsumption architecture (Brooks, 1989). In large-scale exercises, human participants may cooperate with, command or compete against virtual agents.

9.2   Developing a virtual armored company commander

In the following sections we will illustrate the process of building and training an agent to behave as a virtual company commander in a ModSAF-based distributed interactive simulation environment. This agent will receive orders to accomplish certain missions (such as to defend an area against an enemy attack, to move to another position, to attack the enemy, etc.) and will have to generate a plan of actions that will accomplish the mission. For instance, it could receive the mission of defending an area and should decide how to best position its platoons on the ground.

Figure 9.2 shows a simulation scenario corresponding to this mission where the company is defending a valley with its platoons positioned on hills on the two sides of a road. Two enemy companies are approaching the valley along the road. In the actual ModSAF simulation, all of the enemy forces are stopped in the valley, at a cost of approximately 25% of the company defending the area. ModSAF is a real-time, non-deterministic simulation, and there is always variability in the results. However, if the platoons of the defending company are not well placed, then they are invariably destroyed and do not stop the enemy.

Following the methodology described in Chapter 5, a military expert can build a Disciple agent that behaves as a military commander in the ModSAF environment. To illustrate this process, we will consider the case of an armored company commander that commands three or four platoons. To build this agent one needs to first define precisely the tasks to be accomplished by it, as presented in the next section.

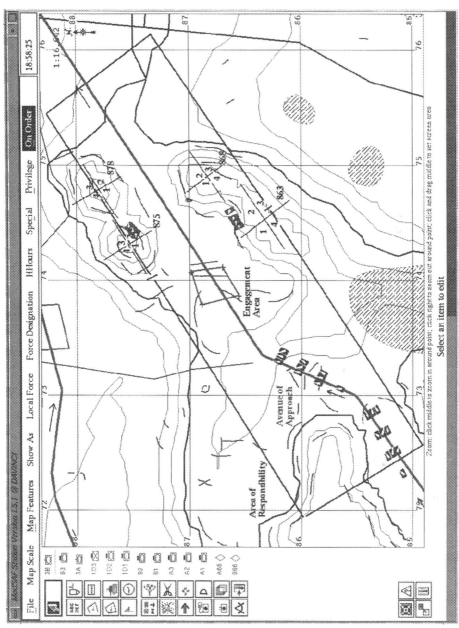

Figure 9.2 ModSAF defensive mission scenario.

9.3   Defining the tasks and the top-level ontology of the agent

As illustrated in the previous section, one task of the company commander is to position its company to defend an area against an enemy attack. To adequately perform this task, the agent has to take into account its area of responsibility (which is the area to defend), the avenue of approach (which is the direction and route from which the enemy is expected to approach), and the engagement area (which is the optimum spot on the avenue of approach for the units to coordinate their fire upon). These three features are drawn on an overlay to the terrain in Figure 9.2. Using these features, the task is represented as follows:

?P1	IS	POSITION-COMPANY	; the task is to position company
	UNIT-ID	?C1	; ?C1
	IN	?C2	; in area of responsibility ?C2
	TO-DESTROY-ENEMY-IN	?E1	; to destroy the enemy in area ?E1
	FOR	?D1	; as part of defensive mission ?D1
?C1	IS	COMPANY	
?C2	IS	COMPANY-AREA-OF-RESPONSIBILITY	
?E1	IS	ENGAGEMENT-AREA	
?D1	IS	DEFEND-AREA-MISSION	

The company's defensive mission requires determination of the best available terrain positions for its platoons, cover and concealment, coordination of fire, ability to retreat, and many other factors. Along with terrain selection, units must be selected to be placed upon the appropriate terrain, since different platoons have different capabilities. Ultimately, the positioning of the company reduces to the positioning of each of its platoons. The task of positioning a tank platoon **?P3** in the location **?L1** is represented as follows:

?P2	IS	POSITION-TANK-PLATOON
	UNIT-ID	?P3
	IN	?L1
?P3	IS	TANK-PLATOON
?L1	IS	LOCATION

Such an analysis of the agent's tasks leads to the identification of the top-level ontology of the agent, a fragment of which is presented in Figure 9.3. This fragment shows only some of the object concepts that form the basis of the semantic network of the agent.

9.4   Developing the semantic network

The next step in building the company commander agent is to populate and extend the initial semantic network. As can be seen from Figure 9.3, the semantic network includes concepts

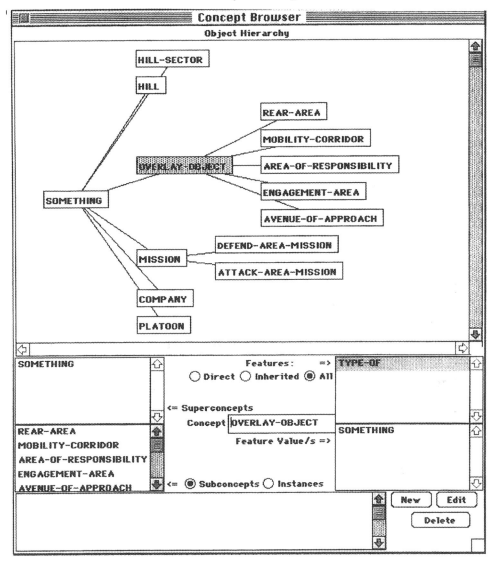

Figure 9.3 A fragment of the top-level ontology of the company commander.

corresponding to terrain, military units, and other military concepts relevant to the tasks which the agent is trained to perform. The semantic network represents information from the ModSAF terrain database at a conceptual level. For instance, Figure 9.4 shows a small portion of the area of Fort Knox, Kentucky, as it is represented in the ModSAF terrain database. The grid overlay represents kilometers. The area shown is about 3 km west to east and a little over 2 km north to south. The curving lines are contour lines, each of which represents an increase in elevation of 10 m.

The ModSAF terrain database contains thousands of objects in the area of interest, but does not contain the kind of objects relevant to the agent. For example, it does not explicitly

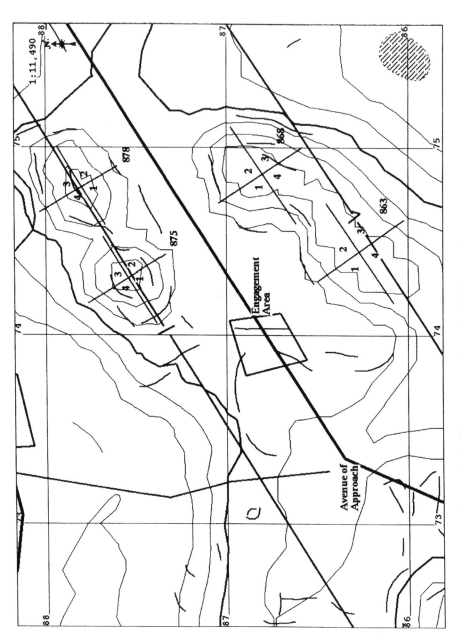

Figure 9.4 Terrain map from ModSAF graphical user interface.

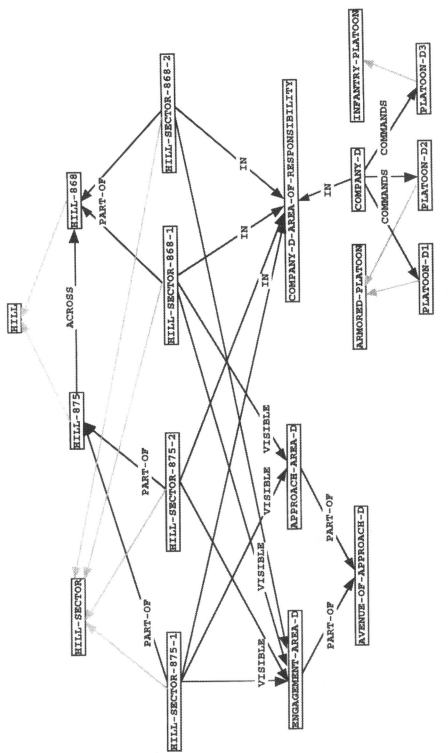

Figure 9.5 Semantic network.

contain objects such as hills, or hill features such as front slope relative to an engagement area, rear slope, or crest. This type of abstract concepts have to be represented in the agent's semantic network, and related to the specific regions of the map that they represent. For instance, the area in Figure 9.4 is represented as containing two hills in the north and two connected hills in the south. A portion of this semantic network is shown in Figure 9.5.

Some nodes in the semantic network represent map regions as objects such as **HILL-SECTOR-875-1** and **HILL-868**, while others represent general concepts like hill-sectors or hills. These concepts are organized in a generalization hierarchy (the gray links in Figure 9.5) which relates a concept to a more general concept. This hierarchy is very important not only for the inheritance of properties but also for learning. The objects are further described in terms of their features (such as **PART-OF**, **IN**, **VISIBLE**). The agent also maintains a correspondence between each concept in the semantic network (e.g., **HILL-SECTOR-875-1**) and the corresponding region on the map that it represents.

In the presented case study, the semantic network was developed to represent several armored companies, each containing three or four platoons. It also contained the symbolic representation of the terrain from the area of responsibility of each of these companies.

9.5   Teaching the agent how to accomplish a defensive mission

In this section we will illustrate how an expert teaches the virtual company commander where to place its units to defend an area from an enemy attack.

9.5.1   Giving the agent an example

The expert initiates the teaching session by showing the agent a specific example of a correct placement. The expert places the three platoons of **COMPANY-D** on the ModSAF map, in order to defend the company's area of responsibility, which is the area shown on the left-hand side of Figure 9.6. This is a screen shot of the ModSAF interface showing a scalable topographic map of the Fort Knox training ground. The expert uses the ModSAF interface as the expert normally would when placing units. An additional code library, libdisciple, was added to the ModSAF application to display examples generated by Disciple. A more detailed discussion of the integration of Disciple and ModSAF is given in Hieb (1996). The right-hand side of Figure 9.6 shows the textual representation of the example mission and also the solution. The textual representation is in terms of the symbolic concepts from the conceptual semantic network representing the map and the forces (see Figure 9.5). As mentioned above, through this network, the agent maintains a correspondence between each concept in the textual representation (e.g., **HILL-SECTOR-868-1**) and the corresponding object (region) on the map.

9.5.2   Helping the agent to understand why the positioning is correct

Next, an explanation-based learning phase is entered where the agent attempts to understand why the indicated solution is correct. As presented in Chapter 4, the expert guides the agent

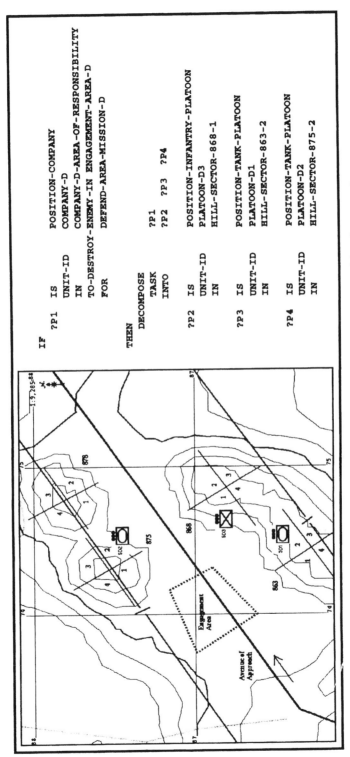

```
IF
  ?P1    IS                  POSITION-COMPANY
         UNIT-ID             COMPANY-D
         IN                  COMPANY-D-AREA-OF-RESPONSIBILITY
         TO-DESTROY-ENEMY-IN ENGAGEMENT-AREA-D
         FOR                 DEFEND-AREA-MISSION-D

THEN
  DECOMPOSE
         TASK                ?P1
         INTO                ?P2 ?P3 ?P4

  ?P2    IS                  POSITION-INFANTRY-PLATOON
         UNIT-ID             PLATOON-D3
         IN                  HILL-SECTOR-868-1

  ?P3    IS                  POSITION-TANK-PLATOON
         UNIT-ID             PLATOON-D1
         IN                  HILL-SECTOR-863-2

  ?P4    IS                  POSITION-TANK-PLATOON
         UNIT-ID             PLATOON-D2
         IN                  HILL-SECTOR-875-2
```

Figure 9.6   Initial placement of **COMPANY-D** indicated by the expert.

to generate plausible explanations from which the expert selects the relevant ones. This strategy is especially important in this domain because, without any guidance, the agent generates a large number of explanations at once, and choosing from them becomes a more difficult process.

One way in which the expert may guide the agent is to point to a relevant object in the example and to ask the agent to generate only plausible explanations involving that object. This is done by clicking on the *Variable Filter* button in the middle pane of the window in Figure 9.7, and then selecting the object from the list of objects. The selected object is **ENGAGEMENT-AREA-D**. The plausible explanations generated by the agent may further be restricted to be of a certain type (association, correlation, property or relation), as indicated in Chapter 4. In this case the agent is asked to generate only explanations that relate **ENGAGE-MENT-AREA-D** to other objects from the example in Figure 9.6. These are the explanations of type associations and are selected by clicking on the corresponding button in the middle pane of the window in Figure 9.7. After providing such guidance, the expert asks the agent to generate plausible explanations, by clicking on the *Generate* button. Figure 9.7 shows the explanations generated by the agent. The expert selects the relevant ones that become marked by an asterisk.

This session continues in this way with the expert pointing successively to other objects from the problem-solving episode (such as **COMPANY-D**, or **PLATOON-D3**) and the agent proposing corresponding plausible explanations from which the expert selects the relevant ones. All the plausible explanations selected are shown in Figure 9.8.

The explanations in Figure 9.8 state that the placement of **COMPANY-D** is correct because this is a mission to defend the company's area of responsibility (**DEFEND-AREA-MISSION-D WITH COMPANY-D** $\land$ **COMPANY-D HAS COMPANY-D-AREA-OF-RESPONSIBILITY**) and the platoons of **COMPANY-D** (**COMPANY-D COMMANDS PLATOON-D1** $\land$ **COMPANY-D COMMANDS PLATOON-D2** $\land$ **COMPANY-D COMMANDS PLATOON-D3**) are positioned on hill sectors of its area of responsibility (**HILL-SECTOR-868-1 IN COMPANY-D-AREA-OF-RESPONSIBILITY** $\land$ **HILL-SECTOR-875-2 IN COMPANY-D-AREA-OF-RESPONSIBILITY** $\land$ **HILL-SECTOR-863-2 IN COMANY-D-AREA-OF-RESPONSIBILITY**), hill sectors from where they can see the engagement area (**HILL-SECTOR-868-1 VISIBLE ENGAGEMENT-AREA-D** $\land$ **HILL-SECTOR-875-2 VISIBLE ENGAGEMENT-AREA-D** $\land$ **HILL-SECTOR-863-2 VISIBLE ENGAGEMENT-AREA-D**). Also, **PLATOON-D3** has weapons classified as light (**PLATOON-D3 WEAPON-CLASSIFICATION light**).

We stress the fact that while it is important to have some explanations of the initial example, there is no requirement that a complete set of explanations be specified. Indeed, the assumption is that this initial explanation set is incomplete and will be completed during experimentation. Consequently, the initial example from Figure 9.6 must always be accessible so that the agent can ask additional questions about this example.

The relevant explanations identified by the expert are used to generate an initial plausible version space (PVS) for a general placement rule to be learned. This PVS rule is indicated in Figure 9.9.

9.5.3 Supervising the agent as it generates other defensive placements

Once the initial PVS rule in Figure 9.9 has been defined, the agent enters an experimentation phase in which it generates other placements for similar defensive missions by analogy

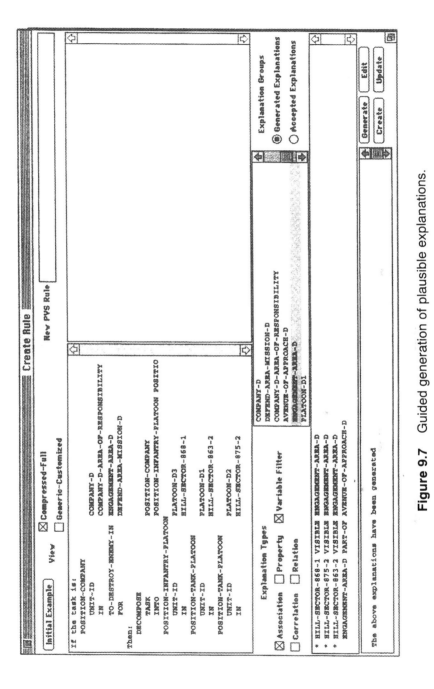

Figure 9.7 Guided generation of plausible explanations.

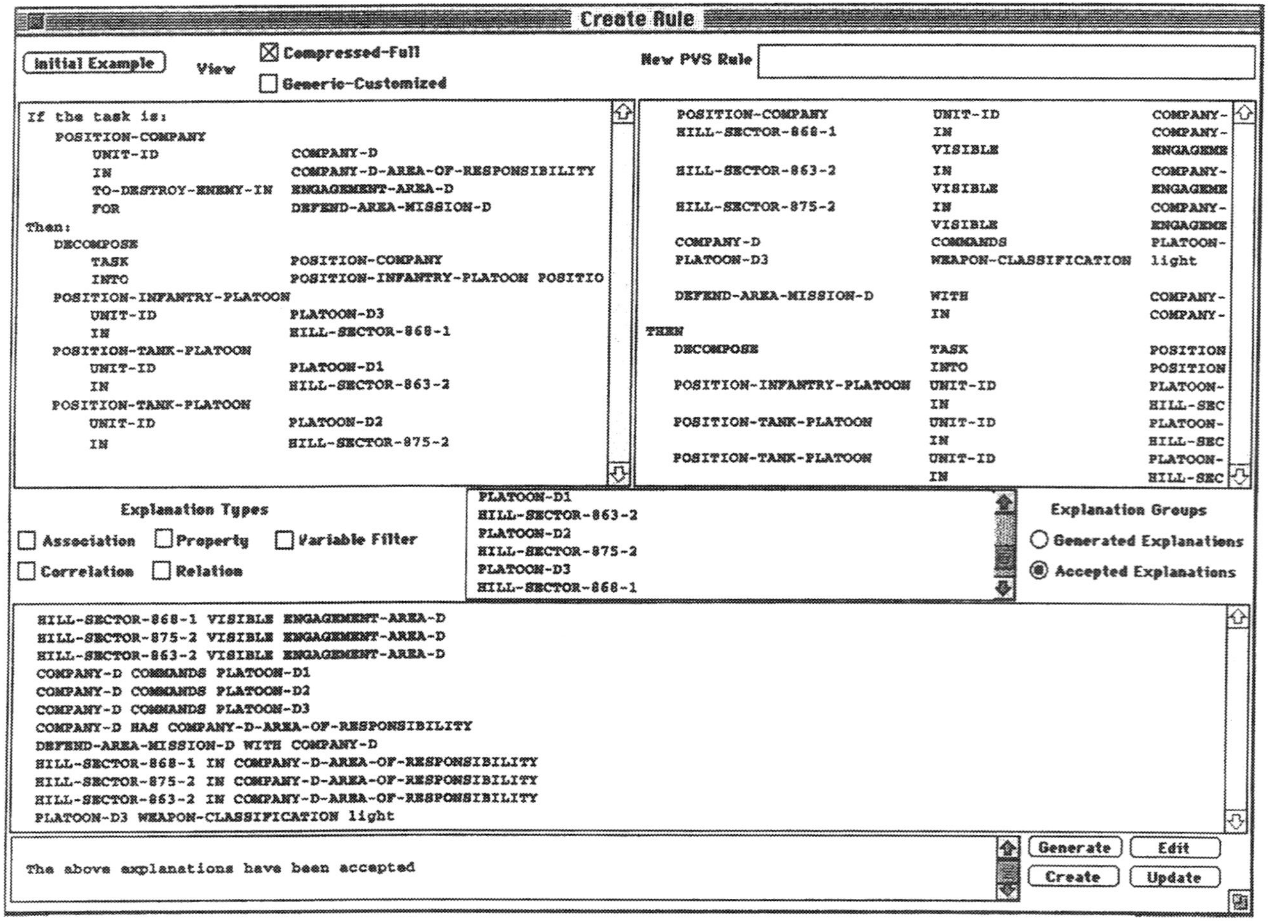

Create Rule

Initial Example   View   ☒ Compressed-Full   ☐ Generic-Customized

New PVS Rule

If the task is:
POSITION-COMPANY
    UNIT-ID   COMPANY-D
    IN   COMPANY-D-AREA-OF-RESPONSIBILITY
TO-DESTROY-ENEMY-IN   ENGAGEMENT-AREA-D
    FOR   DEFEND-AREA-MISSION-D

Then:
DECOMPOSE
    TASK   POSITION-COMPANY
    INTO   POSITION-INFANTRY-PLATOON POSITIO
POSITION-INFANTRY-PLATOON
    UNIT-ID   PLATOON-D3
    IN   HILL-SECTOR-868-1
POSITION-TANK-PLATOON
    UNIT-ID   PLATOON-D1
    IN   HILL-SECTOR-863-2
POSITION-TANK-PLATOON
    UNIT-ID   PLATOON-D2
    IN   HILL-SECTOR-875-2

POSITION-COMPANY   UNIT-ID   COMPANY-
HILL-SECTOR-868-1   IN   COMPANY-
                                        VISIBLE   ENGAGEME
HILL-SECTOR-863-2   IN   COMPANY-
                                        VISIBLE   ENGAGEME
HILL-SECTOR-875-2   IN   COMPANY-
                                        VISIBLE   ENGAGEME
COMPANY-D   COMMANDS   PLATOON-
PLATOON-D3   WEAPON-CLASSIFICATION light

DEFEND-AREA-MISSION-D   WITH   COMPANY-
                                        IN   COMPANY-

THEN
DECOMPOSE   TASK   POSITION
                                        INTO   POSITION
POSITION-INFANTRY-PLATOON   UNIT-ID   PLATOON-
                                        IN   HILL-SEC
POSITION-TANK-PLATOON   UNIT-ID   PLATOON-
                                        IN   HILL-SEC
POSITION-TANK-PLATOON   UNIT-ID   PLATOON-
                                        IN   HILL-SEC

PLATOON-D1
HILL-SECTOR-863-2
PLATOON-D2
HILL-SECTOR-875-2
PLATOON-D3
HILL-SECTOR-868-1

Explanation Types
☐ Association ☐ Property   ☐ Variable Filter
☐ Correlation ☐ Relation

HILL-SECTOR-868-1 VISIBLE ENGAGEMENT-AREA-D
HILL-SECTOR-875-2 VISIBLE ENGAGEMENT-AREA-D
HILL-SECTOR-863-2 VISIBLE ENGAGEMENT-AREA-D
COMPANY-D COMMANDS PLATOON-D1
COMPANY-D COMMANDS PLATOON-D2
COMPANY-D COMMANDS PLATOON-D3
COMPANY-D HAS COMPANY-D-AREA-OF-RESPONSIBILITY
DEFEND-AREA-MISSION-D WITH COMPANY-D
HILL-SECTOR-868-1 IN COMPANY-D-AREA-OF-RESPONSIBILITY
HILL-SECTOR-875-2 IN COMPANY-D-AREA-OF-RESPONSIBILITY
HILL-SECTOR-863-2 IN COMPANY-D-AREA-OF-RESPONSIBILITY
PLATOON-D3 WEAPON-CLASSIFICATION light

The above explanations have been accepted

Explanation Groups
○ Generated Explanations
◉ Accepted Explanations

Generate   Edit
Create   Update

Figure 9.8 Explanations of why the problem-solving episode in Figure 9.6 is correct.

with the initial placement indicated by the expert. Each such placement corresponds to an instance of the plausible upper bound condition, which is not an instance of the plausible lower bound condition. The placement is shown to the expert, who is asked to accept or reject it. The expert can control this experimentation process by 'fixing' some of the parameters of the defensive mission. For instance, it is useful to ask the agent to initially generate only placements of **COMPANY-D** in its area of responsibility. This limits the search space which the agent must deal with.

A domain-specific tool called the example viewer was developed as a user interface for placing ModSAF units in building a ModSAF scenario. The example viewer takes as input a set of unit placements generated as an example and places the units on a ModSAF terrain map. In this way, a subject-matter expert can easily view the unit placements and characterize the placements as good or bad based on his or her expertise.

9.5.3.1   *Rejecting agent's solution and helping it to understand its mistake*

The agent generates a new placement of **COMPANY-D** by matching the plausible upper bound of the rule in Figure 9.9 with the map region in Figure 9.6. It then proposes the placement to the expert on the ModSAF screen, as shown in the left-hand side of Figure 9.10. However, this placement is rejected by the expert.

The placement in Figure 9.10 was generated by analogy with the placement in Figure 9.6 and is consistent with all the explanations of that example. Both these examples are shown in Figure 9.11. The agent has to understand why the generated example is wrong and the initial one is correct. By comparing the two placements in Figure 9.11 the expert can easily explain that the infantry unit in the left-hand side placement is too far away from the area of engagement, while infantry unit in the right-hand side placement is close to the engagement area.

As a result of the additional explanation of the initial example (shown on the right-hand side of Figure 9.11) both the upper and lower bound conditions of the rule in Figure 9.9 are specialized by applying the method from Table 4.10.

The plausible upper bound element

> `?H1   SOMETHING,   IN ?C2,   VISIBLE ?E1`

is specialized to

> `?H1   SOMETHING,   IN ?C2,   VISIBLE ?E1,`
>
> `        DISTANCE-TO-ENGAGEMENT-AREA {close medium far}`

Similarly, the plausible lower bound element

> `?H1   HILL-SECTOR-868-1,   IN ?C2,   VISIBLE ?E1`

is specialized to

> `?H1   HILL-SECTOR-868-1,   IN ?C2,   VISIBLE ?E1,`
>
> `        DISTANCE-TO-ENGAGEMENT-AREA close`

Additionally, the following new pieces of knowledge are added to the semantic network, by applying the method from Table 4.11.

Plausible Upper Bound IF	; IF (Plausible Upper Bound)
?P1 POSITION-COMPANY, UNIT-ID ?C1, IN ?C2, TO-DESTROY-ENEMY-IN ?E1, FOR ?D1	; the task is to position company ?C1 ; in ?C2 to destroy enemy in ?E1 ; for ?D1
?A1 SOMETHING	; and ?A1 is something
?E1 SOMETHING	; and ?E1 is something
?H1 SOMETHING, IN ?C2, VISIBLE ?E1,	; and ?H1 is in ?C2, visible ?E1,
?H2 SOMETHING, IN ?C2, VISIBLE ?E1	; and ?H2 is in ?C2, visible ?E1
?H3 SOMETHING, IN ?C2, VISIBLE ?E1	; and ?H3 is in ?C2, visible ?E1
?C1 SOMETHING, COMMANDS ?P5, COMMANDS ?P6, COMMANDS ?P7	; and ?C1 commands ?P5, ?P6, ; and ?P7
?C2 SOMETHING	; and ?C2 is something
?P5 SOMETHING, WEAPON-CLASSIFICATION {light heavy}	; and ?P5 has weapons ; classified as light or heavy
?P6 SOMETHING	; and ?P6 is something
?P7 SOMETHING	; and ?P7 is something
?D1 SOMETHING, WITH ?C1, IN ?C2	; and ?D1 is with ?C1 in ?C2
Plausible Lower Bound IF	; IF (Plausible Lower Bound)
?P1 POSITION-COMPANY, UNIT-ID ?C1, IN ?C2, TO-DESTROY-ENEMY-IN ?E1, FOR ?D1	; the task is to position company ?C1 ; in ?C2 to destroy enemy in ?E1 ; for ?D1
?A1 AVENUE-OF-APPROACH-D	; and ?A1 is Avenue of Approach D
?E1 ENGAGEMENT-AREA-D	; and ?E1 is Engagement Area D
?H1 HILL-SECTOR-868-1, IN ?C2, VISIBLE ?E1	; and ?H1 is in ?C2, visible ?E1
?H2 HILL-SECTOR-863-2, IN ?C2, VISIBLE ?E1	; and ?H2 is in ?C2, visible ?E1
?H3 HILL-SECTOR-875-2, IN ?C2, VISIBLE ?E1	; and ?H3 is in ?C2, visible ?E1
?C1 COMPANY-D, COMMANDS ?P5, COMMANDS ?P6, COMMANDS ?P7	; and ?C1 commands ?P5, ?P6, ; and ?P7
?C2 COMPANY-D-AREA-OF-RESPONSIBILITY	; and ?C2 is Area of Responsibility D
?P5 PLATOON-D3, WEAPON-CLASSIFICATION light	; and ?P5 is Platoon D3 with ; weapons classified as light
?P6 PLATOON-D1	; and ?P6 is Platoon D1
?P7 PLATOON-D2	; and ?P7 is Platoon D2
?D1 DEFEND-AREA-MISSION, WITH ?C1, IN ?C2	; and ?D1 is with ?C1 in ?C2
THEN	; THEN
DECOMPOSE TASK ?P1, INTO ?P2 ?P3 ?P4	; decompose task ?P1 into ?P2, ?P3, ?P4
?P2 POSITION-INFANTRY-PLATOON, UNIT-ID ?P5, IN ?H1	; ?P2 is the task to position infantry ; platoon ?P5 in ?H1
?P3 POSITION-TANK-PLATOON, UNIT-ID ?P6, IN ?H2	; ?P3 is the task to position tank ; platoon ?P6 in ?H2
?P4 POSITION-TANK-PLATOON, UNIT-ID ?P7, IN ?H3	; ?P4 is the task to position tank ; platoon ?P7 in ?H3

Figure 9.9   Learned PVS rule.

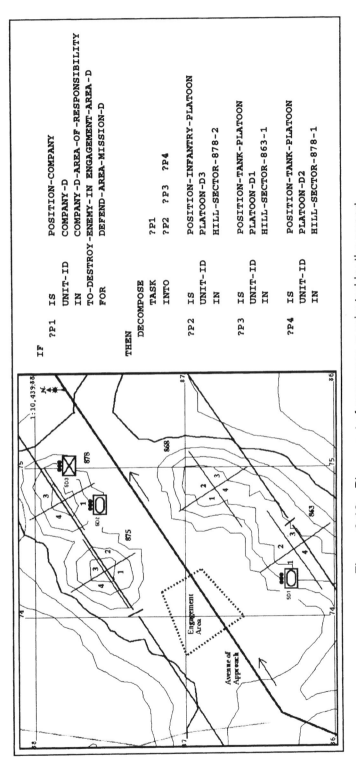

```
IF      ?P1   IS       POSITION-COMPANY
              UNIT-ID  COMPANY-D
              IN       COMPANY-D-AREA-OF-RESPONSIBILITY
              TO-DESTROY-ENEMY-IN ENGAGEMENT-AREA-D
              FOR      DEFEND-AREA-MISSION-D


THEN

        DECOMPOSE
              TASK     ?P1
              INTO     ?P2  ?P3  ?P4

        ?P2   IS       POSITION-INFANTRY-PLATOON
              UNIT-ID  PLATOON-D3
              IN       HILL-SECTOR-878-2

        ?P3   IS       POSITION-TANK-PLATOON
              UNIT-ID  PLATOON-D1
              IN       HILL-SECTOR-863-1

        ?P4   IS       POSITION-TANK-PLATOON
              UNIT-ID  PLATOON-D2
              IN       HILL-SECTOR-878-1
```

Figure 9.10   Placement of **COMPANY-D** rejected by the expert.

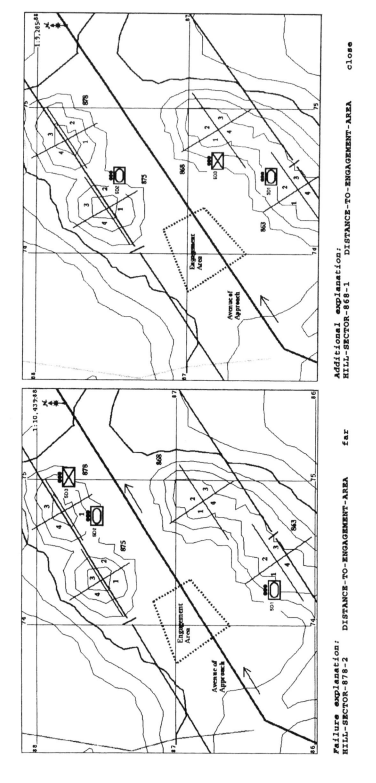

Figure 9.11 Explaining to the agent the mistake made.

HILL-SECTOR-878-2 DISTANCE-TO-ENGAGEMENT-AREA far

HILL-SECTOR-868-1 DISTANCE-TO-ENGAGEMENT-AREA close

9.5.3.2   Confirming the agent's solution

Next, the agent generates the placement in Figure 9.12 which is accepted by the expert. Consequently, the agent makes the following generalizations in the plausible lower-bound condition of the rule:

HILL-SECTOR-868-1, HILL-SECTOR-863-1 → HILL-SECTOR

HILL-SECTOR-863-2, HILL-SECTOR-878-2 → HILL-SECTOR

HILL-SECTOR-875-2, HILL-SECTOR-863-2 → HILL-SECTOR

PLATOON-D1, PLATOON-D2 → ARMORED-PLATOON

These generalizations correspond to the generalization of the positive examples from Figures 9.6 and 9.12 and are performed by climbing the agent's **ISA** hierarchy, part of which is shown in Figure 9.5. As a result of these generalizations, the following part of the plausible lower bound condition of the rule:

?P6 IS   PLATOON-D1

?P7 IS   PLATOON-D2

?H1 IS   HILL-SECTOR-868-1, IN   ?C2, VISIBLE   ?E1,

DISTANCE-TO-ENGAGEMENT-AREA close

?H2 IS   HILL-SECTOR-863-2, IN   ?C2, VISIBLE   ?E1

?H3 IS   HILL-SECTOR-875-2, IN   ?C2, VISIBLE   ?E1

was generalized to

?P6 IS   ARMORED-PLATOON

?P7 IS   ARMORED-PLATOON

?H1 IS   HILL-SECTOR,   IN   ?C2, VISIBLE   ?E1,

DISTANCE-TO-ENGAGEMENT-AREA close

?H2 IS   HILL-SECTOR,   IN   ?C2, VISIBLE   ?E1

?H3 IS   HILL-SECTOR,   IN   ?C2, VISIBLE   ?E1

At this point, all the placement examples for **COMPANY-D** that the agent might generate for **COMPANY-D** are already covered by the plausible lower bound of the PVS rule, and the expert is notified of this situation. Therefore the expert asks the agent to experiment with placing other companies to defend their areas. The agent then generates a new example for

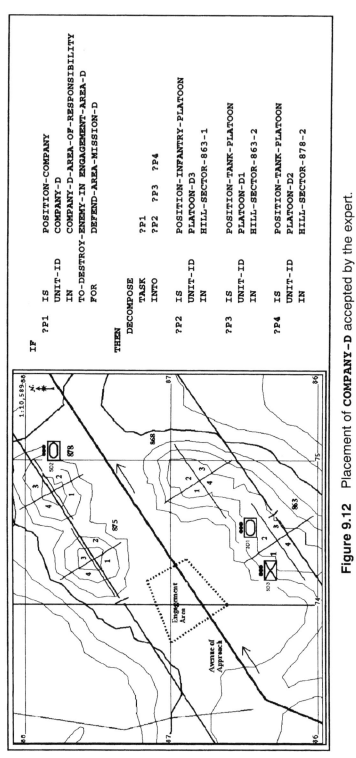

```
IF      ?P1   IS        POSITION-COMPANY
              UNIT-ID   COMPANY-D
              IN        COMPANY-D-AREA-OF-RESPONSIBILITY
              TO-DESTROY-ENEMY-IN ENGAGEMENT-AREA-D
              FOR       DEFEND-AREA-MISSION-D


THEN
       DECOMPOSE
              TASK      ?P1
              INTO      ?P2  ?P3  ?P4


       ?P2   IS        POSITION-INFANTRY-PLATOON
              UNIT-ID   PLATOON-D3
              IN        HILL-SECTOR-863-1

       ?P3   IS        POSITION-TANK-PLATOON
              UNIT-ID   PLATOON-D1
              IN        HILL-SECTOR-863-2

       ?P4   IS        POSITION-TANK-PLATOON
              UNIT-ID   PLATOON-D2
              IN        HILL-SECTOR-878-2
```

Figure 9.12 Placement of **COMPANY-D** accepted by the expert.

the expert to verify, as shown in Figure 9.13. This example comes from a different area on the map, which is in the area of responsibility of **COMPANY-E**.

Because the expert accepted the placement generated for **COMPANY-E**, the agent is able to make a significant reduction in the PVS by generalizing the following concepts from the plausible lower bound condition of the rule and the example in Figure 9.13:

> **COMPANY-D, COMPANY-E → COMPANY**
>
> **ENGAGEMENT-AREA-D, ENGAGEMENT-AREA-E → ENGAGEMENT-AREA**
>
> **AVENUE-OF-APPROACH-D, AVENUE-OF-APPROACH-E**
>   **→ AVENUE-OF-APPROACH**
>
> **DEFEND-AREA-MISSION-D, DEFEND-AREA-MISSION-E**
>   **→ DEFEND-AREA-MISSION**
>
> **PLATOON-D3, PLATOON-E3 → INFANTRY-PLATOON**
>
> **COMPANY-D-AREA-OF-RESPONSIBILITY,**
> **COMPANY-E-AREA-OF-RESPONSIBILITY**
>   **→ AREA-OF-RESPONSIBILITY**

After considering one more area of responsibility (for another company, **COMPANY-F**), the agent learns the rule shown in Figure 9.14.

It is important to stress the fact that, while this rule has been learned from five examples, the agent internally examined over 5000 different placements that were created from the upper bound of the PVS rule. These placements are for the three areas considered so far – for companies D, E and F. The learning process stopped when the rule was refined to the extent that all the placements generated were covered by the lower bound of the rule being learned (i.e., no other placement was both covered by the plausible upper bound and not covered by the plausible lower bound). It is obviously impractical for a human expert to consider this number of solutions individually. However, the expert may continue to verify the learned rule, by examining a testing set consisting of placements covered by the plausible lower bound, by following the method presented in Table 4.13.

9.6   Final remarks

In this chapter we have presented the process for building intelligent agents capable of terrain reasoning, which act as automated commanders within the ModSAF distributed interactive simulation environment. Our research in developing virtual commanders has addressed some difficult issues concerning how to interface with the ModSAF application, creating the initial knowledge base capable of scaling up to such a challenging domain, and defining the tasks to be taught to the ModSAF agent. A more detailed description of this Disciple application is given in Hieb (1996).

The Disciple approach to the development of computer-generated forces offers a number of potential benefits over the traditional approach. In the traditional approach, knowledge and software engineers have to interview military experts and to manualy build the agents to achieve reasonable behavior in a simulation. The Disciple training approach is more natural

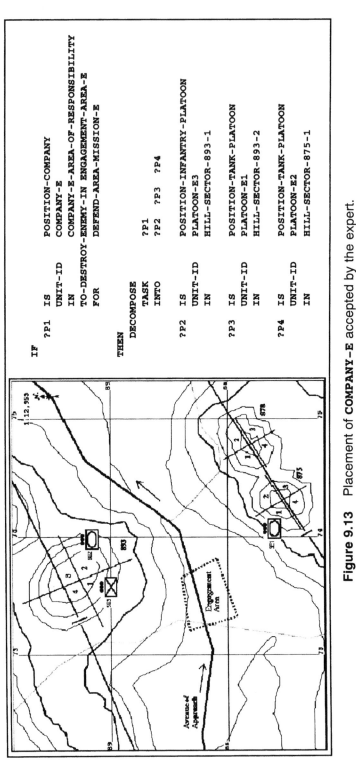

```
IF    ?P1   IS       POSITION-COMPANY
            UNIT-ID  COMPANY-E
            IN       COMPANY-E-AREA-OF-RESPONSIBILITY
            TO-DESTROY-ENEMY-IN ENGAGEMENT-AREA-E
            FOR      DEFEND-AREA-MISSION-E

THEN
      DECOMPOSE
            TASK     ?P1
            INTO     ?P2  ?P3  ?P4

      ?P2   IS       POSITION-INFANTRY-PLATOON
            UNIT-ID  PLATOON-E3
            IN       HILL-SECTOR-893-1

      ?P3   IS       POSITION-TANK-PLATOON
            UNIT-ID  PLATOON-E1
            IN       HILL-SECTOR-893-2

      ?P4   IS       POSITION-TANK-PLATOON
            UNIT-ID  PLATOON-E2
            IN       HILL-SECTOR-875-1
```

Figure 9.13   Placement of **COMPANY-E** accepted by the expert.

```
Plausible  Upper  Bound  IF
    ?P1     POSITION-COMPANY,  UNIT-ID  ?C1,  IN  ?C2,
            TO-DESTROY-ENEMY-IN  ?E1,  FOR  ?D1

    ?A1     SOMETHING

    ?E1     SOMETHING

    ?H1     HILL-SECTOR,  IN  ?C2,  VISIBLE  ?E1,
            DISTANCE-TO-ENGAGEMENT-AREA  {close   medium far}

    ?H2     HILL-SECTOR,  IN  ?C2,  VISIBLE  ?E1

    ?H3     HILL-SECTOR,  IN  ?C2,  VISIBLE  ?E1

    ?C1     SOMETHING,  COMMANDS  ?P5,  COMMANDS  ?P6,  COMMANDS  ?P7

    ?C2     SOMETHING

    ?P5     SOMETHING,   WEAPON-CLASSIFICATION   {light  heavy}

    ?P6     SOMETHING

    ?P7     SOMETHING

    ?D1     SOMETHING,  WITH  ?C1,  IN  ?C2

Plausible  Lower  Bound  IF
    ?P1     POSITION-COMPANY,  UNIT-ID  ?C1,  IN  ?C2,
            TO-DESTROY-ENEMY-IN  ?E1,  FOR  ?D1

    ?A1     AVENUE-OF-APPROACH

    ?E1     ENGAGEMENT-AREA

    ?H1     HILL-SECTOR,  IN  ?C2,  VISIBLE  ?E1

    ?H2     HILL-SECTOR,  IN  ?C2,  VISIBLE  ?E1

    ?H3     HILL-SECTOR,  IN  ?C2,  VISIBLE  ?E1

    ?C1     COMPANY,  COMMANDS  ?P5,  COMMANDS  ?P6,  COMMANDS  ?P7

    ?C2     AREA-OF-RESPONSIBILITY

    ?P5     INFANTRY-PLATOON,  WEAPON-CLASSIFICATION  light

    ?P6     ARMORED-PLATOON

    ?P7     ARMORED-PLATOON

    ?D1     DEFEND-AREA-MISSION,  WITH  ?C1,  IN  ?C2

THEN
    DECOMPOSE  TASK  ?P1,  INTO  ?P2  ?P3  ?P4

    ?P2     POSITION-INFANTRY-PLATOON,  UNIT-ID  ?P5,  IN  ?H1

    ?P3     POSITION-TANK-PLATOON,  UNIT-ID  ?P6,  IN  ?H2

    ?P4     POSITION-TANK-PLATOON,  UNIT-ID  ?P7,  IN  ?H3
```

Figure 9.14   The refined PVS rule.

and significantly simpler, and the role of the knowledge and software engineer is greatly reduced. As has been demonstrated above, the Disciple agent was capable of rapidly learning a complex placement rule for hilly terrain through a very natural interaction with the expert.

Moreover, Disciple's methods of knowledge elicitation and refinement lead to a verified knowledge base.

Tomasz Dybala has contributed to the research on developing a virtual company commander. This research was greatly facilitated through the support of J. Mark Pullen and the Center for Excellence in Command, Control, Communications, and Intelligence at George Mason University. The work on ModSAF applications was sponsored in part by the Defense Modeling and Simulation Office under contract DCA 100-91-C-0033. The research was also aided by various discussions with Andrew Ceranowicz whose insights concerning ModSAF and the development of CGF were very useful. The development of ModSAF was supported by the US Army Simulation, Training and Instrumentation Command.

Selected Bibliography of Machine Learning, Knowledge Acquisition, and Intelligent Agents Research

In addition to the works cited in the book, we have included here several basic titles from the fields of machine learning, knowledge acquisition, and intelligent agents.

Aamodt, A. (1995). Knowledge acquisition and learning by experience – the role of case-specific knowledge. In Tecuci, G. and Kodratoff, Y. (editors), *Machine Learning and Knowledge Acquisition: Integrated Approaches*. Academic Press, London.

Allen, J. (1987). *Natural Language Understanding*, Benjamin/Cummings, Menlo Park, CA.

Allen, J. F., Hendler, J. and Tate, A. (editors) (1990). *Readings in Planning*. Morgan Kaufmann, San Mateo, CA.

American Association for the Advancement of Science (1993). *Benchmarks for Science Literacy: Project 1061*. Oxford University Press, New York.

Anderson, J. R. (1983). *The Architecture of Cognition*. Harvard University Press, Cambridge, MA.

Arens, Y., Knoblock, C. and Shen, W. (1996). Query reformulation for dynamic information integration. *Journal of Intelligent Information Systems*, **6**(2/3), 99–130.

Aronis, J., Buchanan, B. G. and Lee, S. W. (1996). Augmenting medical databases with domain knowledge. In *Proceedings of the AAAI-96 Spring Symposium on AI in Medicine (AIM): Applications of Current Technologies*. Stanford University, CA. AAAI Press, Menlo Park, CA.

Atkeson, C. G. and Shaal, S. (1997). Robot learning from demonstration. In Fisher, D. (editor), *Proceedings of the Fourteenth International Conference on Machine Learning (ICML'97)*, San Francisco. Morgan Kaufmann, San Mateo, CA.

Baclace, P. E. (1992). Competitive agents for information filtering. *Communications of the ACM*, **35**(12), 50.

Baffes, P. T. and Mooney, R. J. (1993). Extending theory refinement to M-of-N rules. *Informatica* (Special Issue on Multistrategy Learning), **17**(4), 387–397.

Bala, J., DeJong, K., and Pachowicz, P. (1994). Multistrategy learning from engineering data by integrating inductive generalization and genetic algorithms. In Michalski, R. S. and Tecuci, G. (editors), *Machine Learning: A Multistrategy Approach*, Vol. 4. Morgan Kaufmann, San Mateo, CA.

Balabanovic, M. and Shoham, Y. (1995). Learning information retrieval agents: experiments with automated web browsing. In *Proceedings of the AAAI Spring Symposium on Information Gathering*, Stanford, CA.

Bareiss, E. R., Porter, B. W. and Murray, K. S. (1989). Supporting start-to-finish development of knowledge bases. *Machine Learning*, **4**, 259–283.

Bareiss, R. (1989). *Exemplar-Based Knowledge Acquisition*, Academic Press, London.

Baroglio, C., Botta, M. and Saitta, L. (1994). WHY: a system that learns using causal models and examples. In Michalski, R. S. and Tecuci, G. (editors), *Machine Learning: A Multistrategy Approach*, Vol. 4. Morgan Kaufmann, San Mateo, CA.

Barr, A., Cohen, P., and Feigenbaum, E. (editors) (1989). *The Handbook of Artificial Intelligence*, Vols I–IV, second printing. Morgan Kaufmann, San Mateo, CA.

Barto, A. G., Sutton, R. S. and Browwer, P. S. (1981). Associative search network: a reinforcement learning associative memory. *Biological Cybernetics*, **40**, 201–211.

Baudin, C., Kedar, S. and Pell, B. (1995). Increasing levels of assistance in refinement of knowledge-based retrieval systems, In Tecuci, G. and Kodratoff, Y. (editors), *Machine Learning and Knowledge Acquisition: Integrated Approaches*. Academic Press, London.

Beale, R. and Wood, A. (1994). Agent-based interaction. In *Proceedings of HCI '94*. Cambridge University Press.

Bell, J. (1995). Changing attitudes. In Wolldridge, M. and Jennings, N. R. (editors), *Intelligent Agents*, pp. 40–56. Springer-Verlag, Berlin.

Bennett, J. S. (1985). ROGET: A knowledge-based system for acquiring the conceptual structure of a diagnostic expert system. *Journal of Automated Reasoning*, **1**, 49–74.

Bergadano, F. and Giordana, A. (1990). Guiding induction with domain theories, In Kodratoff, Y. and Michalski, R. S. (editors), *Machine Learning: An Artificial Intelligence Approach*, Vol. 3. Morgan Kaufmann, San Mateo, CA.

Bergadano, F., Giordana, A. and Saitta, L. (1988). Automated concept acquisition in noisy environments. *IEEE Transactions on Pattern Analysis and Machine Intelligence*, **10**(4), 555–577.

Beyer, B. (1987). *Practical Strategies for the Teaching of Thinking*. Allyn and Bacon, Boston, MA.

Beyer, B. (1988). *Developing a Thinking Skills Program*. Allyn and Bacon, Boston, MA.

Bloom, B. (1956). *Taxonomy of Educational Objectives*. David McKay Co., New York.

Bobrow, D. G. and Winograd, T. (1977). An overview of KRL, a knowledge representation language. *Cognitive Science*, **1**(1), 3–45.

Boose, J. H. (1992). A survey of knowledge acquisition techniques and tools, In Buchanan, B. G. and Wilkins, D. (editors), *Readings in Knowledge Acquisition and Learning: Automating the Construction and the Improvement of Programs*. Morgan Kaufmann, San Mateo, CA.

Boose, J. H. and Bradshaw, J. M. (1988). Expertise transfer and complex problems using Aquinas as a knowledge acquisition workbench for expert systems. *International Journal of Man-Machine Studies*, **26**, 3–28.

Borgida, A., Brachman, R. J., McGuinness, D. L. and Alperine Resnick, L. (1989). CLASSIC: A structural data model for objects. In *Proceedings of the 1989 ACM SIGMOD International Conference on Management of Data*, pp. 58–67, Portland, Oregon.

Borrajo, D. and Veloso, M. (1996). Lazy incremental learning of control knowledge for efficiently obtaining quality plans. *AI Review Journal* (Special Issue on Lazy Learning), **10**, 1–34.

Boy, G. A. (1997). Software agents for cooperative learning. In Bradshaw, J. M. (editor), *Software Agents*, pp. 223–245. AAAI Press, Menlo Park, CA.

Brachman, R. J. (1979). On the epistemological status of semantic networks. In Findler, N. V. (editor), *Associative Networks: Representation and Use of Knowledge by Computers*, pp. 3–50. Academic Press, New York.

Brachman, R. J. and Levesque, H. J. (editors) (1985). *Readings in Knowledge Representation*. Morgan Kaufmann, San Mateo, CA.

Brachman, R. J. and Schmolze, J. G. (1985). An overview of the KL-ONE knowledge representation system. *Cognitive Science*, **9**, 171–216.

Brachman, R. J., McGuinness, D., Patel-Schneider P., Borgida, A. and Resnick, L. (1991). Living with CLASSIC: when and how to use a KL-ONE-like language. In Sowa, J. (editor), *Principles of Semantic Networks*, pp. 401–456. Morgan Kaufmann, San Mateo, CA.

Bradshaw, J. M. (editor) (1997). *Software Agents*. AAAI Press, Menlo Park, CA.

Bradshaw, J. M., Dutfield, S., Beboit, P. and Wooley, J. D. (1997). KAoS: Toward an industrial-strength open agent architecture, In Bradshaw, J. M. (editor), *Software Agents*, pp. 375–418. AAAI Press, Menlo Park, CA.

Bratko, I. (1990). *Prolog Programming for Artificial Intelligence*, second edition. Addison-Wesley, Reading, MA.

Bratko, I. (1992). Qualitative modelling and learning in KARDIO. In Buchanan B. G. and Wilkins, D. (editors), *Readings in Knowledge Acquisition and Learning: Automating the Construction and the Improvement of Programs*. Morgan Kaufmann, San Mateo, CA.

Brennan, S. E. and Ohaeri, J. O. (1994). Effects of message style on users' attributions toward agents. In Plaisant, A. (editor), *CHI '94 Human Factors in Computing Systems Conference Companion*, pp. 281–282. ACM, New York.

Breuker, J. A. and Van de Velde, W. (editors) (1994). *The Common KADS Library for Expertise Modelling*. IOS Press, Amsterdam.

Brooks, R. A. (1989). Engineering approach to building complete, intelligent beings. In *Proceedings of the SPIE – The International Society for Optical Engineering*, **1002**, 618–625.

Brownston, L., Farrell, R., Kant, E. and Martin, N. (1985). *Programming Expert Systems in OPS5: An Introduction to Rule-Based Programming*. Addison-Wesley, Reading, MA.

Brustoloni, J. C. (1991). Autonomous agents: characterization and requirements, *Carnegie Mellon Technical Report CMU-CS-91-204*. Carnegie Mellon University, Pittsburgh.

Buchanan, B. G. and Wilkins, D. C. (editors), (1993). *Readings in Knowledge Acquisition and Learning: Automating the Construction and Improvement of Expert Systems*. Morgan Kaufmann, San Mateo, CA.

Buchanan, B. G., Barstow, D., Bechtal, R., Bennett, J., Clancey, W., Kulikowski, C., Mitchell, T. and Waterman, D. (1983). Constructing an expert system. In Hayes-Roth, F., Waterman, D. and Lenat, D. (editors), *Building Expert Systems*, pp.127–168. Addison-Wesley, Reading, MA.

Burnstein, M. and Smith, D. (1996). ITAS: a portable, interactive transportation scheduling tool using a search engine generated from formal specifications. In Drabble, B. (editor), *Proceedings of the Third International Conference on Artificial Intelligence Planning Systems*. AAAI Press, Menlo Park, CA.

Burstein, M. H. (1986). Concept formation by analogical reasoning and debugging. In Michalski, R., Carbonell, J. and Mitchell, T. (editors), *Machine Learning: An Artificial Intelligence Approach*, Vol. 2, pp. 351–370. Morgan Kaufmann, San Mateo, CA.

Bussmann, S. and Mueller, J. (1993). A communication architecture for cooperating agents. *Computers and Artificial Intelligence*, **12**, 37–53.

Buvac, S. and Fikes, R. A. (1995). Declarative formalization of knowledge translation. In *Proceedings of the ACM CIKM: The 4th International Conference on Information and Knowledge Management*, ACM, New York.

Buvac, S. and McCarthy, J. (1996). Combining planning contexts. In Tate, A. (editor), *Advanced Planning Technology-Technological Achievements of the ARPA/Rome Laboratory Planning Initiative*. AAAI Press, Menlo Park, CA.

Carbonell, J. (1983). Learning by analogy: formulating and generalizing plans from past experience. In Michalski, R., Carbonell, J. and Mitchell, T. (editors), *Machine Learning: An Artificial Intelligence Approach*, pp. 137–162. Tioga Publishing Company, Palo Alto, CA.

Carbonell, J. (1986). Derivational analogy: a theory of reconstructive problem solving and expertise acquisition. In Michalski, R., Carbonell, J. and Mitchell, T. (editors), *Machine Learning: An Artificial Intelligence Approach*, Vol. 2, pp. 371–392. Morgan Kaufmann, San Mateo, CA.

Carbonell, J. G. and Gil, Y. (1990). Learning by experimentation: the operator refinement method. In Kodratoff, Y. and Michalski, R. S. (editors), *Machine Learning: An Artificial Intelligence Approach*, Vol. III, pp. 191–213. Morgan Kaufmann, San Mateo, CA.

Carey, J. R., Liedo, P., Orozco, D. and Vaupel, J. W. (1992). Slowing of mortality rates at older ages in large medflies cohorts. *Science*, **258**, 457–461.

Ceranowicz, A. (1994). ModSAF capabilities. In *Proceedings of the 4th Conference on Computer Generated Forces and Behavior Representation*. Orlando, FL, May.

Chandrasekaran, B. (1983). Towards a taxonomy of problem solving types. *AI Magazine*, **4**, 9–17.

Chandrasekaran, B. (1986). Generic tasks in knowledge-based reasoning: high-level building blocks for expert systems design. *IEEE Expert*, **1**, 23–29.

Chandrasekaran, B. and Johnson, T. R. (1993). Generic tasks and task structures: history, critique and new directions. In David, J.-M., Krivine, J.-P. and Simmons, R. (editors), *Second Generation Expert Systems*, pp. 239–280. Springer, Berlin.

Chang, C.-L. and Lee, R. C.-T. (1973). *Symbolic Logic and Mechanical Theorem Proving.* Academic Press, New York.

Chapman, D. (1987). Planning for conjunctive goals. *Artificial Intelligence,* **32**(3), 333–377.

Cimatt, A. and Serafini, L. (1995). Multi-agent reasoning with belief contexts: the approach and a case study. In Wolldridge, M. and Jennings, N. R. (editors), *Intelligent Agents,* pp. 71–85. Springer-Verlag, Berlin.

Clancey, W. J. (1983). The epistemology of a rule-based expert system: a framework for explanation. *Artificial Intelligence,* **27**, 215–251.

Clancey, W. J. (1984). Classification problem solving, In *Proceedings of the Third National Conference on Artificial Intelligence (AAAI-84),* pp. 49–55. Austin, Texas.

Clancey, W. J. (1985). Heuristic classification. *Artificial Intelligence,* **27**(3), 289–350.

Clark, P. and Porter, B. (1997). Building concept representations from reusable components. In *Proceedings of the Fourteenth National Conference on Artificial Intelligence,* pp. 369–376, Providence, Rhode Island. AAAI/MIT Press, Menlo Park, CA; Cambridge, MA.

Cohen, W. (1991). The generality of overgenerality, In Birnbaum, L. and Collins, G. (editors), *Machine Learning: Proceedings of the Eighth International Workshop,* pp. 490–494, Chicago, IL. Morgan Kaufmann, San Mateo, CA.

Cohen, W. (1995). Text categorization and relational learning. In *Proceedings of the Twelfth International Conference on Machine Learning,* Tahoe City, CA. Morgan Kaufmann, San Mateo, CA.

Collins, A. and Michalski, R. S. (1989). The logic of plausible reasoning: a core theory. *Cognitive Science,* **13**, 1–49.

Coyne, R. D., Rosenman, M. A., Radford, A. D., Balachandran, M. and Gero, J. S. (1990). *Knowledge-Based Design Systems.* Addison-Wesley, Reading, MA.

Cypher, A. (1991). Eager: programming repetitive tasks by example. In Robertson, S. P., Olson, G. M., and Olson, J. S. (editors), *CHI '91 Conference Proceedings,* pp. 33–39. ACM Press, New York.

Cypher, A. (editor), (1993). *Watch What I Do: Programming by Demonstration.* MIT Press, Cambridge, MA.

Danyluk, A. P. (1994). GEMINI: an integration of analytical and empirical learning. In Michalski, R. S. and Tecuci, G. (editors), *Machine Learning: A Multistrategy Learning,* Vol. 4. Morgan Kaufmann, San Mateo, CA.

Davies, T. R. and Russell, S. J. (1987). A logical approach to reasoning by analogy, In *Proceedings of the International Joint Conference on Artificial Intelligence,* pp. 264–270, Milan, Italy. Morgan Kaufmann, San Mateo, CA.

Davis, R. (1979). Interactive transfer of expertise: acquisition of new inference rules. *Artificial Intelligence Approach,* **12**, 121–157.

De Raedt, L. (1993). *Interactive Theory Revision: An Inductive Logic Programming Approach.* Academic Press, Boston.

De Raedt, L. and Bruynooghe, M. (1989). Towards friendly concept-learners. In *Proceedings of IJCAI-89,* Detroit. Morgan Kaufmann, San Mateo, CA.

Dean, T., Allen, J. and Aloimonos, Y. (1995). *Artificial Intelligence: Theory and Practice.* Benjamin Cummings, Menlo Park, CA.

Decker, K., Lesser, V., Nagendra Prasad, M. V. and Wagner, T. (1995). MACRON: an architecture for multi-agent cooperative information gathering. In *Proceedings of the CIKM Workshop on Intelligent Information Agents,* Baltimore, MD.

DeJong, G. and Mooney, R. (1986). Explanation-based learning: an alternative view. *Machine Learning,* **1**, 145–176.

DeJong, K. (1998). *Evolutionary Computation: Theory and Practice,* MIT Press (to appear).

Denning, P. J. and Metcalfe, B. (editors) (1997). *Beyond Calculation: The Next Fifty Years of Computing.* Springer-Verlag, Berlin.

Dongha, P. (1995). Toward a formal model of commitment for resource bounded agents. In Wolldridge, M. and Jennings, N. R. (editors), *Intelligent Agents*, pp. 86–101. Springer-Verlag, Berlin.

Doyle, J. (1979). A truth maintenance system. *Artificial Intelligence*, **12**(2), 231–272.

Doyle, J. and Patil, R. S. (1991). Two theses of knowledge representation: language restrictions, taxonomic classification, and the utility of representation services. *Artificial Intelligence*, **48**(3), 261–297.

Doyle, J., Shoham, Y. and Wellman, M. P. (1991). A logic of relative desire (preliminary report). In Ras, Z. W. and Zemankova, M. (editors), *Methodologies for Intelligent Systems*, 6, Vol. 542 of Lecture Notes in Artificial Intelligence, pp. 16–31. Springer-Verlag, Berlin.

Dunin-Keplicz, B. and Treur, J. (1995). Compositional formal specification of multi-agent systems. In Wolldridge, M. and Jennings, N. R. (editors), *Intelligent Agents*, pp. 102–117. Springer-Verlag, Berlin.

Dybala, T. (1996). Shared expertise model for building interactive learning agents, Ph.D. Dissertation, George Mason University, Fairfax, Virginia.

Dybala, T., Tecuci, G. and Rezazad, H., (1996). The shared expertise model for teaching interactive design assistants. *Journal of Engineering Applications of Artificial Intelligence*, **9**(6), 611–626.

Edmonds, E. A., Candy, L., Jones, R. and Soufi, B. (1994). Support for collaborative design: agents and emergence. *Communications of the ACM*, **37**(7), 41–47.

Ericsson, K. A. and Simon, H. A. (1984). *Protocol Analysis: Verbal Reports as Data*. MIT Press, Cambridge, MA.

Eriksson, H., Puerta, A. R. and Musen, M. A. (1994). Generation of knowledge-acquisition tools from domain ontologies. *International Journal of Human-Computer Studies*, **41**, 425–453.

Eriksson, H., Shahar, Y., Tu, S. W., Puerta, A. R. and Musen, M. A. (1995). Task modeling with reusable problem-solving methods. *Artificial Intelligence*, **79**, 293–326.

Eshelman, L. (1988). MOLE: a knowledge-acquisition system for cover-and-differentiate systems. In Marcus S. (editor), *Automating Knowledge Acquisition for Expert Systems*. Kluwer Academic Publishers, Boston.

Etzioni, O. (1993). Intelligence without robots (a reply to Brooks), *AI Magazine*, **14**(4), 70–77.

Etzioni, O. (1996). Moving up the information food chain: deploying softbots on the World Wide Web, invited talk. In *Proceedings of AAAI-96*, pp. 1322-1326. AAAI Press/MIT Press, Menlo Park, CA.

Etzioni, O. and Weld, D. (1994). A softbot-based interface to the Internet. *Communications of the ACM*, **37**(7), 72–76.

Everett, J. and Forbus, K. (1996). Scaling up logic-based truth maintenance systems via fact garbage collection. In *Proceedings of the 13th National Conference on Artificial Intelligence*, Portland, Oregon. AAAI Press/MIT Press, Menlo Park, CA

Evett, M. P., Hendler, J. A. and Andersen, W. A. (1993). Massively parallel support for effective recognition queries. In *Proceeding of the Eleventh National Conference on Artificial Intelligence* (AAAI-93), pp. 297–302, The AAAI Press/The MIT Press, Cambridge, MA.

Falkenhainer, B. C. and Michalski, R. S. (1990). Integrating quantitative and qualitative discovery in the ABACUS system. In Kodratoff, Y. and Michalski, R. S. (editors), *Machine Learning: An Artificial Intelligence Approach*, Vol. 3, pp. 153–190. Morgan Kaufmann, San Mateo, CA.

Falkenhainer, B., Forbus, K. D., and Gentner, D. (1989). The structure-mapping engine: algorithm and examples. *Artificial Intelligence*, **41**, 1–63.

Farquhar, A., Fikes, R. and Rice, J.(1996). The Ontolingua server: a tool for collaborative ontology construction. In Gaines, B. (editor), *Proceedings of the Knowledge Acquisition Workshop*, Banff, Canada.

Ferguson, R. W. and Forbus, K. D. (1995). Understanding illustrations of physical laws by integrating differences in visual and textual representations. In *Proceedings of the AAAI Fall Symposium on Computational Models for Integrating Language and Vision*, MIT, Cambridge, MA. AAAI Press, Menlo Park, CA.

Fikes, R., Engelmore, R., Farquhar, A. and Pratt, W. (1995). Network-based information brokers. In *Proceedings of the 1995 AAAI Spring Symposium Series on Information Gathering from Distributed, Heterogeneous Environments*, Stanford University, CA. AAAI Press, Menlo Park, CA.

Finin, T. Fritzson, R. and McKay, D. (1992). A language and protocol to support intelligent agent interoperability. In *Proceedings of the CE& CALS Washington '92 Conference*, June.

Finin, T. McKay, D., Fritzson, R. and McEntire, R. (1994). KQML: an information and knowledge exchange protocol. In Fuchi, K. and Yokoi, T. (editors), *Knowledge Building and Knowledge Sharing*, Ohmsha and IOS Press, Japan.

Fisher, D. H. (1987). Knowledge acquisition via incremental conceptual clustering. *Machine Learning*, **2**, 139–172.

Fisher, D. and Langley, P. (1985). Approaches to conceptual clustering, In *Proceedings of the Ninth International Joint Conference on Artificial Intelligence*, pp. 692–697. Morgan Kaufmann, Los Angeles, CA.

Flann, N. and Dietterich, T. (1989). A study of explanation-based methods for inductive learning, *Machine Learning*, **4**, 187–266.

Fontana, L., Debe, C., White, C. and Cates, W. (1993). Multimedia: gateway to higher-order thinking skills in progress. In *Proceedings of the National Convention of the Association for Educational Communications and Technology*, New Orleans. AECT, Washington, DC.

Forbus, K. D. and de Kleer, J. (1993). *Building Problem Solvers*. MIT Press, Cambridge, MA.

Forbus, K. and Gentner, D. (1989). Structural evaluation of analogies: what counts? In *Proceedings of the Eleventh Annual Conference of the Cognitive Science Society*. Erlbaum, Hillside, NJ.

Forbus, K. and Oblinger, D. (1990). Making SME greedy and pragmatic. In *Proceedings of the Twelfth Annual Conference of the Cognitive Science Society*. Erlbaum, Hillside, NJ.

Forbus, K., Gentner, D. and Law, K. (1995). MAC/FAC: a model of similarity-based retrieval. *Cognitive Science*, **19**(2), 141–205.

Forbus, K. D., Ferguson, R. W. and Gentner, D. (1994). Incremental structure mapping. *Proceedings of the Sixteenth Annual Conference of the Cognitive Science Society*. Erlbaum, Hillside, NJ.

Frankiln and Graesser (1996). Is it an agent, or just a program? A taxonomy for autonomous agents. In *Proceedings of the Third International Workshop on Agent Theories, Architectures, and Languages*. Springer-Verlag, Berlin.

Gaines, B. R. (1994). Class library implementation of an open architecture knowledge support system. *International Journal of Human-Computer Studies*, **41**(1/2), 59–107.

Gaines, B. R. and Boose, J. H. (editors) (1988). *Knowledge Acquisition for Knowledge Based Systems*. Academic Press, London.

Gaines, B. R. and Shaw, M. L. G. (1992a). Eliciting knowledge and transferring it effectively to a knowledge-based system, *IEEE Transactions on Knowledge and Data Engineering*, **5**, 4–14.

Gaines, B. R. and Shaw, M. L. G. (1992b). Knowledge acquisition tools based on personal construct psychology. *Knowledge Engineering Review* (Special Issue on Automated Knowledge Acquisition Tools), **8**, 49–85.

Gaines, B. R. and Shaw, M. L. G. (1995). WebMap concept mapping on the Web. In *Proceedings of the Fourth International World Wide Web Conference*, Boston, Massachusetts.

Gammack, J. G. (1987). Different techniques and different aspects on declarative knowledge, In Kidd, A. L. (editor), *Knowledge Acquisition for Expert Systems: A Practical Handbook*, pp. 137–163. Plenum Press, New York.

Genesereth, M. R. (1991). Knowledge interchange format. principles of knowledge representation and reasoning. In *Proceedings of the Second International Conference on Principles of Knowledge Representation and Reasoning*, pp. 599–600. Morgan Kaufmann, Cambridge, MA.

Genesereth, M. R. (1997). An agent-based framework for interoperability, In Bradshaw, J. M. (editor), *Software Agents*, pp. 317–345. AAAI Press, Menlo Park, CA.

Genesereth, M. R. and Ketchpel, S. P. (1994). Software agents, *Communications of the ACM*, **37**(7), 48–53, 147.

Genesereth, M. R. and Nilsson, N. J. (1987). *Logical Foundations of Artificial Intelligence*. Morgan Kaufmann, San Mateo, CA.

Genest, J., Matwin, S. and Plante, B. (1990). Explanation-based learning with incomplete theories: a three-step approach, In Porter B. W. and Mooney, R. J. (editors), *Machine Learning: Proceedings of the Eighth International Workshop*, Austin, TX. Morgan Kaufmann, San Mateo, CA.

Gennari, J. H., Stein, A. R. and Musen, M. A. (1996). Reuse for knowledge-based systems and CORBA components. In *Proceedings of the Tenth Banff Knowledge Acquisition for Knowledge-Bases Systems Workshop*, Banff, CA.

Gennari, J. H., Tu, S. W., Rothenfluh, T. E. and Musen, M. A. (1994). Mapping domains to methods in support of reuse. *International Journal of Human-Computer Studies*, **41**, 399–424.

Gentner, D. (1983). Structure mapping: a theoretical framework for analogy. *Cognitive Science*, 7, 155–170.

Gero, J. S. (1990). Design prototypes: a knowledge representation schema for design. *AI Magazine*, **11**(4), 26–36.

Gil, Y. (1994). Knowledge refinement in a reflective architecture. In *Proceedings of the National Conference on Artificial Intelligence (AAAI-94)*, Seattle, Washington. AAAI Press/MIT Press, Menlo Park, CA.

Gil, Y. and Melz, E. (1996). Explicit representation of problem-solving strategies to support knowledge acquisition. In *Proceedings of the Thirteenth National Conference on Artificial Intelligence*, Portland, Oregon.

Gil, Y. and Paris, C. (1995). Towards method-independent knowledge acquisition, In Tecuci, G. and Kodratoff, Y. (editors), *Machine Learning and Knowledge Acquisition: Integrated Approaches*. Academic Press, London.

Gil, Y. and Swartout, W. (1994). EXPECT: a reflective architecture for knowledge acquisition. In *ARPA/Rome Laboratory Knowledge-based Planning and Scheduling Initiative Workshop Proceedings*, pp. 433–444, Tucson, Arizona, February.

Gil, Y. and Tallis, M. (1997). A script-based approach to modifying knowledge bases. In *Proceedings of the Fourteenth National Conference on Artificial Intelligence*, pp. 377–383, Providence, Rhode Island. AAAI/MIT Press.

Gil, Y., Hoffman, M. and Tate, A. (1994). Domain-specific criteria to direct and evaluate planning systems. In *ARPA/Rome Laboratory Knowledge-based Planning and Scheduling Initiative Workshop Proceedings*, pp. 433–444, Tucson, Arizona, February.

Ginsberg, A. (1988). *Automatic Refinement of Expert System Knowledge Bases*. Pitman, New York.

Ginsberg, A., Weiss, S. and Politakis, P. (1989). Automatic knowledge base refinement for classification systems, *Artificial Intelligence*, **35**, 197–226.

Ginsberg, M. (1993). *Essentials of Artificial Intelligence*. Morgan Kaufmann, San Mateo, CA.

Goldberg, D. E. (1989). *Genetic Algorithms in Search, Optimization and Machine Learning*. Addison-Wesley, Reading, MA.

Gordon, D. F. (1991). An enhancer for reactive plans. In Birnbaum, L. and Collins, G. (editors), *Machine Learning: Proceedings of the Eighth International Workshop*, pp. 505–508. Chicago, IL. Morgan Kaufmann, San Mateo, CA.

Gordon, D. F. and Subramanian, D. (1993). A multistrategy learning scheme for agent knowledge acquisition. *Informatica*, **17**(4), 331–346.

Gordon, D., Daley, R., Shavlik, J., Subramanian, D. and Tecuci, G. (editors) (1995). *Proceedings of the ML'95 Workshop Agent that Learn from Other Agents*, Tahoe City, CA, July.

Graham, P. (1994). *On Lisp: Advanced Techniques for Common Lisp*. Prentice Hall, Englewood Cliffs, NJ.

Graham, P. (1996). *ANSI Common Lisp*. Prentice Hall, Englewood Cliffs, NJ.

Gruber, T. R. (1993a). A translation approach to portable ontology specifications. *Knowledge Acquisition*, **5**(2), 199–220.

Gruber, T. R. (1993b). Toward principles for the design of ontologies used for knowledge sharing. In Guarino, N. and Poli, R. (editors), *Formal Ontology in Conceptual Analysis and Knowledge Representation*. Kluwer Academic, Boston, MA.

Gruber, T. (1996). *Ontolingua Reference Manual*. Stanford University.

Gruber, T. R., Tenenbaum, J. M. and Weber, J. C. (1992). Toward a knowledge medium for collaborative product development. In *Proceedings of the Second International Conference on Artificial Intelligence in Design*, pp. 413–432, Pittsburgh. Kluwer Academic, Boston, MA.

Guarino, N. and Giaretta, P. (1995). Ontologies and knowledge bases: toward a terminological clarification. In Mars, N. J. I. (editor), *Towards Very Large Knowledge Bases*, pp. 25–32. IOS Press, Ohmsha, Tokyo.

Guha, R. V. (1991). Contexts: a formalization and some applications. Ph.D. thesis, Stanford University.

Guha, R. V. and Lenat, D. B. (1990). CYC: a midterm report. *AI Magazine*, **11**(3), 32–59.

Guha, R. V. and Lenat, D. (1994). CYC: Enabling agents to work together. *Communications of the ACM*, **37**(7), 127–142.

Guichard, F. and Ayel, J. (1995). Logical reorganization of DAI systems. In Wolldridge, M. and Jennings, N. R. (editor), *Intelligent Agents*, pp. 118–129. Springer-Verlag, Berlin.

Gunning, D. (1996). *High Performance Knowledge Bases (HPKB)*, DARPA/ISO.

Hass, N. and Hendrix, G. (1983). Learning by being told: acquiring knowledge for information management. In Michalski, R., Carbonell, J. and Mitchell, T. (editors), *Machine Learning: An Artificial Intelligence Approach*, pp. 371–392. Tioga, Palo Alto.

Hayes-Roth, B. (1985). A blackboard architecture for control. *Artificial Intelligence*, **26**, 251–321.

Hayes-Roth, B. (1995). An architecture for adaptive intelligent systems. *Artificial Intelligence* (Special Issue on Agents and Interactivity), **72**, 329–365.

Hayes-Roth, B., Pfleger, K., Lalanda, P., Morignot, P. and Balabanovic, M. (1995). A domain-specific software architecture for adaptive intelligent systems. *IEEE Transactions on Software Engineering*, **21**(4), 288–301.

Hayes-Roth, B., Washington, R., Ash, D., Hewett, R., Collinot, A., Vina, A. and Seiver, A. (1992). Guardian: a prototype intelligent agent for intensive-care monitoring. *Journal of AI in Medicine*, **4**, 165–185.

Hayes-Roth, F., Waterman, D. A. and Lenat, D. B. (editors) (1983). *Building Expert Systems*, Addison-Wesley, Reading, MA.

Hendler, J. A., Stoffel, K. and Taylor, M. (1996). Advances in high performance knowledge representation. *Technical Report CS-TR-3672*, Department of Computer Science, University of Maryland.

Hieb, M. R. (1996). Teaching instructable agents through multistrategy apprenticeship learning. Ph.D. thesis, School of Information Technology and Engineering, George Mason University. Fairfax, Virginia.

Hieb, M. R. and Tecuci, G. (1996). Training an agent interactively through an integration of multistrategy learning and knowledge acquisition. In *Proceedings of the AAAI-96 Spring Symposium: Acquisition, Learning and Demonstration: Automating Tasks for Users*, March, Stanford University, CA. AAAI Press, Menlo Park, CA.

Hieb, M. R., Tecuci, G., Pullen, J. M., Ceranowicz, A. and Hille, D. (1995). A methodology and tool for constructing adaptive command agents for computer generated forces. In *Proceedings of the 5th Conference on Computer Generated Forces and Behavioral Representation*. Orlando, Florida, March.

Hille, D., Hieb, R. M., Tecuci, G. and Pullen, J. M. (1995). Methods for transforming terrain data into abstract models through semantic terrain transformations. In *Proceedings of the 5th Conference on Computer Generated Forces and Behavioral Representation*, Orlando, Florida, March.

Hirsh, H. (1989). Incremental version-space merging: a general framework for concept learning, Doctoral dissertation, Stanford University.

Huffman, S. B. (1994). Instructable autonomous agents, Doctoral dissertation, Department of Computer Science and Engineering. University of Michigan.

Josephson, J. and Josephson, S. (1994). *Abductive Inference*. Cambridge University Press.

Josephson, J. R., Chandrasekaran, B., Smith, J. W. and Tanner, M. C. (1987). A mechanism for forming composite explanatory hypotheses. *IEEE Trans. on Systems, Man and Cybernetics*, **17**, 445–454.

Kaebling, L P. (1990) Learning functions in k-DNF from reinforcement. In *Machine Learning: Proceedings of the Seventh International Conferences*, pp. 162–169. Morgan Kaufmann, Los Altos, CA.

Karp, P. D., Myers, K. and Gruber, T. (1995). The generic frame protocol. In *Proceedings of the International Joint Conference on Artificial Intelligence*, pp. 768–774.

Kedar-Cabelli, S. T. (1987). Issues and case studies in purpose-directed analogy, Ph.D. thesis, Rutgers University, New Brunswick, NJ.

Kelly, G. A. (1955). *The Psychology of Personal Constructs*. Norton, New York.

Khuwaja, R., Desmarais, M. and Cheng, R. (1996). Intelligent guide: combining user knowledge assessment with pedagogical guidance. In *Proceedings of the third International Conference on Intelligent Tutoring Systems*. Montreal, Canada.

Klinker, G. (1988). KNACK: sample-driven knowledge acquisition for reporting systems, In Marcus, S. (editor), *Automating Knowledge Acquisition for Expert Systems*, pp.125–174. Kluwer, New York.

Knoblock, C. A. and Ambite, J-L. (1997). Agents for information gathering, In Bradshaw, J. M. (editor), *Software Agents*, pp. 347–373, AAAI Press, Menlo Park, CA.

Kodratoff, Y. (1988). *Introduction to Machine Learning*. Pitman, New York.

Kodratoff, Y. and Ganascia, J-G. (1986). Improving the generalization step in learning, In Michalski, R., Carbonell, J. and Mitchell, T. (editors), *Machine Learning: An Artificial Intelligence Approach*, Vol. 2, pp. 215–244. Morgan Kaufmann, San Mateo, CA.

Kodratoff, Y. and Michalski, R. S. (editors) (1990). *Machine Learning: An Artificial Intelligence Approach*, Vol. III. Morgan Kaufmann, San Mateo, CA.

Kodratoff, Y. and Tecuci, G. (1986). Rule learning in DISCIPLE. In *Proceedings of the First European Working Session on Learning*, Orsay, France, February.

Kodratoff, Y. and Tecuci, G. (1987). DISCIPLE1: a learning apprentice system for weak theory domains. In *Proceedings of the 10th International Joint Conference on Artificial Intelligence*, Milano, Italy, August. Morgan Kaufmann, San Mateo, CA.

Kodratoff, Y. and Tecuci, G. (1988a). Techniques of design and DISCIPLE learning apprentice. *International Journal of Expert Systems: Research and Applications*, **1**(1), 39–66. Reprinted In Buchanan B. G. and Wilkins, D. (editors) (1993). *Readings in Knowledge Acquisition and Learning: Automating the Construction and the Improvement of Programs*. Morgan Kaufmann, San Mateo, CA.

Kodratoff, Y. and Tecuci, G. (1988b). Learning based on conceptual distance. *IEEE Transactions on Pattern Analysis and Machine Intelligence*, **10**, 897–909.

Kolodner, J. (1993). *Case-Based Reasoning*. Morgan Kaufmann, San Mateo, CA.

Koppel, M., Segre, A. and Feldman, R. (1995). An integrated framework for knowledge representation and theory revision, In Tecuci, G. and Kodratoff, Y. (editors), *Machine Learning and Knowledge Acquisition: Integrated Approaches*, pp. 95–113. Academic Press, London.

Kowalski, R. (1979). *Logic for Problem Solving*. Elsevier/North-Holland, Amsterdam, London, New York.

LaFrance, M. (1987). The knowledge acquisition grid: a method for training knowledge engineers. *International Journal of Man-Machine Studies*, **27**, 245–255.

Laird, J. E., Newell, A. and Rosenbloom, P. S. (1987). SOAR: an architecture for general intelligence. *Artificial Intelligence*, **33**(1), 1–64.

Laird, J. E., Rosenbloom, P. S. and Newell, A. (1986). Chunking in SOAR: the anatomy of a general learning mechanism, *Machine Learning*, **1**(1), 11–46.

Langley, P. (1996). *Elements of Machine Learning*. Morgan Kaufmann, San Francisco, CA.

Langley, P. and Simon, H. A. (1995). Applications of machine learning and rule induction. *Communications of the ACM*, **38**(11), 55–64.

Langley, P., Simon H. A., Bradshow G. L. and Zytkow, J. M. (1987). *Scientific Discovery: Computational Explorations of the Creative Processes*. MIT Press, Cambridge, MA.

Langley, P., Simon, H. A. and Bradshow, G. L. (1990). Heuristics for empirical discovery. In *Readings in Machine Learning*, pp. 356–372. Morgan Kaufmann, San Mateo, CA.

Lashkari, Y., Metral, M. and Maes, M. (1994). Collaborative interface agents. In *Proceedings of AAAI '94 Conference*, Seattle, Washington, August.

Laurel, B. (1997). Interface agents: metaphors with characters. In Bradshaw, J. M. (editor), *Software Agents*, pp. 67–77. AAAI Press, Menlo Park, CA.

Laurel, B. (editor) (1990a). *The Art of Human Computer Interface Design*. Addison Wesley, Reading, MA.

Laurel, B. (1990b). Interface agents: metaphors with character. In Laurel, B. (editor), *The Art of Human-Computer Interface Design*, pp. 355–365. Addison-Wesley, Reading, MA.

Laurel, B., Oren, T. and Don, A. (1990). Issues in multimedia interface design: media integration and interface agents. In *CHI '90 Conference Proceedings*, pp. 133–139, ACM Press, New York.

Lebowitz, M. (1986). Integrated learning: controlling explanation. *Cognitive Science*, **10**, 219–240.

Lee, O. and Tecuci, G. (1993). MTLS: an inference-based multistrategy learning system. In *Proceedings of InfoScience '93*, October 21–22, Seoul, Korea.

Lee, S. W. (1995). A relational database interface for the RL machine learning program. Master's thesis. *Intelligent Systems Laboratory Report NO. ISL-95-17*. Department of Computer Science, University of Pittsburgh, Pittsburgh, PA.

Lee, S. W., Fischthal, S. and Wnek, J. (1997). A multistrategy learning approach to flexible knowledge organization and discovery. In Whitehall, B. (editor), *Proceedings of the Fourteenth National Conference on Artificial Intelligence workshop on Artificial Intelligence and Knowledge Management*, pp. 15–24, Providence, Rhode Island. AAAI Press, Menlo Park, CA.

Lenat, D. (1995a). Artificial intelligence. *Scientific American*, September.

Lenat, D. (1995b). CYC: a large-scale investment in knowledge infrastructure. *Communications of the ACM*, **38**(11), 33–38.

Lenat, D. B. and Guha, R. V. (1990). *Building Large Knowledge-Based Systems: Representation and Inference in the CYC Project*. Addison-Wesley, Reading, MA.

Levy, A. Y., Rajaraman, A. and Ordille, J. J. (1996). Query-answering algorithms for information agents, In *Proceedings of AAAI-96*, pp. 40–47, August, Portland, Oregon. AAAI Press/ MIT Press, Menlo Park, CA.

Liang, Y., Golshan, A. and Tecuci, G. (1995). Apprenticeship learning of domain models. In *Proceedings of the Seventh International Conference on Software Engineering and Knowledge Engineering (SEKE'95)*, Rockville, Maryland, June.

Lieberman, H. (1994). A user interface for knowledge acquisition from video. In *Proceedings of the Twelfth Conference on Artificial Intelligence*, Vol. 1, pp. 527–534, Seattle, WA, August. AAAI Press/MIT Press, Menlo Park, CA.

Lieberman, H. (1995). Letizia: an agent that assists Web browsing. In *Proceedings of the Fourteenth International Joint Conference on Artificial Intelligence*, pp. 924–929, Montreal, Quebec. Morgan Kaufmann, San Mateo, CA.

Luger, G. F. and Stubblefield, W. A. (1993). *Artificial Intelligence, Structures and Strategies for Complex Problem Solving*, second edition. Benjamin/Cummings.

MacGregor, R. (1990). The evolving technology of classification-based representation systems. In Sowa, J. (editor), *Principles of Semantic Networks: Explorations in the Representation of Knowledge*. Morgan Kaufmann, San Mateo, CA.

MacGregor, R. and Bates, R. (1987). The loom knowledge representation language. In *Proceedings of the Knowledge-Based Systems Workshop*, St. Louis, Missouri, April.

MacGregor, R. and Burstein, M. H. (1991). Using a description classifier to enhance knowledge representation. *IEEE Expert*, **6**(3), 41–46.

Maes, P. (1994). Agents that reduce work and information overload. *Communications of the ACM*, **37**(7), 30–40.

Maes, P. (1995a). Artificial life meets entertainment: life like autonomous agents. *Communications of the ACM*, **38**(11), 108–114.

Maes, P. (1995b). Intelligent software. *Scientific American*, **273**(3), 84–86.

Maes, P. and Kozierok, R. (1993). Learning interface agents, In *Proceedings of AAAI-93*, Washington DC, pp. 459–465.

Maes, P., Darrell, T., Blumberg, B., and Pentland, A. (1994). ALIVE: artificial life interactive video environment. In *Proceedings of the Twelfth National Conference on Artificial Intelligence (AAAI-94)*, p. 1506, Seattle, Washington. AAAI Press, Menlo Park, CA.

Mahadevan, S. (1989). Using determinations in explanation-based learning: a solution to incomplete theory problem, In Segre, A. (editor), *Proceedings of the Sixth International Workshop on Machine Learning*, Ithaca, NY. Morgan Kaufman, San Mateo, CA.

Mallery, J. C. (1994). A common LISP hypermedia server, In *Proceedings of the First International Conference on The World-Wide Web*, CERN, Geneva, May 25.

Malone, T. W., Grant, K. R. and Lai, K-Y. (1997). Agents for information sharing and coordination: a history and some reflections, In Bradshaw, J. M. (editor), *Software Agents*, pp. 109–143. AAAI Press, Menlo Park, CA.

Marcus, S. (editor) (1988). *Automating Knowledge Acquisition for Expert Systems*. Kluwer, Boston, MA.

Marcus, S. (editor) (1989). *Machine Learning* (Special Issue on Knowledge Acquisition), **4**(3–4).

Marcus, S. and McDermott, J. (1989). SALT: a knowledge acquisition language for propose-and-revise systems. *Artificial Intelligence*, **39**, 1–37.

Maulsby, D. (1994). Instructable agents. PhD thesis. Department of Computer Science. University of Calgary.

Maulsby, D. and Witten, I. H. (1995). Learning to describe data in actions. In *Proceedings of ICML-95 Workshop on Learning from Examples vs. Programming by Demonstration*, University of California, Santa Cruz, CA. Morgan Kaufmann, Los Altos, CA.

McCarthy, J. (1968). Programs with common sense. In Minsky, M. L. (editor), *Semantic Information Processing*, pp. 403–418. MIT Press, Cambridge, MA.

McCarthy, J. (1984). Some expert systems need common sense. In Pagels, H. (editor), *Computer Culture: The Scientific, Intellectual and Social Impact of the Computer*, Vol. 426. Annals of the New York Academy of Science.

McCarthy, J. (1987). Generality in artificial intelligence. *Communications of the ACM*, **30**(12), 1029–1035.

McCarthy, J. (1990). *Formalizing Common Sense: Papers by John McCarthy*. Ablex, Norwood, NJ.

McCarthy, J. (1993). Notes on formalizing context. In *Proceedings of Thirteenth International Joint Conference on Artificial Intelligence (IJCAI '93)*, pp. 81–98. Morgan Kaufmann, Los Altos, CA.

McCarthy, J. and Hayes, P. I. (1969). Some philosophical problems from the standpoint of artificial intelligence. In Meltzer, B., Michie, D. and Swann, M. (editors), *Machine Intelligence* 4, pp. 463–502. Edinburgh University Press, Edinburgh, Scotland.

McDermott, J. (1982). R1: A rule-based configurer of computer systems. *Artificial Intelligence*, **19**(1), 39–88.

McDermott, J. (1988). Preliminary steps toward a taxonomy of problem solving methods, In Marcus, S. (editor), *Automating Knowledge Acquisition for Expert Systems*. Kluwer, New York.

McDermott, D. and Doyle, J. (1980). Non-monotonic logic-I. *Artificial Intelligence*, **13**, 41–72.

Michalski, R. S. (1983). A theory and methodology of inductive learning. In Michalski, R. S., Carbonell, J. G. and Mitchell, T. M. (editors), *Machine Learning: An Artificial Intelligence Approach*, Vol. 1. Tioga Publishing Co., Palo Alto, CA.

Michalski, R. S. (1993). Inferential theory of learning as a conceptual framework for multistrategy learning. *Machine Learning* (Special Issue on Multistrategy Learning), **11**(2–3), 111–151.

Michalski, R. S. and Stepp, B. (1983). Learning from observation: conceptual clustering, In Michalski, R. S., Carbonell, J. G. and Mitchell, T. M. (editors), *Machine Learning: An Artificial Intelligence Approach*, Vol. I, pp. 331–364. Tioga Publishing Co., Palo Alto, CA.

Michalski, R. S. and Tecuci, G. (editors) (1991). *Proceedings of the First International Workshop on Multistrategy Approach*, Harpers Ferry, WV, November 7–9, Center for Artificial Intelligence, George Mason University.

Michalski, R. S. and Tecuci, G. (editors) (1993). *Proceedings of the Second International Workshop on Multistrategy Approach*, Harpers Ferry, WV, May 26–29, Center for Artificial Intelligence, George Mason University.

Michalski, R. S. and Tecuci, G. (editors) (1994). *Machine Learning: A Multistrategy Approach*, Vol. 4, Morgan Kaufmann, San Mateo, CA.

Michalski, R. S., Carbonell, J. G. and Mitchell, T. M. (editors) (1983, 1986). *Machine Learning: An Artificial Intelligence Approach*, Vol. I–II. Morgan Kaufmann, San Mateo, CA.

Miller, C. A. and Levi, K. R. (1995). A machine learning approach to knowledge-based software engineering, In Tecuci, G. and Kodratoff, Y. (editors), *Machine Learning and Knowledge Acquisition: Integrated Approaches*. Academic Press, London.

Miller, D. C. (1994). Network performance requirements for real-time distributed simulation. In Plattner, B. and Kiers, J. (editors), *Proceedings of INET '94/JENC5*, pp. 241-1–241-7, Internet Society (isoc@isoc.org).

Miller, G. A. (1995). WordNet: a lexical database for English, *Communications of the ACM*, **38**, 11.

Minsky, M. (1975). A framework for representing knowledge. In Winston, P. H. (editor), *The Psychology of Computer Vision*, pp. 211–277. McGraw-Hill, New York.

Minsky, M. (1985). *The Society of Mind*. Simon & Schuster, New York.

Minsky, M. and Riecken, D. (1994). A conversation with Marvin Minsky about agents. *Communications of the ACM* (Special Issue on Intelligent Agents), **37**(7), 22–29.

Minton, S. (1988). Quantitative results concerning the utility of explanation-based learning. In *Proceedings of the Seventh National Conference on Artificial Intelligence (AAAI-88)*, St. Paul, Minnesota. Morgan Kaufmann, San Mateo, CA.

Minton, S. (1990). Quantitative results concerning the utility of explanation-based learning, *Artificial Intelligence*, **42**, 363–392.

Mitchell, T. M. (1978). Version spaces: an approach to concept learning. Doctoral dissertation, Stanford University.

Mitchell, T. M. (1982). Generalization as search. *Artificial Intelligence*, **18**(2), 203–226.

Mitchell, T. M. (1997). *Machine Learning*. McGraw-Hill, New York.

Mitchell, T. M., Allen, J., Chalasani, P., Cheng, J., Etzioni, O., Ringuette, M. and Schlimmer, J. C. (1990). Theo: a framework for self-improving systems. In VanLehn, K. (editor), *Architectures for Intelligence*. Erlbaum, Hillsdale, NJ.

Mitchell, T. M., Caruana, R., Freitag, D., McDermott, J. and Zabowski, D. (1994). Experience with a learning personal assistant, *Communications of the ACM*, **37**, 81–91.

Mitchell, T. M., Keller, R. M., Kedar-Cabelli, S. T. (1986). Explanation-based generalization: a unifying view, *Machine Learning*, **1**, 47–80.

Mitchell, T. M., Mahadevan, S., and Steinberg, L. I. (1985). LEAP: a learning apprentice system for VLSI design, In Kodratoff, Y. and Michalski, R. S. (editors), *Machine Learning*, Vol. III. Morgan Kaufmann, San Mateo, CA.

Mitchell, T. M., Steinberg, L. I. and Shulman, J. S. (1984). A knowledge-based approach to design. In *Proceeding of the IEEE Workshop on Principles of Knowledge-Based Systems*, pp. 27–34, Denver, CO, December. IEEE Computer Society Press.

Mitchell, T. M., Utgoff, P. E. and Banerji, R. (1983). Learning by experimentation: acquiring and refining problem-solving heuristics, In Michalski, R. S., Carbonell, J. G. and Mitchell, T. M.

(editors), *Machine Learning: An Artificial Intelligence Approach*, Vol. 1, pp. 163–190. Tioga Publishing Co., Palo Alto, CA.

Mooney, R. and Bennet, S. (1986) A domain independent explanation based generalizer. In *Proceedings AAAI-86*, pp. 551–555, Philadelphia.

Mooney, R. and Ourston, D. (1994). A multistrategy approach to theory refinement, In Michalski, R. S. and Tecuci, G. (editors), *Machine Learning: An Multistrategy Approach*, Vol. 4. Morgan Kaufmann, San Mateo, CA.

Morik, K. (1989). Sloppy modeling, In Morik, K. (editor), *Knowledge Representation and Organization in Machine Learning*, pp. 107–134. Springer-Verlag, Berlin.

Morik, K. (1994). Balanced cooperative modeling. In Michalski, R. S. and Tecuci, G. (editors), *Machine Learning: A Multistrategy Approach*, Vol. 4. Morgan Kaufmann. San Mateo, CA.

Morik, K., Wrobel, S., Kietz, J-U. and Emde, W. (1993). *Knowledge Acquisition and Machine Learning: Theory, Methods and Applications*. Academic Press, London.

Mostow, D. J. (1983). Machine transformation of advice into a heuristic search procedure. In Michalski, R. S., Carbonell, J. G. and Mitchell, T. M. (editors), *Machine Learning: An Artificial Intelligence Approach*, Vol. I. Morgan Kaufmann, San Mateo, CA.

Mowbray, T. J. and Zahavi, R. (1995). *The ESSENTIAL CORBA: System Integration Using Distributed Objects*. John Wiley and Object Management Group.

Muggleton, S. (editor) (1992). *Inductive Logic Programming*. Academic Press, London.

Muggleton, S. and Buntine, W. (1988). Machine invention of first order predicates by inverting resolution. In *Proceedings of the Fifth International Conference on Machine Learning*, Ann Arbor, Michigan, pp. 339-352. Morgan Kaufmann, San Mateo, CA.

Musen, M. A. (1992). Overcoming the limitations of role-limiting methods. *Knowledge Acquisition*, **4**(2), 165–170.

Musen, M. A. (1993). Automated support for building and extending expert models, In Buchanan, B. G. and Wilkins, D. C. (editors), *Readings in Knowledge Acquisition and Learning*. Morgan Kaufmann, San Mateo, CA.

Musen, M. A. and Tu, S. W. (1993). Problem-solving models for generation of task-specific knowledge acquisition tools. In Cuena, J. (editor), *Knowledge-Oriented Software Design*. Elsevier, Amsterdam.

Naisbitt, J. (1990). *Megatrends 2000*. William Morrow, New York.

National Academy of Sciences (1996). *National Science Education Standards*. National Academy Press, Washington DC.

National Committee on Science Education Standards and Assessment (1996). *National Science Education Standards: 1996*. National Academy Press, Washington DC.

Neches, R., Fikes, R., Finin, T., Gruber, T., Patil, R., Senator, T. and Swartout, W. (1991). Enabling technology for knowledge sharing. *AI Magazine*, **12**(3), 36–56.

Nedellec, C. (1991). A smallest generalization step strategy, In Birnbaum, L. and Collins, G. (editors), *Machine Learning: Proceedings of the Eighth International Workshop*, pp. 529–533, Chicago, IL. Morgan Kaufmann, San Mateo, CA.

Negoita, C. (1985). *Expert Systems and Fuzzy Systems*. Benjamin/Cummings, Menlo Park, CA.

Negroponte, N. (1997). Agents: from direct manipulation to delegation, In Bradshaw, J. M. (editor), *Software Agents*, pp. 57–66. AAAI Press, Menlo Park, CA.

Newell, A. (1982). The knowledge level. *Artificial Intelligence*, **18**(1), 87–127.

Nilsson, N. J. (1980). *Principles of Artificial Intelligence*. Morgan Kaufmann, San Mateo, CA.

Norman, D. (1994). How might people interact with agents. *Communications of the ACM*, **37**(7), 68–71.

O'Keefe, R. and O'Leary D. (1989). *Verifying and Validating Expert Systems*. Tutorial, IJCAI-89. In *The Eleventh International Joint Conference on Artificial Intelligence*, Detroit, MI. Morgan Kaufmann, Los Altos, CA.

O'Rorke, P., Morris, S. and Schulenburg, D. (1990). Theory formation by abduction: a case study based on the chemical revolution, In Shrager, J. and Langley, P. (editors), *Computational Models of Scientific Discovery and Theory Formation*. Morgan Kaufmann, San Mateo, CA.

Paley, S. M., Lowrance, J. D. and Karp, P. D. (1997). A generic knowledge-base browser and editor. In *Proceedings of the Fourteenth National Conference on Artificial Intelligence*, pp. 1045–1051, Providence, Rhode Island. AAAI Press/MIT Press, Menlo Park, CA.

Pazzani, M. J. (1988). Integrating explanation-based and empirical learning methods in OCCAM, In *Proceedings of the Third European Working Session on Learning*, pp. 147–166. Glasgow, Scotland.

Pazzani, M. J. (1990). *Creating a Memory of Causal Relationships: an Integration of Empirical and Explanation-Based Learning Methods*. Lawrence Erlbaum, NJ.

Pazzani, M. J., Muramatsu, J. and Billsus, D. (1996). Syskill & Webert: identifying interesting Web sites. In *Proceedings of the Thirteenth National Conference on Artificial Intelligence*, pp. 54–59, Portland, OR. AAAI Press/MIT Press, Menlo Park, CA.

Plotkin, G. D. (1970). A note on inductive generalization. In Meltzer, B. and Michie, D. (editors), *Machine Intelligence 5*, pp. 165–179. Edinburgh University Press, Edinburgh.

Porter, B. W., Bareiss, R. and Holte, R. C. (1990). Concept learning and heuristic classification in weak-theory domains. In Shavlik, J. W. and Dietterich, T. G. (editors), *Readings in Machine Learning*, pp. 710–746. Morgan Kaufmann, San Mateo, CA.

Prieditis, A. E. (editor) (1988). *Analogica*. Kluwer, Boston, MA.

Puerta, A. R., Egar, J. W., Tu, S. W. and Musen, M. A. (1992). A multiple-method knowledge-acquisition shell for the automatic generation of knowledge-acquisition tools. *Knowledge Acquisition*, **4**, 171–196.

Pullen, J. M. (1994). Networking for distributed virtual simulation. In Plattner, B. and Kiers, J. (editors), *Proceedings of INET '94/JENC5*, pp. 243-1–243-7, Internet Society (isoc@isoc.org).

Quillian, M. R. (1968). Semantic memory, In Minsky, M. (editor), *Semantic Information Processing*, pp. 227–270. MIT Press, Cambridge, MA.

Quinlan, J. R. (1986a). Induction of decision trees. *Machine Learning*, **1**, 81–106.

Quinlan, J. R. (1986b). Learning efficient classification procedures and their application to chess end games. In Michalski, R. S., Carbonell, J. G. and Mitchell, T. M. (editors), *Machine Learning: An Artificial Intelligence Approach*, Vol. 1, pp. 463–482. Tioga Publishing Co., Palo Alto, CA.

Quinlan, J. R. (1990). Learning logical definitions from relations. *Machine Learning*, **5**, 239–266.

Quinlan, J. R. (1993). *C4.5: Programs for Machine Learning*. Morgan Kaufmann, San Mateo, CA.

Ram, A. and Cox, M. (1994). Introspective reasoning using meta-explanations for multistrategy learning, In Michalski, R. S. and Tecuci, G. (editors), *Machine Learning: A Multistrategy Approach*, Vol. 4, pp. 349–377. Morgan Kaufmann, San Mateo, CA.

Reich, Y. (1994). Macro and micro perspectives of multistrategy learning. In Michalski, R. S. and Tecuci, G. (editors), *Machine Learning: A Multistrategy Approach*, Vol. 4, pp. 379–407. Morgan Kaufmann, San Mateo, CA.

Rice, J., Farquhar, A., Piernot, P. and Gruber, T. (1996). Using the Web instead of a window system. In *Conference on Human Factors in Computing Systems (CHI96)*, pp. 103–110, Vancouver, Canada. Addison Wesley, Reading, MA.

Rich, E. and Knight, K. (1991). *Introduction to Artificial Intelligence*. McGraw-Hill, New York.

Riecken, D. (1994a). A conversation with Marvin Minsky about agents. *Communications of the ACM*, **37**(7), 22–29.

Riecken, D. (1994). M: an architecture of integrated agents. *Communications of the ACM*, **37**(7), 107–116.

Riecken, D. (1994). Intelligent agents, *Communications of the ACM* (Special Issue on Intelligent Agents), **37**(7), 18–21.

Rouveirol, C. and Puget, J.-F. (1989). A simple and general solution for inverting resolution. In *Proceedings of the European Working Session on Learning*, pp. 201–210, Porto, Portugal. Pitman, New York.

Rumelhart, D. E. and McClelland, J. L. (editors) (1986). *Parallel Distributed Processing*. MIT Press, Cambridge, MA.

Ruspini, E. H. (1991). On the semantics of fuzzy logic. *International Journal of Approximate Reasoning*, **5**, 45–88.

Russell, S. (1989). *The Use of Knowledge in Analogy and Induction*. Morgan Kaufmann, San Mateo, CA.

Russell, S. and Norvig, P. (1995). *Artificial Intelligence, A Modern Approach*. Prentice Hall, Englewood Cliffs, NJ.

Rychener, M. D. (1983). The instructable production system: a retrospective analysis. In Michalski, R. S., Carbonell, J. G. and Mitchell, T. M. (editors), *Machine Learning: An Artificial Intelligence Approach*, Vol. I, pp. 429–459. Morgan Kaufmann, San Mateo, CA.

Sacerdoti, E. D. (1977). A *Structure for Plans and Behavior*. Elsevier/North-Halland, Amsterdam, London, New York.

Saitta, L. and Botta, M. (1993). Multistrategy learning and theory revision. *Machine Learning* (Special Issue on Multistrategy Learning), **11**, 153–172.

Sammut, C. and Banerji, R.B. (1986). Learning concepts by asking questions, In Michalski, R. S., Carbonell, J. G. and Mitchell, T. M. (editors), *Machine Learning: An Artificial Intelligence Approach*, Vol. 2, pp. 167–191. Morgan-Kaufmann, San Mateo, CA.

Schlimmer, J. C. and Hermens, L. A. (1993). Software agents: completing patterns and constructing user interfaces. *Journal of AI Research*, **1**, 61–89.

Schmalhofer, F. and Tschaitschian, B. (1995). Cooperative knowledge evolution for complex domains. In Tecuci, G. and Kodratoff, Y. (editors), *Machine Learning and Knowledge Acquisition: Integrated Approaches*. Academic Press, London.

Schreiber, A., Wielinga, B. and Breuker, J. (1993). *KADS: A Principled Approach to Knowledge-Based Development*. Academic Press, London.

Selker, T. (1994). Coach: a teaching agent that learns. *Communications of the ACM*, **37**(7), 92–99.

Shavlik, J. W. and Dietterich, T. (editors) (1990). *Readings in Machine Learning*. Morgan Kaufmann, San Mateo, CA.

Shavlik, J. W. and Towell, G. G. (1990). An approach to combining explanation-based and neural learning algorithms. In Shavlik, J. W. and Dietterich, T. (editors), *Readings in Machine Learning*. Morgan Kaufmann, San Mateo, CA.

Shaw, M. L. G. and Gaines, B. R. (1987). An interactive knowledge elicitation technique using personal construct technology. In Kidd, A. L. (editor), *Knowledge Acquisition for Expert Systems: A Practical Handbook*, pp. 109–136. Plenum Press, New York.

Shaw, M. L. G. and Gaines, B. R. (1988). KITTEN: knowledge initiation and transfer tools for experts and novices. In Boose J. and Gaines B. (editors), *Knowledge Acquisition Tools for Expert Systems*. Academic Press, London.

Sheth, B. and Maes, P. (1993). Evolving agents for personalized information filtering. In *Proceedings of the Ninth IEEE Conference on Artificial Intelligence for Applications*, Washington, DC, pp. 345–352. IEEE Computer Society Press.

Shneiderman, B. (1995). Looking for the bright side of user interface agents. *Interactions*, **2**(1), 13–15.

Shneiderman, B. (1997). Direct manipulation versus agents: paths to predictable, controllable, and comprehensible interfaces. In Bradshaw, J. M. (editor), *Software Agents*, pp. 97–106. AAAI Press, Menlo Park, CA.

Shoham, Y. (1997). An overview of agent-oriented programming, In Bradshaw, J. M. (editor), *Software Agents*, pp. 271–290, AAAI Press, Menlo Park, CA.

Shortliffe, E. H., Buchanan, B. G. and Feigenbaum, E. A. (1979). Knowledge engineering for medical decision making: a review of computer-based clinical decision aides. In *Proceedings of IEEE*, **67**(9), 1207–1224.

Smith, D. C., Cypher, A. and Spohrer, J. (1994). KidSim: programming agents without a programming language. *Communications of the ACM*, **37**(7), 55–67.

Soderland, S., Fisher, D., Aseltine, J. and Lehnert, W. (1995). Crystal: Inducing a conceptual dictionary. In *Proceedings of the Fourteenth International Joint Conference on Artificial Intelligence*, Montreal, Canada. Morgan Kaufmann, Los Altos, CA.

Sowa, J. F. (1984). *Conceptual Structures: Information Processing in Mind and Machine*. Addison-Wesley, Reading, MA.

Sridharan, N. and Bresina, J. (1983). A mechanism for the management of partial and indefinite descriptions. *Technical Report CBM-TR-134*, Rutgers University.

Sriram, D. and Tong, C. (1991). Artificial intelligence and engineering design. *Tutorial Presented on the Ninth National Conference on Artificial Intelligence, July 14, Anaheim, CA*. MIT Press, Cambridge, MA.

Steele, G. (1990). *Common LISP: The Language*, second edition. Digital Press, Bedford, MA.

Stefik, M. (1980). Planning with constraints (MOLGEN: Part 1). *Artificial Intelligence*, **14**(2), 111–139.

Stoffel, K., Taylor, M. and Hendler, J. A. (1997). Efficient management of very large ontologies. In *Proceedings of American Association for Artificial Intelligence Conference* (AAAI-97). The AAAI Press/The MIT Press, Cambridge, MA.

Stone, P. and Veloso, M. (1997). Using decision tree confidence factors for multiagent control. In *Proceedings of AAAI-97 Workshop on Multiagent Learning*. AAAI Press, Menlo Park, CA.

Sutton, R. S. (1988). Learning to predict by the methods of temporal differences. *Machine Learning*, **3**, 9–44.

Swartout, W. R. and Gil, Y. (1995). EXPECT: explicit representations for flexible acquisition. In *Proceedings of the Ninth Knowledge Acquisition for Knowledge-Based Systems Workshop*, Banff, Alberta.

Swartout, W. R. and Moore, J. D. (1993). Explanation in second-generation expert systems. In David, J.-M., Krivine, J.-P. and Simmons, R. (editors), *Second Generation Expert Systems*. Springer-Verlag, Berlin.

Swartout, W. R., Neches, R. and Patil, R. (1993). Knowledge sharing: prospects and challenges. In *Proceedings of the International Conference on Building and Sharing of Very Large-Scale Knowledge Bases '93*, Tokyo, Japan.

Swartout, W. R., Paris, C. L. and Moore, J. D. (1991). Design for explainable expert systems. *IEEE Expert*, **6**(3), 58–64.

Swartout, W. R., Patil, R., Knight, K. and Russ, T. (1996). Toward distributed use of large-scale ontologies. In *Proceedings of the Banff Knowledge Acquisition Workshop*, November, Banff, Canada.

Sycara, K. and Miyashita, K. (1995). Learning control knowledge through case-based acquisition of user optimization preferences in ill-structured domains. In Tecuci, G. and Kodratoff, Y. (editors), *Machine Learning and Knowledge Acquisition: Integrated Approaches*. Academic Press, London.

Tambe, M., Johnson, W., L., Jones, R. M., Koss, F., Laird, J. E., Rosenbloom, P. S. and Schwamb, K. (1995). Intelligent agents for interactive simulation environments. *AI Magazine*, **16**(1), 15–39.

Tanimoto, S. (1990). *The Elements of Artificial Intelligence*. Computer Science Press, Rockville, MD.

Tate, A. (1977). Generating project networks. In *Proceedings of IJCAI-77*, Cambridge, MA, pp. 888–893. Morgan Kaufmann, Los Altos, CA.

Tate, A. (1996). Towards a plan ontology. *Journal of the Italian AI Association (AIIA)*, **9**(1), 19–26.

Tecuci, G. (1981). H-graphs and their applications to pattern recognition. *Journal of the Polytechnic Institute of Bucharest*, **43**(3), 23–34.

Tecuci, G. (1984). Learning hierarchical descriptions from examples. *Computers and Artificial Intelligence*, **3**(3), 211–222.

Tecuci, G. (1988). Disciple: a theory, methodology and system for learning expert knowledge. Doctoral dissertation, University of Paris South.

Tecuci, G. (1991). A multistrategy learning approach to domain modeling and knowledge acquisition. In Kodratoff, Y. (editor), *Proceedings of the European Conference on Machine Learning, Machine Learning EWSL-91*, pp. 14-32, Porto, March. Springer-Verlag, Berlin.

Tecuci, G. (1992a). Cooperation in knowledge base refinement. In Sleeman, D. and Edwards, P. (editors), *Machine Learning: Proceedings of the Ninth International Conference (ML92)*, pp. 445–450. Morgan Kaufmann, San Mateo, CA.

Tecuci, G. (1992b). Automating knowledge acquisition as extending, updating, and improving a knowledge base. *IEEE Trans. on Systems, Man and Cybernetics*, **22**, 1444–1460.

Tecuci, G. (1993a). Plausible justification trees: a framework for the deep and dynamic integration of learning strategies. *Machine Learning* (Special Issue on Multistrategy Learning), **11**, 237–261.

Tecuci, G. (guest editor) (1993b). *Informatica* (Special Issue on Multistrategy Learning), **17**(4), December.

Tecuci, G. (1994). An inference-based framework for multistrategy learning, In Michalski, R. S. and Tecuci, G. (editors), *Machine Learning: A Multistrategy Approach*, Vol. 4, pp. 107–138. Morgan Kaufmann, San Mateo, CA.

Tecuci, G. (1995). Building knowledge bases through multistrategy learning and knowledge acquisition. In Tecuci, G. and Kodratoff, Y. (editors), *Machine Learning and Knowledge Acquisition: Integrated Approaches*, pp. 13–50. Academic Press, London.

Tecuci, G. and Duff, D. (1994). A framework for knowledge base refinement through multistrategy learning and knowledge acquisition. *Knowledge Acquisition Journal*, **6**(2), 137–162.

Tecuci, G. and Hieb, M. (1994). Consistency-driven knowledge elicitation: using a learning-oriented knowledge representation that supports knowledge elicitation in NeoDISCIPLE. *Knowledge Acquisition Journal*, **6**(1), 23–46.

Tecuci, G. and Hieb, M. H. (1996). Teaching intelligent agents: the disciple approach. *International Journal of Human-Computer Interaction*, **8**(3), 259–285.

Tecuci, G. and Kodratoff, Y. (1989). Multistrategy learning in non-homogeneous domain theories. In Segre, A. (editor), *Proceedings of the 6th International Conference on Machine Learning*, Cornell University, Ithaca, New York, June. Morgan Kaufmann, San Mateo, CA.

Tecuci, G. and Kodratoff, Y. (1990). Apprenticeship learning in imperfect theory domains. In Kodratoff, Y. and Michalski, R. S. (editors), *Machine Learning: An Artificial Intelligence Approach*, Vol. 3, pp. 514–551. Morgan Kaufmann, San Mateo, CA.

Tecuci, G. and Kodratoff, Y. (editors) (1995). *Machine Learning and Knowledge Acquisition: Integrated Approaches*. Academic Press, London.

Tecuci, G. and Michalski, R. S. (1991a). A method for multistrategy task-adaptive learning based on plausible justifications. In Birnbaum, L. and Collins, G. (editors), *Machine Learning: Proceedings of the Eighth International Workshop*, pp. 549–553, Chicago, IL. Morgan Kaufmann, San Mateo, CA.

Tecuci, G. and Michalski, R. S. (1991b). Input understanding as a basis for multistrategy task-adaptive learning. In *Proceedings of the International Symposium on Methodologies for Intelligent Systems*, pp. 419–428, Charlotte, NC. Springer-Verlag, Berlin.

Tecuci, G., Hieb, M. R. and Dybala, T. (1995). Teaching an automated agent to monitor the electrical power system of an orbital satellite. *Telematics and Informatics*, **12**(3–4), 229–245.

Tecuci, G., Kedar, S. and Kodratoff, Y. (editors), (1993). *Proceedings of the IJCAI-93 Workshop on Machine Learning and Knowledge Acquisition: Common Issues, Contrasting Methods, and Integrated Approaches*, Chambery, France.

Tecuci, G., Kedar, S. and Kodratoff, Y. (guest editors) (1994). *Knowledge Acquisition* (Special Issue on the Integration of Machine Learning and Knowledge Acquisition), **6**(2).

Tecuci, G., Kodratoff, Y., Bodnaru, Z. and Brunet, T. (1987). DISCIPLE: an expert and learning system. In Moralee, D. S. (editor), *Research and Development in Expert Systems IV*, pp. 234–245. Cambridge University Press.

Tecuci, G., Mândutianu, D. and Voinea, S. (1983). A hierarchical system for robot programming. *Computers and Artificial Intelligence*, **2**(2), 167–188.

Touretzky, D. S. (1986). *The Mathematics of Inheritance Systems*. Pitman and Morgan Kaufmann, London and San Mateo, CA.

Towell, G. G. and Shavlik, J. W. (1994). Refining symbolic knowledge using neural networks. In Michalski, R. S. and Tecuci, G. (editors), *Machine Learning: A Multistrategy Approach*, Vol. 4, pp. 405–429. Morgan Kaufmann, San Mateo, CA.

Tu, S. W., Eriksson, H., Gennari, J. H., Shahar, Y. and Musen, M. A. (1995). Ontology-based configuration of problem-solving methods and generation of knowledge-acquisition tools: applications of PROTEGE-II to protocol-based decision support. *Artificial Intelligence in Medicine*, 7, 257–289.

Valente, A., van de Velde, W. and Breuker, J. (1994). Common KADS library for expertise modelling. In Breuker, J. and van de Velde, W. (editors), *Common KADS Expertise Modeling Library*, Chapter 3, pp. 31–56. IOS Press, Amsterdam.

Veloso, M. (1994). *Planning and Learning by Analogical Reasoning*. Springer Verlag, Berlin.

Veloso, M. (1996). Flexible strategy learning: analogical replay of problem solving episodes. In Leake, D. (editor), *Case-Based Reasoning: Experiences, Lessons, and Future Directions*. AAAI Press/The MIT Press, Menlo Park, CA.

Veloso, M. and Aamodt, A. (1995). *Case-Based Reasoning Research and Development*. Springer-Verlag, Berlin.

Veloso, M. M. and Carbonell, J. G. (1993). Derivational analogy in PRODIGY: automating case acquisition, storage and utilization. *Machine Learning*, 10, 249–278.

Veloso, M. and Carbonell, J. G. (1994). Case-based reasoning in PRODIGY. In Michalski, R. S. and Tecuci, G. (editors), *Machine Learning: A Multistrategy Approach*,Vol. 4, pp. 523–548. Morgan Kaufmann, San Mateo, CA.

Veloso, M., Carbonell, J. G., Perez, M. A., Borrajo, D., Fink, E. and Blythe, J. (1995). Integrating planning and learning: the PRODIGY architecture. *Journal of Experimental and Theoretical Artificial Intelligence*, 7(1), 81–120.

Waterman, D. and Hayes-Roth, F. (1978). *Pattern-Directed Inference Systems*. Academic Press, New York.

Watkins, C. J. (1989). Models of delayed reinforcement learning. PhD thesis. Psychology Department, Cambridge University, Cambridge, UK.

Whitehall, B. L. (1990). Knowledge-based learning: integration of deductive and inductive leaning for knowledge base completion. PhD thesis, University of Illinois at Urbana-Champaign.

Whitehall, B. L. and Lu, S. C-Y. (1994). Theory completion using knowledge-based learning. In Michalski, R. S. and Tecuci, G. (editors), *Machine Learning: A Multistrategy Approach*, Vol. 4, pp. 165–187. Morgan Kaufmann, San Mateo, CA.

Widmer, G. (1994). Learning with a qualitative domain theory by means of plausible explanations. In Michalski, R. S. and Tecuci, G. (editors), *Machine Learning: A Multistrategy Approach*, Vol. 4, pp. 635–655. Morgan Kaufmann, San Mateo, CA.

Wiederhold, G. (1992). The roles of artificial intelligence in information systems. *Journal of Intelligent Information Systems*, 11(1), 35–56.

Wiederhold, G. and Genesereth, M. (1997). The conceptual basis for mediation services. *IEEE Expert*, 12(5), 38–47.

Wiederhold, G., Wegner, P. and Ceri, S. (1992). Toward megaprogramming. *Communications of the ACM*, 33(11), 89–99.

Wielinga, B. J., Schreiber, A. T. and Breuker, J. A. (1992). KADS: a modelling approach to knowledge engineering. *Knowledge Acquisition*, 4(1), 5–53.

Wielinga, B. J., Van de Velde, W., Schreiber, A. T. and Akkermans, J. M. (1993). Towards a unification of knowledge modelling approaches. In David, J. M., Krivine, J. P. and Simmons, R. (editors), *Second Generation Expert Systems*, pp. 299–335. Springer-Verlag, Berlin.

Wilensky, R. (1986). *Common LISPcraft*. Norton and Company, New York.

Wilkins, D. C. (1990). Knowledge base refinement as improving an incorrect and incomplete domain theory. In Kodratoff, Y. and Michalski, R. S. (editors), *Machine Learning: An Artificial Intelligence Approach*, Vol. 3, pp. 493–513. Morgan Kaufmann, San Mateo, CA.

Wilkins, D. C., Clancey, W. J. and Buchanan, B. G. (1986). *An Overview of the Odysseus Learning Apprentice*. Kluwer Academic Press, New York.

Wilkins, D. E. (1988). *Practical Planning: Extending the Classical AI Planning Paradigm*. Morgan Kaufmann, San Mateo, CA.

Wille, R. (1992). Concept lattices and conceptual knowledge systems. In Lehmann, F. (editor), *Semantic Networks in Artificial Intelligence*, pp. 493–516. Pergamon Press, Oxford.

Winston, P. H. (ed.) (1975). *The Psychology of Computer Vision*. McGraw-Hill, New York.

Winston, P. H. (1980). Learning and reasoning by analogy, *Communications of the ACM*, **23**, 689–703.

Winston, P. H. (1986). Learning by augmenting rules and accumulating censors. In Michalski, R. S., Carbonell, J. G. and Mitchell, T. M. (editors), *Machine Learning: An Artificial Intelligence Approach*, Vol. 2, pp. 45–61. Morgan-Kaufmann, San Mateo, CA.

Winston, P. H. (1993). *Artificial Intelligence*, third edition. Addison-Wesley, Reading, MA.

Winston, P. H. and Horn, B. K. P. (1989). *LISP*. Addison-Wesley, Reading, MA.

Wisniewski, E. J. and Medin, D. J. (1991). Is it a pocket or a purse? Tightly coupled theory and data driven learning. In Birnbaum, L. and Collins, G. (editors), *Machine Learning: Proceedings of the Eighth International Workshop*, pp. 564–568, Chicago, IL. Morgan Kaufmann, San Mateo, CA.

Wnek, J. and Hieb, M. (1994). Bibliography of multistrategy learning research. In Michalski, R. S. and Tecuci, G. (editors), *Machine Learning: A Multistrategy Approach*, Vol. 4, pp. 657–729. Morgan Kaufmann, San Mateo, CA.

Wooldridge, M. and Jennings, N. R. (1995a). Agent theories, architectures, and languages: a survey. In Wooldridge, M. and Jennings, N. R. (editors), *Intelligent Agents*, pp. 1–39. Springer-Verlag, Berlin.

Wooldridge, M. J. and Jennings, N. R. (editors) (1995b). *Intelligent Agents*. Springer-Verlag, Berlin.

Wrobel, S. (1989). Demand-driven concept formation. In Morik, K. (editor), *Knowledge Representation and Organization in Machine Learning*, pp. 289–319. Springer-Verlag, New York.

Index

Printed and bound by CPI Group (UK) Ltd, Croydon, CR0 4YY

09/10/2024

01042679-0001